"十三五"高等职业教育核心课程规划教材·食品类

食品安全与质量控制

主　编　刘　皓　侯　婷
副主编　魏　玮　马长路　路冠茹　魏纪平　吕春晖
　　　　范兆军　安　娜　〔澳〕Oscar Jenkins
主　审　王　芃　王立晖
参　编　揣玉多　傅　维　李晓阳　张轶斌　刘　晨

西安交通大学出版社
XI'AN JIAOTONG UNIVERSITY PRESS

图书在版编目(CIP)数据

食品安全与质量控制 / 刘皓，侯婷主编. —西安：西安交通大学出版社，2019.5(2021.12重印)
 ISBN 978-7-5693-1122-8

Ⅰ. ①食… Ⅱ. ①刘… ②侯… Ⅲ. ①食品安全 ②食品-质量控制 Ⅳ. ①TS201.6 ②TS207.7

中国版本图书馆 CIP 数据核字(2019)第 041963 号

书　　名	食品安全与质量控制
主　　编	刘　皓　侯　婷
副 主 编	魏　玮　马长路　路冠茹　魏纪平　吕春晖　范兆军　安　娜　〔澳〕Oscar Jenkins
策划编辑	曹　昳　陈　昕
责任编辑	陈　昕　于睿哲
出版发行	西安交通大学出版社 (西安市兴庆南路1号　邮政编码 710048)
网　　址	http://www.xjtupress.com
电　　话	(029)82668357　82667874(发行中心) (029)82668315(总编办)
传　　真	(029)82668280
印　　刷	西安日报社印务中心
开　　本	787 mm×1092 mm　1/16　印张 23.25　字数 574千字
版次印次	2019年5月第1版　2021年12月第3次印刷
书　　号	ISBN 978-7-5693-1122-8
定　　价	42.00元

如发现印装质量问题，请与本社发行中心联系调换。

订购热线：(029)82665248　(029)82665249
投稿热线：(029)82669097　QQ：8377981
本书同时开发了数字教材，如需了解更多丰富的数字化教学资源，请联系以下信箱：lg_book@163.com

版权所有　侵权必究

前　言

"民以食为天,食以安为先"。食品生产工业化和食品消费社会化使得食品安全事件的影响范围急剧扩大。食品质量安全状况已成为一个国家或地区经济发展水平和人民生活质量的重要标志。随着经济的全球化,世界各国之间食品贸易日益增加,食品安全也就成为影响国家农业和食品工业竞争力的关键因素之一。

本教材是针对食品相关专业的学生开设的专业基础课程而编写的,主要以食品安全科学理论、管理法规和控制措施为指导思想,在食品安全风险分析的基础上,围绕食品供应过程,详细阐述了有关食品安全与质量控制的基本概念、各类食品质量安全因素的来源及控制措施,强调食品加工过程的ISO9000、危害分析与关键控制点(HACCP)等体系对食品安全控制的重要性。

本教材的编写源于高等职业教育食品专业群国际化教学的要求,由于食品企业生产过程中所需的安全管理体系大多为国际通行的,因此利用双语教材培养具备较高国际语言能力和行业国际视野的食品安全管理人才尤为重要,而在"食品安全与质量控制"课程的教学工作中一直苦于没有一本内容和深度较适合的国际化教材。为此,特组织从事此课程教学的主干教师共同编写了此教材。全书由刘皓、侯婷担任主编,魏玮、马长路、路冠茹、魏纪平、吕春晖、范兆军、安娜和澳大利亚外籍教师Oscar Jenkins等担任副主编,揣玉多、傅维、李晓阳、张轶斌、刘晨参编。在上述各位老师负责编写、翻译和轮流审稿的基础上,由王芃、王立晖主审。

本教材在编写过程中,得到了编者所在院校和相关企业的指导、帮助和支持,在此深表谢意! 由于编者水平有限,时间仓促,加之食品安全与质量控制方面的内容仍在不断发展变化之中,书中疏漏和不妥之处在所难免,恳请各位读者批评指正。

编　者

2018年12月

目　录

项目1　食品安全质量控制与监管 ……………………………………………… (1)
 任务1　食品安全和食品质量 …………………………………………………… (1)
 任务2　食品安全管理和食品质量控制的重要性 ……………………………… (2)
 任务3　食品行业监管体系 ……………………………………………………… (3)
 任务4　中国国家认证认可监督管理委员会职责及相关文件 ……………… (10)

项目2　食品法律法规和标准体系 …………………………………………… (20)
 任务1　我国食品法律法规体系 ……………………………………………… (20)
 任务2　食品法律法规 …………………………………………………………… (22)
 任务3　我国食品标准体系 ……………………………………………………… (31)
 任务4　食品安全标准 …………………………………………………………… (46)
 任务5　食品安全企业标准的编写及备案 …………………………………… (48)

项目3　食品企业生产许可证申办 …………………………………………… (58)
 任务1　食品生产许可制度 ……………………………………………………… (58)
 任务2　食品生产许可的申请 …………………………………………………… (61)
 任务3　食品生产许可的审查 …………………………………………………… (69)

项目4　食品质量安全管理常用工具应用方法 …………………………… (91)
 任务1　食品质量安全数据及随机变量 ……………………………………… (91)
 任务2　分层法的应用 …………………………………………………………… (99)
 任务3　调查表法的应用 ……………………………………………………… (103)
 任务4　排列图法的应用 ……………………………………………………… (110)
 任务5　因果图法的应用 ……………………………………………………… (118)
 任务6　对策表法的应用 ……………………………………………………… (123)
 任务7　直方图法的应用 ……………………………………………………… (126)

项目5　ISO9000质量管理体系的建立与认证 …………………………… (133)
 任务1　质量管理体系 ………………………………………………………… (133)
 任务2　质量管理认证标准 …………………………………………………… (136)
 任务3　质量管理体系的建立 ………………………………………………… (171)
 任务4　内审和管理评审 ……………………………………………………… (183)

项目 6　食品安全管理基础知识 …………………………………………… (190)
任务 1　食品安全危害及控制手段 ……………………………………… (190)
任务 2　良好的操作规范(GMP) ………………………………………… (201)
任务 3　卫生标准操作程序(SSOP) ……………………………………… (226)
任务 4　依据卫生标准操作程序要求编写卫生标准操作规范 …………… (248)

项目 7　危害分析与关键控制点(HACCP)体系的建立与认证 …………… (249)
任务 1　危害分析与关键控制点体系 ……………………………………… (249)
任务 2　危害分析与关键控制点体系认证对食品企业的价值 …………… (253)
任务 3　危害分析与关键控制点体系认证标准 …………………………… (255)
任务 4　危害分析与关键控制点体系建立方法 …………………………… (277)
任务 5　危害分析与关键控制点体系文件编写方法 ……………………… (298)
任务 6　内审和管理评审 …………………………………………………… (300)

项目 8　ISO22000 食品安全管理体系的建立与认证 …………………… (303)
任务 1　ISO22000 食品安全管理体系概况 ……………………………… (303)
任务 2　ISO22000:2005 标准内容 ………………………………………… (305)
任务 3　危害分析与关键控制点体系、食品安全管理体系认证实施规则比较
　　　　　　　　　　　　　　　　　　　　　　　　　　　　………………… (309)

项目 9　绿色食品与有机食品的认证 ……………………………………… (312)
任务 1　绿色食品认证 ……………………………………………………… (312)
任务 2　有机食品认证 ……………………………………………………… (339)

项目1 食品安全质量控制与监管

项目概述

通过对食品安全管理与食品质量控制的学习,可以对食品行业的价值与现状有很好的了解,从而掌握食品安全监管体系。

任务1 食品安全和食品质量

1.1.1 基础知识

1. 食品

食品指各种供人食用或者饮用的成品和原料,以及按照传统既是食品又是中药材的物品,但是不包括以治疗为目的的物品。(2015年4月24日第十二届全国人民代表大会常务委员会第十四次会议修订的《中华人民共和国食品安全法》中"食品"的含义)

2. 食品安全

食品安全指食品无毒、无害,符合应当有的营养要求,对人体健康不造成任何急性、亚急性或者慢性危害。

3. 食品质量

食品质量特性可分为内在(固有)食品质量特性和外在(非固有)食品质量特性。内在食品质量特性包括:①食品本身的安全性与健康性;②感官品质与货架期;③产品的可靠性与便利性。外在食品质量特性包括:①生产系统特性;②环境特性;③市场特性。

Ⅰ. Food

Food refers to all kinds of finished products and raw materials for people to eat or drink, as well as items that are traditionally used as food and Chinese herbal medicines, but does not include articles for the purpose of treatment. (the 14th meeting of the Standing Committee of the 12th National People's Congress on April 24th 2015 amended the meaning of "food" in the *Food Safety Law of the People's Republic of China*)

Ⅱ. Food Safety

Food safety means that food is non-toxic, harmless and meets the nutritional requirements, which does not cause any acute, subacute or chronic harm to human health.

Ⅲ. Food Quality

Food quality characteristics can be divided into internal (inherent) food quality characteristics and external (non-inherent) food quality characteristics. The characteristics of internal food quality include: a. the safety and health of food itself; b. the sensory quality and the shelf life; c. the reliability and convenience of products. The characteristics of external food quality include: a. the production system characteristics; b. the environmental characteristics; c. the marketing character.

1.1.2　实训任务

实训组织：对学生进行分组，每个组参照"基础知识"中的内容并利用网络资源，分析食品质量与食品安全的区别，填写表1-1。

表1-1　食品质量与食品安全的区别

序号	项目	区　　别
1	食品质量	
2	食品安全	

任务2　食品安全管理和食品质量控制的重要性

1.2.1　基础知识

1. 国家层面

食品质量安全管理涉及民生问题，只有食品质量安全得到保障，人民才能安居乐业，社会才能稳定团结。

Ⅰ. National Level

Food quality and safety management involves the livelihood of the people. Only when food quality and safety is guaranteed can people live and work in peace and contentment whilst society remains stable and unified.

2. 企业层面

食品质量安全管理是企业第一要务，只有企业食品质量安全没有问题，企业才可能生存盈利，否则最终会面临倒闭。

Ⅱ. Enterprise Level

Food quality and safety management is the first priority of enterprises, only if there is no problem in food quality and safety, enterprises will be able to survive and make profits, otherwise, they will face closure eventually.

1.2.2　实训任务

实训组织：对学生进行分组，每个组参照"基础知识"中的内容并利用网络资源，分析食品质量安全控制对于国家和企业的价值，填写表1-2。

表1-2　食品质量安全控制对于国家和企业的价值

序号	项目	食品质量安全控制的价值
1	国家层面	
2	企业层面	

任务3 食品行业监管体系

1.3.1 基础知识

1. 中国食品安全监管体系

第一阶段：2003年以前，以卫生部门为主进行监管。

第一个时期：20世纪50年代到60年代。

1953年政务院批准建立各级卫生防疫站，各级卫生行政部门在防疫站内设立食品卫生监督机构，负责食品卫生监督管理工作。

1953年卫生部颁布《清凉饮食物管理暂行办法》。

1964年国务院转发了卫生部、商业部等五部委制定的《食品卫生管理试行条例》。

第二个时期：20世纪70年代到80年代。

1978年国务院批准由卫生部牵头，会同其他有关部门组成"全国食品卫生领导小组"。

1979年国务院正式颁发《中华人民共和国食品卫生管理条例》。

1982年五届全国人大常委会第二十五次会议审议通过《中华人民共和国食品卫生法（试行）》。

第三个时期：20世纪90年代至

Ⅰ. China's Food Safety Regulatory System

Phase one: before 2003, the department of health was the primary supervision.

The first period: 1950s to 1960s.

In 1953, the government administration council approved the establishment of sanitation and antiepidemic stations at all levels, and the health administrative departments at all levels set up the agency for food hygiene supervision in the epidemic prevention stations to be responsible for the supervision and administration of food hygiene.

In 1953, the Ministry of Health promulgated the *Interim Measures for the Management of Cool Diets*.

In 1964, the State Council forwarded the *Trial Regulations on Food Hygiene Management* formulated by the Ministry of Health and the Ministry of Commerce and other five ministries and commissions.

The second period: 1970s to 1980s.

In 1978, the State Council approved the establishment of the "National Food Hygiene Leading Group" under the leadership of the Ministry of Health and in conjunction with other relevant departments.

In 1979, the State Council formally issued the *Regulations on the Management of Food Hygiene in People's Republic of China*.

In 1982, the 25th meeting of the Standing Committee of the 5th National People's Congress deliberated and adopted the *Food Hygiene Law of the People's Republic of China (Trial)*.

The third period: from 1990s to 2003.

2003年。

1993年国务院机构改革撤销了轻工部,食品企业在体制上正式与轻工业主管部门分离,食品生产经营方式发生了较大变化。

1995年八届人大常委会第十六次会议审议通过了《中华人民共和国食品卫生法》。

《食品卫生法》确定了卫生部门食品卫生执法主体地位,废除了原有政企合一体制下主管部门的管理职权,明确规定国家实行食品卫生监督制度。

第二阶段:2003年以后,分段监管与综合协调相结合。

第一个时期:2003年到2008年,由食品药品监管局负责综合协调。

2003年,国务院机构改革在原国家药品监督管理局的基础上组建了国家食品药品监督管理总局。

2004年9月,国务院印发《关于进一步加强食品安全工作的决定》(国发〔2004〕23号),启动修订《食品卫生法》。

2004年12月中央编办印发《关于进一步明确食品安全部门职责分工有关问题的通知》(中央编办发〔2004〕35号)。

国家食品药品监督管理总局各部门职责如图1-1所示。

In 1993, the State Council's institutional reform abolished the Ministry of Light Industry. The food enterprises were formally separated from the competent department of light industry in the system, and the mode of food production and operation changed greatly.

In 1995, the 16th meeting of the Standing Committee of the 8th National People's Congress deliberated and adopted the *Food Hygiene Law of the People's Republic of China*.

The *Food Hygiene Law* determined the main law enforcement status of food hygiene in the Department of Health, abolished the administrative authority of the competent department under the original system of integration of government and enterprises, and clearly stipulated that the country shall implement the food hygiene supervision system.

Phase two: after 2003, the combination of subsection supervision and comprehensive coordination.

The first period: from 2003 to 2008, the Food and Drug Administration was responsible for comprehensive coordination.

In 2003, the institutional reform of the State Council established the China Food and Drug Administration on the basis of the former National Medical Products Administration.

In September 2004, the State Council issued the *Decision on Further Strengthening of Food Safety*(〔2004〕No. 23 issued by China), which initiated the revision of the "Food Hygiene Law".

In December 2004, State Commission Office for Public Sector Reform(SCOPSR) issued the *Notice on Further Clarifying the Issues Related to the Division of Responsibilities of the Food Safety Department*(〔2004〕No. 35 issued by SCOPSR).

The responsibilities of various departments of the China Food and Drug Administration are shown in Figure 1-1.

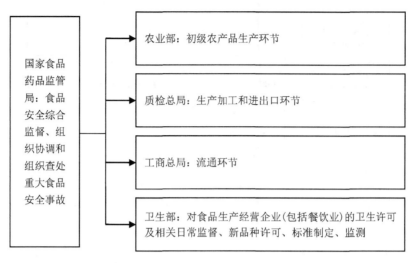

图1-1　国家食品药品监督管理总局各部门职责

第二个时期:2008年到2010年,由卫生部负责综合协调。

2008年,国务院机构改革将食品安全综合协调和组织查处重大食品安全事故职责由国家食品药品监管局划入卫生部,并将该局调整为卫生部管理的国家局,具体如图1-2所示。

2009年2月,《食品安全法》发布,从法律上明确了分段监管和综合协调相结合的体制,并规定国务院成立食品安全委员会作为高层次议事协调机构。

第三个时期:从2010年至2011年,食品安全委员会办公室承办国务院食品安全委员会交办的综合协调任务。

2010年2月6日,国务院印发《关于设立国务院食品安全委员会的通知》(国发〔2010〕6号)。

2010年12月6日,中央编办印发

The second period: from 2008 to 2010, the Ministry of Health was responsible for comprehensive coordination.

In 2008, the State Council institutional reform assigned the responsibility of comprehensive coordination and organization of handling of major food safety accidents from the China Food and Drug Administration to the Ministry of Health, and adjusted this bureau to the State Bureau under the Ministry of Health, as shown in Figure 1-2.

In February 2009, the *Food Safety Law* was issued, which legally defined the system of the combination of subsection supervision and comprehensive coordination, and stipulated that the State Council set up the Food Safety Committee as a high-level agency for negotiation and coordination.

The third period: from 2010 to 2011, the Office of the Food Safety Committee undertook the comprehensive coordination task assigned by the Food Safety Committee of the State Council.

In February 6, 2010, the State Council issued the *Notice on the Establishment of the Food Safety Committee of the State Council* (〔2010〕No. 6 issued by China).

In December 6, 2010, SCOPSR issued the

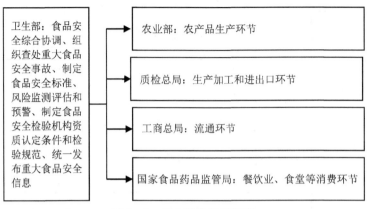

图 1-2 卫生部各部门职责

《关于国务院食品安全委员会办公室机构设置的通知》(中央编办发〔2010〕202号)。

食品安全委员会定位为国务院食品安全的高层次议事协调机构,有19个部门参加。主要职责为分析食品安全形势,研究部署、统筹指导食品安全工作,提出食品安全监管的重大政策措施,督促落实食品安全监管责任,具体如图1-3和图1-4所示。

Notice on the Establishment of the Office of the Food Safety Committee of the State Council (〔2010〕No. 202 issued by SCOPSR).

The Food Safety Committee has been designated as the high-level agency for negotiation and coordination of food safety under the State Council, with 19 participating departments. Its main duties are to analyze the situation of food safety, to study and plan and guide the work of food safety, to put forward major policy measures for food safety supervision, and to urge the implementation of the responsibility of food safety supervision, as shown in Figure 1-3 and Figure 1-4.

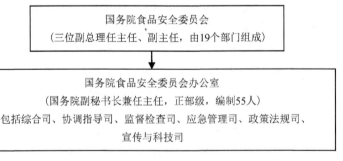

图 1-3 国务院食品安全委员会结构图

第四个时期:从2011年至2013年,将卫生部综合协调、牵头组织食品安全重大事故调查、统一发布重大食品安全信息的职责划入国务院食品安全委员会办公室。

The fourth period: from 2011 to 2013, the comprehensive coordination and leading organization of the Ministry of Health for the investigation of major food safety accidents and the unified release of major food safety information were assigned to

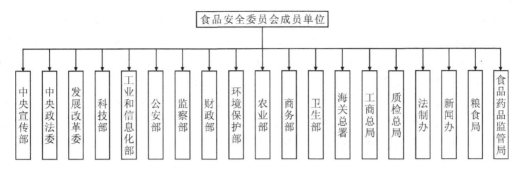

图1-4 食品安全委员会成员单位

2011年11月9日,中央编办印发《关于国务院食品安全委员会办公室机构编制和职责调整有关问题的批复》(中央编办复字〔2011〕216号),决定将卫生部综合协调、牵头组织食品安全重大事故调查、统一发布重大食品安全信息的职责划入国务院食品安全委员会办公室。

三项职能划归食品安全委员会后,卫生部保留三项职能,包括食品安全标准的制定、食品安全风险检测评估、对检验机构资质条件的认定。卫生部承担的各项食品安全职责是食品安全的基础性工作,是食品安全监管的重要技术依据。在国家层面,实行分段监管为主、品种监管为辅和综合协调相结合的体制;在地方政府层面,实行地方政府负总责下的部门分段监管和综合协调相结合的体制。

the Office of the Food Safety Committee of the State Council.

In November 9, 2011, SCOPSR issued a *Written Reply on Issues Related to the Organization Establishment and Responsibilities Adjustment of the Office of the Food Safety Commission of the State Council* (〔2011〕No. 216 printed by the SCOPSR), which decided that the comprehensive coordination and leading organization of the Ministry of Health for the investigation of major food safety accidents and the unified release of major food safety information were assigned to the Office of the Food Safety Committee of the State Council.

After the three functions were assigned to the Food Safety Committee, the Ministry of Health reserved three functions, including the formulation of food safety standards, the assessment of food safety risks, and the determination of the qualification conditions of inspection agencies. The responsibility of food safety undertaken by the Ministry of Health is the fundamental job of food safety and the important technical basis of food safety supervision. At the national level, implementing the system based on subsection supervision as the main, the various supervision as the supplementary and the comprehensive coordination. At the level of local government, implementing a system of a combination of sectoral supervision and comprehensive coordination under the overall responsibility of the local government.

第五个时期：从 2013 年至 2018 年 3 月，我国监管机制主体进一步集中，形成以农业、食药为监管主体，卫生与计划生育委员会为科技支撑，食品安全委员会为综合协调的两段式的监管新格局，大大减少了链条中的空白点、盲点，降低了成本。国家风险评估中心的成立和统一的食品安全标准，为建立在风险评估科学基础上的防御性食品安全体系打下了基础，风险评估及标准制定职能与食品安全监管部门分开，避免了过去既是裁判员又是运动员的尴尬局面。检验检测机构职能转变，检验检测体系"去部门化""管办分离"，实现资源共享，建立法人治理结构，形成统一的检测机构。

The fifth period: from 2013 to March 2018, the main body of supervision mechanism in our country has been further concentrated, forming a new pattern of two stages of supervision, which takes agriculture, food and medicine as the main body of supervision, and the Health and Family Planning Commission is the support of science and technology, and the Food Safety Committee is the comprehensive coordination. This greatly reduces the blind spot in the chain and the cost. The establishment of the National Risk Assessment Center and the unified food safety standards have laid the foundation for a defensive food safety system based on the scientific basis of risk assessment. The separation of risk assessment and standard-setting functions from food safety regulators avoid the embarrassment of being both referees and athletes in the past. The functions of the inspection and testing organization are changed, the inspection and testing system is "removed the departmentalization" and "separated the government regulation from management", so as to achieve resource sharing, establish corporate governance structure, and form a unified inspection agency.

2018 年 3 月中共中央印发《深化党和国家机构改革方案》："为完善市场监管体制，推动实施质量强国战略，营造诚实守信、公平竞争的市场环境，进一步推进市场监管综合执法、加强产品质量安全监管，让人民群众买得放心、用得放心、吃得放心，将国家工商行政管理总局的职责，国家质量监督检验检疫总局的职责，国家食品药品监督管理总局的职责，国家发展和改革委员会的价格监督检查与反垄断执法职责，商务部的经营者集中反垄断执法以及国务院反垄断委员会办公室等职责整合，组建国家市场监督管理总局，作为国务院直属机构。"如图 1-5 所示。

2. 食品企业案例分析

2013 年 12 月，某职业学院毕业生小马经过 3 轮面试顺利应聘中粮集团法规部，其工作职责：①掌握中国最新法律法规动态；②对公司生产部、品管部进行法律法规培训；③负责指导公司食品生产许可（SC）、ISO9001、ISO22000 等相关认证的法律法规支持工作；④与监管部门沟通对中粮集团的监管要求并予以传达。

结合以上案例，请同学们讨论小马需要掌握哪些中国食品安全监管知识，并填写在表 1-3 中。

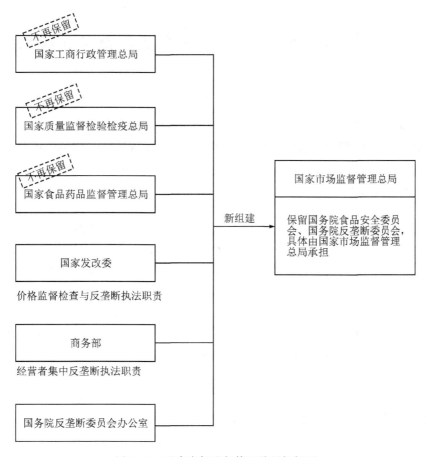

图 1-5 国家市场监督管理总局框架图

表 1-3 小马需要掌握的食品安全监管知识

序号	内　容
1	
2	

1.3.2　实训任务

实训组织:某食品企业要设立安全监管体系,请你按照食品行业监管体系的要求,给企业制定安全监管体系。

实训成果:食品企业安全监管体系的制定。

实训评价:由企业质量负责人或主讲教师进行评价。

任务4　中国国家认证认可监督管理委员会职责及相关文件

1.4.1　基础知识

1. 学习国家认证认可监督管理委员会在食品认证监管中的职责和权限

1）了解国家认证认可监督管理委员会的机构设置

中国国家认证认可监督管理委员会(中华人民共和国国家认证认可监督管理局)，Certification and Accreditation Administration of the People's Republic of China(简称CNCA)，成立于2001年，是国务院决定组建并授权，履行行政管理职能，统一管理、监督和综合协调全国认证认可工作的主管机构。

国家认证认可监督管理委员会（以下简称认监委）下设多个部门、直属单位和管理单位。

(1)下设部门。

①认可监管部。负责拟定认可制度、认证人员注册制度、管理体系认证制度、人员认证制度及其规则和工作规划、计划；研究拟定对认可机构和认证人员注册机构的监督管理的规定，负责认可机构授权和监督管理工作；研究拟定对认证机构、人员认证、认证咨询机构和认证培训机构的资质审核制度以及从业资格审批制度、规定、程序、规划和监督管理的规定，并组织实施；负责对认证认可行业自律组织的管理和指导等。

Ⅰ. Study the Duties and Powers of the Certification and Accreditation Administration of the People's Republic of China in Food Certification and Supervision

A. Understand the Structural Establishment of the Certification and Accreditation Administration of the People's Republic of China

Certification and Accreditation Administration of the People's Republic of China (hereinafter referred to as the CNCA), was established in 2001. It is the competent organization established and authorized by the decision of the State Council to perform the functions of administration, unified management, supervision, and comprehensive coordination of the work of national certification and accreditation.

The CNCA shall set up a number of departments, the units directly under and the management unit.

1. Subordinate department.

(1) Department of Accreditation and Supervision. Be responsible for drawing up the accreditation system, the registration system for certifying personnel, the certification system of the management system, the personnel certification system and its rules, work programme and plans; study and formulate provisions for the supervision and administration of accreditation body and registration authority of certified personnel, responsible for the authorization, supervision and administration of accreditation body; study and formulate regulations on the qualification examination system of the certification body, personnel certification, certification advisory body and certification training organization, as well as examination and approval system, regulations, procedures, planning and supervision and

②认证监管部。研究拟定强制性产品认证与安全质量许可制度的建立、规划、计划,并组织实施和监督管理;负责起草强制性产品认证与安全质量许可制度的产品目录、认证标志管理办法和合格评定程序;负责组织确定承担强制性认证任务的认证机构、检查机构和实验室,并监督检查;研究拟定自愿性产品认证制度的建立、规划、计划,并组织实施和监督管理;负责对产品认证活动和认证结果的监督检查,负责协调强制性产品认证行政执法检查工作中的技术性政策问题等。

③注册管理部。研究拟定进出口食品、化妆品生产、加工、储存等企业的卫生注册登记,以及陶瓷出口质量许可的工作规章、制度和工作规划、计划;负责出口食品、化妆品生产、加工、储存等企业的卫生注册登记和陶瓷出口质量许可的评审、发证和监督管理工作,负责相关重大问题和质量事故的调查处理工作;负责统一办理向境外推荐企业注册和组织接待境外主管部门来华检查工作;负责进口食品、化妆品生产、加工、储存等企业的注册评审和监督管理工作;负责卫生注册评

administration of the qualification of employment, and organize the implementation; be responsible for the management and guidance of self-regulatory organizations in certification and accreditation industry; etc.

(2) Department of Certification and Supervision. Study and formulate the establishment, programming, planning, organization and implementation as well as supervision and administration of the compulsory product certification and safety quality licensing system; be responsible for drafting the product catalogue, certification mark regulation and conformity assessment procedure of compulsory product certification and safety quality licensing systems; be responsible for organizing certification bodies, inspection bodies and laboratories that undertake compulsory certification tasks, and supervise and inspect; study and formulate the establishment, programming and planning of voluntary product certification system, and organize and implement as well as supervise and administrate; be responsible for supervision and inspection of product certification activities and certification results, responsible for coordinating technical policy issues in the administrative enforcement inspection of compulsory product certification, etc.

(3) Registration Management Department. Study and formulate the sanitary registration for the production, processing, storage, and other enterprises of imported and exported food and cosmetics, as well as the working rules, regulations, work programming and plans for the quality licensing of ceramic exports; be responsible for the sanitary registration of the export food, cosmetics production, processing, storage and other enterprises and the evaluation, certification, supervision and administration of the quality licensing of ceramic exports, responsible for the investigation and handling of related major problems and quality accidents; be

审员的管理工作;研究拟定食品和农产品认证规划和实施计划,管理和协调食品和农产品认证体系建设工作等。

④其他部门。认监委下设的其他部门还有政策与法律事务部、实验室与监测监管部、国际合作部、科技与标准管理部等。

(2)直属单位和管理单位。国家认监委有三个直属单位,分别是机关服务中心、信息中心、认证认可技术研究所。管理单位有中国认证认可协会、中国合格评定国家认可中心、中国信息安全认证中心、中国检验认证(集团)有限公司、中国检验有限公司(香港)等。

2)了解国家认证认可监督管理委员会的职责

作为统一管理、监督和综合协调全国认证认可工作的主管机构,国家认监委的主要职能有以下几点。

(1)研究起草并贯彻执行国家认证认可、安全质量许可、卫生注册和合

responsible for the unified registration of overseas referral enterprises and the organization and reception of overseas competent authorities for inspection in China; be responsible for the registration, evaluation and supervision and administration of imported food, cosmetics production, processing and storage enterprises; be responsible for the management of health registration assessor; study and formulate the certification planning and starting plan for the certification of food and agricultural products, manage and coordinate the construction of certification systems for food and agricultural products; and so on.

(4)Other departments. Other departments under the CNCA include Policy and Legal Affairs Department, Laboratory and Monitoring Supervision Department, International Cooperation Department, Science, Technology and Standards Management Department, and so on.

2. Units directly under and the management unit. CNCA has three units directly under, which are the agency service center, information center, certification and accreditation technology research institute respectively. Management units include China Certification and Accreditation Association, China National Accreditation Center for Qualification Assessment, China Information Security Certification Center, China Inspection Certification (Group) Co., Ltd., China Inspection Company Limited (Hong Kong), etc.

B. Understand the Responsibilities of the Certification and Accreditation Administration of the People's Republic of China

As the competent body for the unified management, supervision and integrated coordination of the national accreditation and certification, the main functions of the CNCA are as follows.

1. Study, draft and implement laws, regulations and rules on national certification and accreditation,

格评定方面的法律、法规和规章,制定、发布并组织实施认证认可和合格评定的监督管理制度、规定。

(2)研究提出并组织实施国家认证认可和合格评定工作的方针政策、制度和工作规则,协调并指导全国认证认可工作,监督管理相关的认可机构和人员注册机构。

(3)研究拟定国家实施强制性认证与安全质量许可制度的产品目录,制定并发布认证标志(标识)、合格评定程序和技术规则,组织实施强制性认证与安全质量许可工作。

(4)负责进出口食品和化妆品生产、加工单位卫生注册登记的评审和注册等工作,办理注册通报和向国外推荐事宜。

(5)依法监督和规范认证市场,监督管理自愿性认证、认证咨询与培训等中介服务和技术评价行为;根据有关规定,负责认证、认证咨询、培训机构和从事认证业务的检验机构(包括中外合资、合作机构和外商独资机构)的资质审批和监督;依法监督管理外国(地区)相关机构在境内的活动;受理有关认证认可的投诉和申诉,并组织查处;依法规范和监督市场认证行为,指导和推动认证中介服务组织的改革。

safety and quality licensing, hygiene registration and conformity assessment, and formulate, issue and organize supervision and administration systems and regulations for certification, accreditation and conformity assessment.

2. Study and organize the implementation of the principles, policies, systems and working rules for the work of national certification and accreditation as well as conformity assessment, coordinate and guide the national certification and accreditation work. Supervise and manage the relevant accreditation body and personnel registration authority.

3. Study and draw up the product catalogue for the implementation of the compulsory certification and safety quality licensing system by the state, formulate and issue certification mark (identification), conformity assessment procedures and technical rules, organize and implement the compulsory certification and safety quality licensing.

4. Responsible for the evaluation and registration of sanitary registration of imported and exported food and cosmetics production and processing units, and handle registration, notification and recommendations to foreign countries.

5. Supervise and standardize the certification market according to law, supervise and administer intermediary services and technical evaluation activities such as voluntary certification, certification consultation and training; according to relevant regulations, responsible for the approval and supervision of qualification of certification, certification consulting, training institutions and inspection institutions (including sino-foreign joint ventures, cooperative institutions and wholly foreign-owned institutions) engaged in certification business; supervise and administer the activities of relevant foreign (regional) institutions within the territory according to law; accept complaints and appeals about certification and accreditation, and organize

（6）管理相关校准、检测、检验实验室技术能力的评审和资格认定工作，组织实施对出入境检验检疫实验室和产品质量监督检验实验室的评审、计量认证、注册和资格认定工作；负责对承担强制性认证和安全质量许可的认证机构及承担相关认证检测业务的实验室、检验机构的审批；负责对从事相关校准、检测、检定、检查、检验检疫和鉴定等机构（包括中外合资、合作机构和外商独资机构）技术能力的资质审核。

（7）管理和协调以政府名义参加的认证认可和合格评定的国际合作活动，代表国家参加国际认可论坛（IAF）、太平洋认可合作组织（PAC）、国际人员认证协会（IPC）、国际实验室认可合作组织（ILAC）、亚太实验室认可合作组织（APLAC）等国际或区域性组织以及国际标准化组织（ISO）和国际电工委员会（IEC）的合格评定活动，签署与合格评定有关的协议、协定和议定书；归口协调和监督以非政府组织名义参加的国际或区域性合格评定组织的活动；负责国际标准化组织和国际电工委员会中国国家委员会的合格评定工作；负责认证认可、合格评定等国际活动的外事审批。

the investigation and punishment; standardize and supervise market certification in accordance with the law, guide and promote the reform of certification intermediary service organizations.

6. Manage the assessment and qualification of the technical competence of the related calibration, testing and inspection laboratories, organize and implement the evaluation, metrology and certification, registration and qualification of entry-exit inspection and quarantine laboratories and product quality supervision and inspection laboratories; responsible for the examination and approval of the certification bodies that undertake compulsory certification and safety and quality license, and the laboratories and inspection agencies that undertake the relevant certification and testing business; responsible for the qualification audit of the technical capability of the institutions （including sino-foreign joint ventures, cooperative institutions and wholly foreign-owned institutions） engaged in relevant calibration, testing, verification, inspection, inspection and quarantine and appraisal.

7. Manage and coordinate international cooperation activities for certification and accreditation and conformity assessment in the name of the government, represent countries to attend the conformity assessment activities organised by ISO and IEC, such as IAF, PAC, IPC, ILAC, APLAC and so on, sign the agreements, conventions and protocols related to conformity assessment; coordinate and supervise the activities of international or regional conformity assessment organizations participated in the name of non-governmental organizations; responsible for the conformity assessment of National Committee of China in ISO and IEC; responsible for the examination and approval of foreign affairs for international activities such as certification, accreditation and conformity assessment.

(8)负责与认证认可有关的国际准则、指南和标准的研究和宣传贯彻工作;管理认证认可与相关的合格评定的信息统计,承办世界贸易组织/技术性贸易壁垒协定、实施卫生与植物卫生措施协定中有关认证认可的通报和咨询工作。

8. Responsible for the research and promotion of international standards, guidelines and criterion related to certification and accreditation; manage information statistics on certification and accreditation and related conformity assessment, undertake World Trade Organization/technical barriers to trade agreement, and implement notification and consultation on certification and accreditation in sanitary and phytosanitary measures agreement.

(9)配合国家有关主管部门,研究拟订认证认可收费办法,并对收费办法的执行情况进行监督检查。

9. Cooperate with the relevant competent departments of the state to study and formulate the charge measures for certification and accreditation, supervise and inspect the implementation of the charge measures.

2. 学习国家认证认可监督管理委员会的相关管理文件

1)《认证证书和认证标志管理办法》

《认证证书和认证标志管理办法》于2004年4月30日由国家质量监督检验检疫总局公布,自2004年8月1日起施行。

该办法中所称的认证证书是指产品、服务、管理体系通过认证所获得的证明性文件。认证证书包括产品认证证书、服务认证证书和管理体系认证证书。认证标志是指证明产品、服务、管理体系通过认证的专有符号、图案或者符号、图案以及文字的组合。认证标志包括产品认证标志、服务认证标志和管理体系认证标志。

国家认证认可监督管理委员会(即国家认监委)依法负责认证证书和认证标志的管理、监督和综合协调工作。地方质量技术监督部门和各地出入境检验检疫机构(统称地方认证监督管理部门)按照各自职责分工,依法负责所辖区域内的认证证书和认证标志的监督检查工作。

(1)对认证证书的管理。认证机构应当按照认证基本规范、认证规则从事认证活动,对认证合格的,应当在规定的时限内向认证委托人出具认证证书。

产品认证证书包括以下基本内容:委托人名称、地址;产品名称、型号、规格,需要时对产品功能、特征的描述;产品商标、制造商名称、地址;产品生产厂名称、地址;认证依据的标准、技术要求;认证模式;证书编号;发证机构、发证日期和有效期;其他需要说明的内容。

服务认证证书包括以下基本内容:获得认证的组织名称、地址;获得认证的服务所覆盖的业务范围;认证依据的标准、技术要求;认证证书编号;发证机构、发证日期和有效期;其他需要说明的内容。

管理体系认证证书包括以下基本内容:获得认证的组织名称、地址;获得认证的组织的管理体系所覆盖的业务范围;认证依据的标准、技术要求;证书编号;发证机构、发证日期和有效期;其他需要说明的内容。

该办法中明确要求,获得认证的组织应当在广告、宣传等活动中正确使用认证证书和有关信息,不得利用产品认证证书和相关文字、符号误导公众认为其服务、管理体系通过认证;不得利用服务认证证书和相关文字、符号误导公众认为其产品、管理体系通过认证;不得利用管理体系认证证书和相关文字、符号,误导公众认为其产品、服务通过认证。获得认证的产品、服务、管理体系发生重大变化时,获得认证的组织和个人应当向认证机构申请变更,未经变更或者经认证机构调查发现不符合认证要求的,不得继续使用该认证证书。

(2) 对认证标志的管理。认证标志分为强制性认证标志和自愿性认证标志。强制性认证标志和国家统一的自愿性认证标志属于国家专有认证标志。自愿性认证标志包括国家统一的自愿性认证标志和认证机构自行制定的认证标志。认证机构自行制定的认证标志是指认证机构专有的认证标志。

该办法中规定,强制性认证标志和国家统一的自愿性认证标志的制定和使用,由国家认监委依法规定,并予以公布。认证机构自行制定的认证标志的式样(包括使用的符号)、文字和名称,需满足不得与强制性认证标志、国家统一的自愿性认证标志或者已经国家认监委备案的认证机构自行制定的认证标志相同或者近似,不得妨碍社会管理秩序等要求。

该办法中要求认证机构建立认证标志管理制度,明确认证标志使用者的权利和义务,对获得认证的组织使用认证标志的情况实施有效跟踪调查,发现其认证的产品、服务、管理体系不能符合认证要求的,应当及时作出暂停或者停止其使用认证标志的决定,并予以公布。

(3) 监督检查。国家认监委组织地方认证监督管理部门对认证证书和认证标志的使用情况实施监督检查,对伪造、冒用、转让和非法买卖认证证书和认证标志的违法行为依法予以查处。

国家认监委对认证机构的认证证书和认证标志管理情况实施监督检查。认证机构应当对其认证证书和认证标志的管理情况向国家认监委提供年度报告。年度报告中应当包括其对获证组织使用认证证书和认证标志的跟踪调查情况。

任何单位和个人对伪造、冒用、转让和非法买卖认证证书和认证标志等违法、违规行为可以向国家认监委或者地方认证监督管理部门举报。

2)《有机产品认证管理办法》

《有机产品认证管理办法》于2004年9月27日由国家质量监督检验检疫总局公布,自2005年4月1日起施行。

该办法中所称的有机产品,是指生产、加工、销售过程符合有机产品国家标准的供人类消费、动物食用的产品。有机产品认证,是指认证机构按照有机产品国家标准和本办法的规定对有机产品生产和加工过程进行评价的活动。

国家认证认可监督管理委员会(以下简称国家认监委)负责有机产品认证活动的统一管理、综合协调和监督工作。地方质量技术监督部门和各地出入境检验检疫机构(以下统称地方认证监督管理部门)按照各自职责依法对所辖区域内有机产品认证活动实施监督检查。

(1) 机构管理。该办法中规定,有机产品认证机构应当依法设立,具有《中华人民共和国认证认可条例》规定的基本条件和从事有机产品认证的技术能力,并取得国家认监委确定的认可机构(以下简称认可机构)的认可后,方可从事有机产品认证活动。境外有机产品认证机构在中国境内开展有机产品认证活动的,应当符合《中华人民共和国认证认可条例》和其他有关法律、行政法规以及本办法的有关规定。

从事有机产品认证的检查员应当经认可机构注册后,方可从事有机产品认证活动。

从事与有机产品认证有关的产地（基地）环境检测、产品样品检测活动的机构（以下简称有机产品检测机构）应当具备相应的检测条件和能力，并通过计量认证或者取得实验室认可。国家认监委定期公布符合规定的有机产品认证机构和有机产品检测机构的名录。不在目录所列范围之内的认证机构和产品检测机构，不得从事有机产品的认证和相关检测活动。

（2）认证实施。该办法中规定，有机产品认证机构，应当公开有机产品认证依据的标准、认证基本规范、规则和收费标准等信息。

有机产品生产、加工单位和个人或者其代理人（以下统称申请人），可以自愿向有机产品认证机构提出有机产品认证申请。有机产品认证机构应当自收到申请人书面申请之日起10日内，完成申请材料的审核，并作出是否受理的决定；对不予受理的，应当书面通知申请人，并说明理由。有机产品认证机构受理有机产品认证后，应当按照有机产品认证基本规范、规则规定的程序实施认证活动，并按照相关标准或者技术规范的要求及时作出认证结论。

有机产品认证机构应当按照规定对获证单位和个人、获证产品进行有效跟踪检查，保证认证结论能够持续符合认证要求。生产、加工、销售有机产品的单位及个人和有机产品认证机构，应当采取有效措施，按照认证证书确定的产品范围和数量销售有机产品，保证有机产品的生产和销售数量的一致性。

该办法中明确规定，有机产品认证机构不得对有机配料含量（指重量或者液体体积，不包括水和盐）低于95%的加工产品进行有机认证。

（3）认证证书和标志。国家认监委规定有机产品认证证书的基本格式和有机产品认证标志的式样。

有机产品认证证书有效期为一年。获得有机产品认证证书的单位或者个人，在有机产品认证证书有效期内，发生信息变更或其他情况的，可向有机产品认证机构办理变更或重新申请手续，较为严重的，认证机构可作出暂停、撤销认证证书的决定。

3）《食品检验机构资质认定管理办法》

《食品检验机构资质认定管理办法》于2010年7月22日由国家质量监督检验检疫总局公布，自2010年11月1日起施行。

该办法中所称的食品检验机构资质认定，是指依法对食品检验机构的基本条件和能力，是否符合食品安全法律法规的规定以及相关标准或者技术规范要求实施的评价和认定活动。对向社会出具具有证明作用的数据和结果的食品检验机构开展资质认定活动应当遵守本办法。

国家质检总局统一管理食品检验机构资质认定工作。国家认监委负责食品检验机构资质认定实施、监督管理和综合协调工作。各省级质量技术监督部门按照职责分工，负责所辖区域内食品检验机构资质认定实施和监督检查工作。

（1）资质认定条件与程序。食品检验机构应当按照国家有关认证认可的规定依法取得资质认定后，方可从事食品检验活动。

该办法中规定，食品检验机构资质认定程序为：申请资质认定的食品检验机构（以下简称申请人），应当向国家认监委或者省级质量监督部门（以下统称资质认定部门）提出书面申请；资质认定部门应当对申请人提交的申请材料进行书面审查，并自收到材料之日起5日内作出受理或者不予受理的书面决定；申请材料不齐全或者不符合法定形式的，应当一次性告知申请人需要补正的全部内容；资质认定部门应当自受理申请之日起6个月内，对申请人完成技术评审工作;资质认定部门应当自技术评审完结之日起 20 日内,对技术评审结果进行审查,并作出是否批准的决定。决定批准的,向申请人颁发资质认定证书,并准许其使用资质认定标志;不予批准的,应当书面告知申请人,并说明理由。

食品检验机构资质认定证书有效期为 3 年。食品检验机构需要延续依法取得的资质认定的有效期的,应当在资质认定证书有效期届满前 6 个月,向资质认定部门提出复查换证申请。

(2)技术评审。国家认监委根据国家有关法律法规、国务院卫生行政部门规定的资质认定条件和相关国家标准的规定,制定食品检验机构资质认定评审准则。

资质认定部门应当按照评审准则的要求,组成技术评审组,对申请人的基本条件、管理体系和检验能力等资质条件的符合性情况进行技术评审。技术评审组对申请人的检验能力进行评审时,应当审查确认申请人具备相关能力验证、比对试验、测量审核的证明;需要进行现场试验的,应当按照评审准则的要求进行考核。

(3)监督管理。国家质检总局统一监督管理食品检验机构的相关检验活动。国家认监委负责组织对取得资质认定的食品检验机构进行监督检查,发现食品检验机构有违法违规行为的,应当予以查处,涉及国务院有关部门职责的,及时通报有关部门并协调处理。

国家认监委对省级质量监督部门实施的食品检验机构资质认定工作进行监督、指导。省级质量监督部门组织地(市)、县级质量监督部门对所辖区域内的食品检验机构进行监督检查或者专项监督检查,地(市)、县级质量监督部门应当对所辖区域内的食品检验机构进行日常监督。各直属出入境检验检疫局对所属食品检验机构进行日常监督管理。

该办法中规定,食品检验实行食品检验机构与检验人负责制。食品检验机构应当依据法律法规、检验规范的相关规定及委托检验合同的约定出具食品检验报告。食品检验报告应当加盖食品检验机构公章,并有检验人(授权签字人)的签名或者盖章。食品检验机构和检验人对出具的食品检验报告负责。

该办法中明确指出,任何单位和个人对食品检验机构的检验活动中的违法违规行为,有权向资质认定部门举报,资质认定部门应及时调查处理,构成犯罪的,依法追究刑事责任。

4)查询国家认监委相关管理文件的方法

作为统一管理、监督和综合协调全国认证认可工作的主管机构,国家认监委研究起草并贯彻执行国家认证认可、安全质量许可、卫生注册和合格评定方面的法律、法规和规章,制定、发布并组织实施认证认可和合格评定的监督管理制度、规定,相关信息均在国家认监委网站上进行发布。

在国家认监委的网站 http://www.cnca.gov.cn 上,可查找认监委制定、施行的所有法律和行政法规、部门规章、行政规范性文件。

1.4.2 实训任务

实训组织:对学生进行分组,每个组参照"基础知识"中的内容并利用网络资源,对以下案例进行分析。

案例 1 2012 年 12 月 19 日,国家认监委在网站上发布公告,世界认证服务(中国)有限公司未经国家认监委批准,擅自在中国境内从事认证活动并颁发质量管理体系认证证书。其行为违反了《中华人民共和国认证认可条例》的规定,属非法认证,所颁发认证证书均属无效。这已是 2012 年度国家认监委通报的第 6 家认证机构。认监委在公告中提醒社会各界,应选择经国家批准的合法认证机构提供认证服务。一旦发现非法从事认证活动的机构,可向所在地出入境检验检疫局、质量技术监督局或国家认监委举报。

结合上述案例,查阅国家认监委的性质、职能及其相关信息,并尝试回答以下问题。

1. 国家认监委各部门的职能是什么?
2. 认证申请人发现类似问题应该如何处置?

案例2 一家茶叶生产企业申请有机产品认证。

该企业是否需要了解相关认证认可的法律法规?如果你是这家茶叶生产企业的管理者,如何查询、搜集相关法规信息?

思考题

1. 国家食品药品监督管理总局的职责、权限、机构改革和职能转变情况有哪些内容?
2. 食品安全监督管理的现状是怎样的?
3. 卫生部在食品监管中的职责和权限有哪些?
4. 我国食品标准体系存在的问题有哪些?
5. 认监委对于食品工业发展的作用有哪些?
6. 国家认监委的其他管理文件还有哪些?
7. 如何实现认监委信息的有效利用?

拓展学习网站

1. 国家卫生健康委员会(http://www.nhc.gov.cn)
2. 国家市场监督管理总局(http://samr.saic.gov.cn)
3. 食品伙伴网(http://www.foodmate.net)

项目 2　食品法律法规和标准体系

项目概述

本项目将学习食品法律法规与标准,分析食品安全案例,编写食品产品企业标准和查找与食品企业相关的法律法规和标准。

任务1　我国食品法律法规体系

2.1.1　基础知识

1. 食品法律法规

食品法律法规指的是由国家制定的适用于食品从农田到餐桌各个环节的一整套法律规定,其中食品法律和由职能部门制定的规章是食品生产、销售企业必须强制执行的,而有些标准、规范为推荐内容。食品法律法规是国家对食品进行有效监督管理的基础。我国目前已基本形成了由国家基本法律、行政法规和部门规章构成的食品法律法规体系,我国食品法律法规框架如图2-1所示。

Ⅰ. Food Laws and Regulations

Food laws and regulations refer to a set of laws and regulations formulated by the state that apply to all aspects of food from farmland to dining table, in which food laws and regulations formulated by functional departments must be enforced by food production and marketing enterprises, while some standards and specifications are recommended. Food laws and regulations are the basis for the effective supervision and management of food. At present, China has basically formed a food laws and regulations system composed of the basic laws, administrative regulations and departmental regulations of the state. The framework of food laws and regulations in China is shown in Figure 2-1.

图 2-1　我国食品法律法规框架

1) 食品法律

法律是由全国人民代表大会和全国人民代表大会常务委员会依据特定的立法程序制定的有关的规范性法律文件。

食品法律包括《中华人民共和国食品安全法》《中华人民共和国产品质量法》《中华人民共和国农产品质量安全法》《中华人民共和国进出口商品检验法》《中华人民共和国国境卫生检疫法》《中华人民共和国动物防疫法》《中华人民共和国进出境动植物检疫法》《中华人民共和国消费者权益保护法》《中华人民共和国标准化法》等。

2) 食品行政法规

行政法规分国务院制定的行政法规和地方性行政法规两类。行政法规是对法律的补充,在完善的情况下它的法律效力仅次于法律。

食品行政法规包括《中华人民共和国食品安全法实施条例》《中华人民共和国进出口商品检验法实施条例》《乳品质量安全监督管理条例》《农业转基因生物安全管理条例》《农药管理条例》《兽药管理条例》《中华人民共和国工业产品生产许可证管理条例》《中华人民共和国认证认可条例》《饲料和饲料添加剂管理条例》《粮食流通管理条例》《国务院关于加强食品等产品安全监督管理的特别规定》等。

3) 食品部门规章

部门规章是国务院各部门、各委员会、审计署等根据法律和行政法规的规定和国务院的决定,在本部门的权限范围内制定和发布的调整本部门范围内的行政管理关系,并不得与宪法、法律和行政法规相抵触的规范性文件,主要形式是命令、指示、规定等。食品的部门规章分为国务院各行政部门制定的部门规章和地方人民政府制定的规章。

食品部门规章包括《食品安全抽样检验管理办法》《食品安全国家标准管理办法》《进出口食品安全管理办法》《出入境检验检疫报检企业管理办法》《流通环节食品安全监督管理办法》《食品召回管理办法》《新食品原料安全性审查管理办法》《食品检验机构资质认定管理办法》《产品质量监督抽查管理办法》《食品添加剂新品种管理办法》《农产品质量安全监测管理办法》《食品生产许可管理办法》《进出口乳品检验检疫监督管理办法》《餐饮服务食品安全监督管理办法》《农业转基因生物安全评价管理办法》《无公害农产品管理办法》《食品添加剂生产监督管理规定》《水产养殖质量安全管理规定》等。

4) 其他规范性文件

规范性文件是指除政府规章外,行政机关及法律、法规授权的具有管理公共事务职能的组织,在法定职权范围内依照法定程序制定并公开发布的针对不特定的多数人和特定事项,涉及或者影响公民、法人或者其他组织权利义务,在本行政区域或其管理范围内具有普遍约束力,在一定时间内相对稳定、能够反复适用的行政措施、决定、命令等行政规范文件的总称。

食品规范性文件包括《国务院关于进一步加强食品安全工作的决定》《国务院关于加强食品安全工作的决定》《国务院关于加强产品质量和食品安全工作的通知》《国务院关于地方改革完善食品药品监督管理体制的指导意见》《国务院办公厅关于进一步加强乳品质量安全工作的通知》《国务院办公厅关于印发国家食品安全监管体系"十二五"规划的通知》《国务院办公厅关于加强地沟油整治和餐厨废弃物管理的意见》《国务院办公厅关于严厉打击食品非法添加行为切实加强食品添加剂监管的通知》《国务院办公厅转发食品药品监管总局等部门

关于进一步加强婴幼儿配方乳粉质量安全工作意见的通知》等。

2. 食品法规文献检索

1) 国内食品法规的检索

(1) 检索工具。选择合适的检索工具，如《中华人民共和国食品监督管理实用法规手册》(中国食品工业协会编辑)、《中华人民共和国法规汇编》(中国法制出版社出版)等书目检索工具，利用手工检索办法从中找到有关的食品法规。

(2) 网站检索。登录国内的专业网站检索食品法规，主要有国家市场监督管理总局(http://samr.saic.gov.cn)、万方数据库(http://www.wanfangdata.com.cn)、食品伙伴网(http://www.foodmate.net)等。

2) 国外食品法规的检索

(1) 检索工具。选择合适的检索工具，如《欧洲共同体法规目录》《最新国内外食品管理制度规范与政策法规实用手册》等书目检索工具，利用手工检索办法从中找到有关的食品法规。

(2) 网站检索。登录国外的专业网站检索食品法规，主要有德国标准学会(http://www.din.de)、法国标准化协会(http://catafnor.afnor.fr)、日本工业标准调查会(http://www.jisc.go.jp)、美国国家标准系统网络(http://www.nssn.org)等。

2.1.2　实训任务

查找乳制品企业的相关法律法规与标准。

实训组织：对学生进行分组，每个组参照"基础知识"中的内容并利用网络资源，查找和列出涉及乳制品企业的法律法规(表2-1)。

表2-1　乳制品企业适用的法律法规清单

序号	法律法规	发布单位	实施日期

任务2　食品法律法规

2.2.1　基础知识

1. 学习法律

以《食品安全法》为例。

1)《食品安全法》立法的意义

现行的《食品安全法》于2015年10月1日正式施行。这部被誉为"史上最严"的食品安全法典的实施，对规范食品生产经营活动，重塑我国食品

Ⅰ. Learning Laws

Take the *Food Safety Law* as an example.

A. Significance of Legislation on *Food Safety Law*

The current *Food Safety Law* was formally implemented in October 1st 2015. The implementation of the "most strict" food safety code in history has played an active role in standardizing food

安全公信力,开启食品安全监管新阶段发挥了积极作用。民以食为天,食以安为先。食品安全问题一直是公众最关心的话题之一。然而,近年来食品安全问题时有发生,三聚氰胺、苏丹红、地沟油等每一起事件都牵动着公众的神经。面对乱象丛生的食品安全格局和执法实践中暴露出的诸多问题,《食品安全法》以食品生产经营者为"第一责任人"的角色定位更加凸显,食品安全社会共治的思路进一步得到展现,具体民事责任和刑事责任也更合乎国情、更具震慑力。

(1)保障食品安全,保证公众身体健康和生命安全。通过实施《食品安全法》,建立以食品安全标准为基础的科学管理制度,理顺食品监管体制,明确各监管部门的职责,确立食品生产经营者是保证食品安全第一责任人的义务,可以从法律制度上更好地解决我国当前食品安全工作中存在的主要问题,防控食品污染以及食品中有害因素对人体健康的危害,预防和控制食源性疾病的发生,切实保障食品安全,保证公众身体健康和生命安全。

(2)促进我国食品工业和食品贸易发展。通过实施《食品安全法》,可以更加严格地规范食品生产经营行为,促使食品生产者依照法律、法规和食品安全标准从事生产经营活动,在食品生产经营活动中重质量、重服务、

production and operation activities, weighing up our country food safety credibility, opening food safety supervision new stage. Hunger breeds discontentment, safety is the supremacy for food. Food safety has been one of the most concerned topics among the public. However, food safety problems have occurred frequently in recent years, melamine, Sudan, illegal cooking oil and other incidents have affected public nerves. Faced with a variety of problems exposed in food safety pattern and law enforcement practice, *Food Safety Law* has become more prominent with food production operators as "first responsible person". The idea of food safety governed by society has been further demonstrated. Concrete civil liability and criminal liability are more suitable for national conditions and more deterrent.

1. Ensure food safety, ensure public health and life safety. Through implementation of the *Food Safety Law* we can establish the scientific management system based on food safety standards, rationalize the food regulatory system, clear the responsibilities of regulatory authorities, establish that food producers are the obligations to ensure the first responsibility of food safety, so as to solve the main problems better in the current food safety work in China from the legal system, prevent and control the substances that harm the health of the human body, such as food pollution and the harmful factors in food, prevent and control the occurrence of foodborne diseases, ensure food safety, ensure public health and life safety.

2. Promote development of food industry and food trade in China. Through the implementation of the *Food Safety Law*, food production and business activities can be regulated more strictly, and food producers can be urged to engage in production and business activities in accordance

重自律,对社会和公众负责,以良好的质量、可靠的信誉推动食品产业规模不断扩大,继续发展,从而极大地促进我国食品行业的发展,同时可以树立重视和保障食品安全的良好国际形象,有利于推动我国对外食品贸易的发展。

(3)加强社会领域立法,完善我国食品安全法律制度。实施《食品安全法》,在法律框架内解决食品安全问题,着眼于以人为本,关注民生,切实解决人民群众最关心、最直接、最现实的利益问题,促进社会的和谐稳定,维护广大人民群众根本利益的需要。同时,《食品安全法》与《农产品质量安全法》《农业法》《动物防疫法》《产品质量法》等法律、法规相配套,有利于进一步完善我国食品安全法律制度,为我国社会主义市场经济的健康发展提供法律保障。

2)《食品安全法》的内容体系

《中华人民共和国食品安全法》共分10章154条,主要包括总则、食品安全风险监测和评估、食品安全标准、食品生产经营、食品检验、食品进出口、食品安全事故处置、监督管理、法

with laws, regulations and food safety standards. In the food production and operation activities, it attaches importance to quality, service and self-discipline. It is responsible to society and the public, and promotes the scale of the food industry to expand and continue to develop with good quality and reliable reputation. Thus, it greatly promote the development of the food industry in China, at the same time it can set up a good international image of attaching importance to and ensuring food safety, which is conducive to promoting the development of China's foreign food trade.

3. Strengthen legislation in society and perfect food safety law system in China. Through the implementation of the *Food Safety Law*, the issue of food safety should be solved within the framework of the law, with a view to putting people first, paying attention to the people's livelihood, effectively solving the most concerned, direct and realistic interests of the people, and promoting social harmony and stability, safeguarding the fundamental interests of the broad masses of the people. At the same time, the *Food Safety Law* is complementary to the following laws and regulations, such as *The Law on Quality Safety of Agricultural Products*, *The Law on Agriculture*, *The Law on Animal Immunization*, *The Law on Quality of Products*, and so on, which is conductive to the further improvement of China's food safety legal system and provide legal protection for the healthy development of the socialist market economy in China.

B. The Content System of the *Food Safety Law*

The *Food Safety Law of the People's Republic of China* is divided into 10 chapters and 154 articles, which mainly include general principles, food safety risk monitoring and assessment, food safety standards, food production

律责任、附则。

第 1 章,总则,包括第 1 条至第 13 条,共 13 条。总则是整部法律的纲领性规定,是法律的灵魂,分别为立法目的、调整范围、工作方针、生产经营者社会责任、部门及地方政府职责、评议考核制度、部门沟通配合、协会责任、宣传教育、举报、表彰与奖励等内容,明确指出食品安全工作的基本原则是预防为主、风险管理、全程控制、社会共治。

第 2 章,食品安全风险监测和评估,包括第 14 条至第 23 条,共 10 条,主要是食品安全风险监测、风险评估、风险警示、风险交流的规定和要求。

第 3 章,食品安全标准,包括第 24 条至第 32 条,共 9 条,分别规定了食品安全标准制定原则、制定内容、制定主体及程序和标准的公布、跟踪评价。

第 4 章,食品生产经营,包括第 33 条至第 83 条,共 51 条。该章占到全法条款的三分之一,分一般规定、生产经营过程控制、标签说明书和广告、特殊食品四节,主要规定的是食品生产经营者在生产经营过程中必须遵守的各项义务要求。食品生产经营者是保障食品安全最直接、最重要、最关键的因素,对食品安全负的是第一位的主体责任。现行《食品安全法》在这一部分增加了食品安全风险自查、全程追

and operation, food inspection, imported and exported food, food safety accident disposal, supervision and management, legal responsibility, and supplementary provisions.

Chapter 1, general principles, including articles 1 to 13, a total of 13 articles. The general principles are the programmatic provisions of the whole law and the soul of the law. Respectively for legislative purposes, adjustment scope, work policy, production and operator social responsibility, departments and local government responsibilities, evaluation and assessment system, departmental communication and coordination, association responsibility, publicity and education, reporting, recognition and reward, and so on. It is clearly pointed out that the basic principles of food safety work are prevention, risk management, whole process control and social co-treatment.

Chapter 2, food safety risk monitoring and assessment, including articles 14 to 23, a total of 10 articles, which are mainly the provisions and requirements of food safety risk monitoring, risk assessment, risk warning and risk communication.

Chapter 3, food safety standards, including articles 24 to 32, a total of 9 articles. It respectively stipulates the principles, contents, subjects, procedures and standards for the formulation of food safety standards as well as the publication and tracking evaluation of standards.

Chapter 4, food production and operation, including articles 33 to 83, a total of 51 articles. This chapter accounts for 1/3 of the provisions of the whole law, divided into four sections of general provisions, production and business process control, labeling instructions and advertising as well as special food. The main provisions are the food production operators in the process of production and operation must comply with the obligations of the requirements. Food producers and operators are the most direct, most important

溯、责任约谈等20多项制度,这些重要制度的创新是全面贯彻落实党中央、国务院提出的"四个最严"要求的具体体现与制度保障。

and key factors to ensure food safety, and bear the primary responsibility for food safety. In this part, the current *Food Safety Law* has added more than 20 systems such as self-inspection of food safety risks, traceability of the whole process, and responsibility interviews. The innovation of these important systems is the concrete embodiment and system guarantee of carrying out the "four strictest" requirements put forward by the Party Central Committee and the State Council in an all-round way.

第5章,食品检验,包括第84条至第90条,共7条,分别对食品检验机构、检验人的资质和职责规定,监督抽验、复检、委托检验等进行了规定。

Chapter 5, food inspection, including articles 84 to 90, a total of 7 articles, respectively stipulates on the qualifications and responsibilities of food inspection institutions and inspectors, supervision sampling, re-inspection, commissioned inspection, etc.

第6章,食品进出口,包括第91条至第101条,共11条,主要明确了进出口食品安全的监督管理部门及进出口食品的监管要求。

Chapter 6, imported and exported food, including articles 91 to 101, a total of 11 articles. It mainly clarifies the supervision and management department of imported and exported food safety and the supervision requirements of imported and exported food.

第7章,食品安全事故处置,包括第102条至第108条,共7条,主要对食品安全事故应急预案、应急处置、报告、通报,以及事故责任的调查进行了规定。

Chapter 7, food safety accident disposal, including articles 102 to 108, a total of 7 articles. It mainly stipulates on the food safety accident emergency plan, emergency disposal, reporting, notification, and accident responsibility investigation.

第8章,监督管理,包括第109条至第121条,共13条,规定了食品安全监督管理的职责内容。

Chapter 8, supervision and administration, including articles 109 to 121, for a total of 13 articles. It stipulates the content of the responsibility of food safety supervision and administration.

第9章,法律责任,包括第122条至第149条,共28条,主要规定的是食品生产经营者、政府、监管部门以及风险监测、风险评估、检验、认证等机构和人员违反本法规定所应承担的法律责任,包括行政责任、刑事责任及民事责任。

Chapter 9, legal responsibility, including articles 122 to 149, a total of 28 articles. The main provisions are the legal responsibility of the following institutions and personnel for violating the provisions of this Law, such as food production and business operators, governments, regulatory departments, risk monitoring, risk assessment,

第 10 章,附则,包括第 150 条至第 154 条,共 5 条,对《食品安全法》相关术语和实施时间进行了规定。

2. 学习行政法规

以《食品安全法实施条例》为例。

1)《食品安全法实施条例》制定的意义

2016 年修订的《食品安全法实施条例》是依据《食品安全法》制定的,是针对《食品安全法》具体实施的安排和要求,其内容是建立在《食品安全法》的内容之上的,也是对《食品安全法》细节的补充和说明,是《食品安全法》的具体细化。

2)《食品安全法实施条例》的内容体系

《食品安全法实施条例》共分 10 章 64 条,主要包括总则、食品安全风险监测和评估、食品安全标准、食品生产经营、食品检验、食品进出口、食品安全事故处置、监督管理、法律责任、附则。

第 1 章,总则,包括第 1 条至第 4 条,共 4 条,阐明了条例制定的依据,县级以上地方人民政府应当履行食品安全法规定的职责,食品生产经营者和食品安全监督管理部门的责任。

第 2 章,食品安全风险监测和评估,包括第 5 条至第 14 条,共 10 条。该章对国家食品安全风险监测计划、监测方案、监测工作以及食品安全风险评估的具体实施进行了具体说明。

第 3 章,食品安全标准,包括第 15 条至第 19 条,共 5 条。该章对各级卫生行政部门对食品安全国家标准和企业标准的工作责任进行了具体说明。

第 4 章,食品生产经营,包括第 20 条至第 33 条,共 14 条。该章对食品生产经营者依据《食品安全法》应尽的义务进行了补充说明。

第 5 章,食品检验,包括第 34 条至第 35 条,共 2 条。该章对复检的申请和费用承担进行了补充说明。

第 6 章,食品进出口,包括第 36 条至第 42 条,共 7 条。该章对进口食品的报检和放行、出入境检验检疫部门的职责以及进口的食品添加剂的标签进行了补充说明。

第 7 章,食品安全事故处置,包括第 43 条至第 46 条,共 4 条。该章对食品安全事故的应急处理与调查进行了补充说明。

第 8 章,监督管理,包括第 47 条至第 54 条,共 8 条。该章对县级以上地方人民政府对食品的监管职责,卫生行政、质量监督、工商行政管理和国家食品药品监督管理部门的监督管理工作进行了补充说明。

第 9 章,法律责任,包括第 55 条至第 61 条,共 7 条。该章补充说明了食品生产经营中违法行为的处罚规定。

第 10 章,附则,包括第 62 条至第 64 条,共 3 条,对该条例相关术语和实施时间进行了规定。

inspection, certification, etc., including administrative liability, criminal liability and civil liability.

Chapter 10, the supplementary provisions, including articles 150 to 154, a total of 5 articles. The relevant terms and time of implementation of the *Food Safety Law* are regulated.

3. 学习部门规章

以《食品生产许可管理办法》为例。

1)《食品生产许可管理办法》制定的意义

现行《食品生产许可管理办法》于2015年10月1日起实施,是《食品安全法》的配套规章,在规范企业必备生产条件、督促企业加强生产过程控制、落实食品安全主体责任,以及改善食品安全总体水平,乃至推动食品工业健康持续发展等方面都发挥了积极而重要的作用。

作为《食品安全法》的配套规章,该办法是全面贯彻新食品安全法的一项重要举措,是适应监管体制改革的必然要求。按照国务院的统一部署,各地食品安全监管职能调整和体制改革相继到位。在新的监管体制下,食品安全监管需要加强源头监管,通过实施生产许可,督促企业完善管理制度,提高环境、卫生保障能力,提升装备、设施水平,保证食品安全。通过事前把关,将不能保证质量安全的生产者淘汰出局。同时该办法从许可申请、现场核查、换发证书等多个方面体现了便民惠民的原则,解决了企业反映强烈的问题。

2)《食品生产许可管理办法》的内容体系

《食品生产许可管理办法》共分8章62条,主要包括总则,申请与受理,审查与决定,许可证管理,变更、延续、补办与注销,监督检查,法律责任以及附则。

第1章,总则,包括第1条至第9条,共9条,对立法目的、适用范围、立法原则及各级食品药品监督管理部门职责等内容作了规定。

第2章,申请与受理,包括第10条至第19条,共10条,对申请食品生产许可的条件以及受理等内容作了规定。

第3章,审查与决定,包括第20条至第26条,共7条,对审查申请时间、现场核查方法等作出规定。

第4章,许可证管理,包括第27条至第31条,共5条,对食品生产许可证的内容、编号、日常监管及保管作出规定。

第5章,变更、延续、补办与注销,包括第32条至第43条,共12条,对食品生产许可证变更、延续、补办与注销作出规定。

第6章,监督检查,包括第44条至第49条,共6条,对各级食品药品监督管理部门的监督检查管理工作等作出规定。

第7章,法律责任,包括第50条至第56条,共7条,对法律责任、处罚措施等内容作出规定。

第8章,附则,包括第57条至第62条,共6条,对该办法的适用范围和实施时间等内容进行规定。

4. 学习规范性文件

以《食品药品监管总局关于贯彻实施〈食品生产许可管理办法〉的通知》为例。

食品药品监管总局关于贯彻实施《食品生产许可管理办法》的通知

食药监食监一〔2015〕225号

2015年9月30日发布

各省、自治区、直辖市食品药品监督管理局,新疆生产建设兵团食品药品监督管理局:

新修订的《食品生产许可管理办法》(国家食品药品监督管理总局令第16号,以下简称《办法》)将于2015年10月1日正式实施。为指导地方各级食品药品监督管理部门认真贯彻执行食品(含食品添加剂,下同)生产许可制度,现就《办法》实施的有关事项通知如下。

一、关于《办法》与原有规章制度的关系问题

(一)《办法》实施后,食品生产许可申请、受理、审查、决定和证书的发放、变更、延续、补办、注销,以及食品药品监督管理部门开展食品生产许可工作监督检查等,严格按照《办法》的规定执行。

(二)在新的生产许可审查通则、细则修订出台前,原有的生产许可审查通则和细则继续有效,但是有关申请材料、许可程序、许可时限、发证检验等内容与《办法》不一致的,应当以《办法》规定为准。

二、关于食品生产许可审批权限下放

省级食品药品监督管理部门要按照国务院简政放权职能转变工作部署和要求,结合食品药品监管体制改革的工作实际,综合衡量基层监管机构、人员、许可审批和现场核查能力等方面因素,合理划分并公布省、市、县级食品生产许可管理权限。同时,要在全面考核基层监管部门食品生产许可能力建设、熟练掌握和执行《办法》的基础上,逐步下放食品生产许可审批权限,保证"放得下、接得住、管得好",实现食品生产许可审批权限的平稳移交。

三、关于"一企一证"的实施

(一)食品生产许可实行"一企一证",对具有生产场所和设备设施并取得营业执照的一个食品生产者,从事食品生产活动,仅发放一张食品生产许可证。

(二)食品生产者应当按照省级食品药品监督管理部门确定的食品生产许可管理权限,向有关食品药品监督管理部门提交生产许可、变更、延续申请。有关食品药品监督管理部门受理申请后,应当按照《办法》的规定,组织审查、作出决定。

(三)食品生产者生产多个类别食品的,应当按照省级食品药品监督管理部门确定的食品生产许可管理权限,向省、市或者县级食品药品监督管理部门一并提出申请。其中,许可事项非受理部门审批权限的,受理部门应当及时告知有相应审批权限的食品药品监督管理部门,组织联合审查,按照规定时限作出决定,由受理申请的食品药品监督管理部门根据决定颁发食品生产许可证,并在副本中注明许可生产的食品类别。

四、关于旧版食品生产许可证变更及延续

(一)已获证食品生产者于2015年10月1日前提出延续申请但未完成现场核查,且申请人声明生产条件未发生变化的,可以不再实施现场核查,经审核申请材料符合要求的,予以换发新版食品生产许可证。

(二)持有旧版食品生产许可证的生产者申请变更或者延续许可,应当向原有关许可机关提出申请,经审查符合要求的,一律换发新版食品生产许可证。持有多张旧版食品生产许可证的,按照"一企一证"的原则,可以一并申请,换发一张新版食品生产许可证;也可以分别申请,其生产的食品类别在已换发的新版食品生产许可证副本上予以变更。换发新证后,持有的原许可证予以注销。新证书副本上应当一一标注原生产许可证编号。

五、关于食品生产许可证编号标注及"QS"标志

(一)新获证及换证食品生产者,应当在食品包装或者标签上标注新的食品生产许可证编号,不再标注"QS"标志。食品生产者存有的带有"QS"标志的包装和标签,可以继续使用完为止。2018年10月1日起,食品生产者生产的食品不得再使用原包装、标签和"QS"标志。

(二)使用原包装、标签、标志的食品,在保质期内可以继续销售。

六、关于食品生产许可证编号的食品类别编码

食品生产许可证编号中食品类别编码具体为:第1位数字代表食品、食品添加剂生产许可识别码,"1"代表食品,"2"代表食品添加剂;第2、3位数字代表食品、食品添加剂类别编号,其中食品类别编号按照《办法》第十一条所列食品类别顺序依次标识。食品添加剂类别编号标识为:"01"代表食品添加剂,"02"代表食品用香精,"03"代表复配食品添加剂。

七、关于食品检验报告的核查

现场核查时,除首次申请许可或申请增加食品类别需提供试制食品检验合格报告外,不再要求食品生产者提供检验报告。试制食品检验可由生产者自行检验,或者委托有资质的食品检验机构检验。

八、关于产业政策的执行

食品生产许可申请人应当遵守国家产业政策。申请项目属于《产业结构调整指导目录》中限制类的,按照《国务院关于发布实施〈促进产业结构调整暂行规定〉的决定》(国发〔2005〕40号),不得办理相关食品生产许可手续。地方性法规、规章或者省、自治区、直辖市人民政府有关文件对贯彻执行产业政策另有规定的,还应当遵守其规定。

九、关于落实工作保障

(一)《办法》实施后,食品生产许可将不再收取审查费,为保障许可审查工作的顺利实施,地方各级食品药品监督管理部门要积极争取地方财政支持,将实施食品生产许可所需经费列入本级行政机关的预算。

(二)地方各级食品药品监督管理部门要高度重视食品生产许可工作,积极落实各项保障措施,确保食品生产许可工作落到实处。要配备证书打印、二维码赋码、档案保存等设备设施,保障工作有序开展。各省级食品药品监督管理部门可以根据本地实际情况,自行确定食品生产许可证的启用时间。

(三)地方各级食品药品监督管理部门要严格遵守《办法》规定的食品生产许可申请、审批工作程序。各省级食品药品监督管理部门要编制并公布食品生产许可法律依据、条件、程序、时限、申请书示范文本等资料,方便企业办理许可事项。

(四)要大力加强食品生产许可信息化系统建设,实现食品生产许可审批的公开化、透明化,提升审批工作效率。

十、关于加强行政许可监督

各级食品药品监督管理部门要采取有效措施,加强对实施行政许可的监督检查。要把行政许可公开、办理程序、审批时限、廉政要求等的执行情况作为重点内容进行督查。对发现违规收费、违规实施行政许可的,要坚决予以纠正。要建立健全对违法和不当的行政许可决定的申诉等制度,及时发现、纠正违法实施行政许可的行为,依法追究相关责任人员责任。

各省级食品药品监管部门应将本通知执行情况,及时上报国家食品药品监督管理总局。

食品药品监管总局

2015年9月30日

2.2.2 实训任务

食品法律法规案例分析。

实训组织:对学生进行分组,每个组参照"基础知识"的内容并利用网络资源,进行相关的食品违法案例分析。

请阐述下述案例中违反了《食品安全法》中的哪些条款,应该如何处理违规企业?

案例1 2013年1月4日,石门县工商局查实,某公司销售超过保质期的糕点。当事人于2012年12月在自己经营的超市加工房加工糕点,然后在超市进行销售。2013年1月4日,执法人员在其超市货架上发现保质期至2012年12月23日和2013年1月2日的糕点,现场查获过期糕点货值152元。至案发时止,销售额3072元。

案例2 岳阳市工商局岳阳楼分局根据举报查实,当事人陆某在未取得食品流通许可证的情况下,从2012年5月起,在某市场从事冷冻食品经营。至案发时止,当事人经营冷冻食品货值1945元,获利213元。

案例3 衡阳市工商局根据抽检结果查实,衡阳某超市销售的两种橄榄油,标签不符合食品安全标准。2012年1月至2012年12月,当事人经销的这两种橄榄油合计货值60480元,获利5437.4元。

任务3 我国食品标准体系

2.3.1 基础知识

1. 食品标准

1)标准化和标准

(1)标准化。中国国家标准GB/T 20000.1—2014《标准化工作指南 第1部分:标准化和相关活动的通用术语》中对"标准化"的定义是:"为了在既定范围内获得最佳秩序,促进共同效益,对现实问题或潜在问题确立共同使用和重复使用的条款以及编制、发布和应用文件的活动。"标准化活动确立的条款,可形成标准化文件,包括标准和其他标准化文件。标准化的主要效益在于为了产品、过程或服务的预期目的改进它们的适用性,促进贸易、交流以及技术合作。标准化可以有一个或更多特定目的,以使产品、过程或服务适合其用途。这些目的可能包括但不限于品种控制、可用性、兼容

Ⅰ. **Food Standards**

A. Standardization and Standard

1. Standardization. The definition of "standardization" in the Chinese national standard GB/T 20000.1—2014 *Guidelines for Standardization Work Part 1: Generic Terminology for Standardization and Related Activities* is that, "In order to get the best order and promote common benefits within a given range, establish the terms of common use and reuse of real or potential problems, as well as the activities for the preparation, publication and application of documents". Provisions established by standardization activities may form standardized documents, including standards and other standardization documents. The main benefits of standard-ization are to improve their applicability, to facilitate trade, exchange and technical cooperation for the intended purposes of products, processes or

性、互换性、健康、安全、环境保护、产品防护、相互理解、经济绩效、贸易。这些目的可能相互重叠。

对食品生产企业来说,标准化是组织现代化生产的重要手段,是质量管理的重要组成部分,有利于提高产品质量和生产效率。对国家来说,标准化是国家经济建设和社会发展的重要基础工作,搞好标准化工作,对于加快发展国民经济,提高劳动生产率,有效利用资源,保护环境,维护人民身体健康都有重要作用。在当前全球经济一体化的世界格局下,标准化的重要意义在于改进产品、过程和服务的实用性,防止贸易壁垒,并促进各国的科学、技术、文化的交流与合作。

(2)标准。中国国家标准 GB/T 20000.1—2014《标准化工作指南 第1部分:标准化和相关活动的通用术语》中对"标准"的定义是:"通过标准化活动,按照规定的程序经协商一致制定,为各种活动或其结果提供规则、指南或特性,供共同使用和重复使用的文件。"其中,"规定的程序"是指制定标准的机构颁布的标准制定程序。标准宜以科学、技术和经验的综合成果为基础。"协商一致"是指普遍同意,即有关重要利益相关方对于实质

services. Standardization can have one or more specific purposes for making a product, process, or service fit for its use. These objectives may include but are not limited to various control, availability, compatibility, interchangeability, health, safety, environmental protection, product protection, mutual understanding, economic performance, trade. These aims may overlap.

As a food production enterprise, standardization is an important means of organizing modern production and an important part of quality management, which is conducive to improving product quality and production efficiency. As a country, standardization is an important foundation for national economic construction and social development. Doing well the standardization work plays an important role in speeding up the development of the national economy, improving labor productivity, effectively utilizing resources, protecting the environment and safeguarding the health of the people. In the current world situation of global economic integration, the significance of standardization lies in improving the practicability of products, processes and services, preventing trade barriers, and promoting scientific, technological and cultural exchanges and cooperation among countries.

2. Standard. The definition of "standard" in the Chinese national standard GB/T 20000.1—2014 *Guidelines for Standardization Work Part 1: Generic Terminology for Standardization and Related Activities* is that, "the documents for common use and reuse through standardized activities, according to the prescribed procedures, it is formulated by consensus, providing rules, guidelines or characteristics for various activities or results", in which "the prescribed procedure" refers to the standard-setting procedure promulgated by the standard-setting body. The

性问题没有坚持反对意见,同时按照程序考虑了有关各方的观点并且协调了所有争议。协商一致并不意味着全体一致同意。

2)食品标准的作用

食品标准是食品行业的技术规范,在食品生产经营中具有极其重要的作用,具体体现在以下几个方面。

(1)保证食品的卫生质量。食品是供人食用的特殊商品,食品质量特别是卫生质量关系到消费者的生命安全。食品标准在制定过程中充分考虑到在食品生产销售过程中可能存在的和潜在有害因素,并通过一系列标准的具体内容,对这些因素进行有效的控制,从而使符合食品标准的食品都可以防止被污染有毒有害物质,保证食品的卫生质量。

(2)国家管理食品行业的依据。国家为了保证食品质量、宏观调控食品行业的产业结构和发展方向、规范稳定食品市场,就要对食品企业进行有效管理。例如对生产设施、卫生状况、产品质量进行检查等,这些检查就是以相关的食品标准为依据的。

(3)食品企业科学管理的基础。食品企业只有通过试验方法、检验规

standard should be based on the combined results of science, technology and experience. "Consensus" means that there is general agreement that important stakeholders do not insist on opposing substantive issues, taking into account the views of the parties concerned and coordinating all disputes in accordance with the procedure. Consensus does not mean the unanimous agreement of the whole.

B. The Role of Food Standards

Food standard is the technical standard of the food industry, which plays an extremely important role in the food production and operation, which is embodied in the following aspects.

1. Ensure the hygienic quality of food. Food is a special commodity for human consumption. The quality of food, especially the quality of hygiene is related to the life and safety of consumers. Food standards take fully into account the possible and potential harmful factors in the course of food production and marketing. Through the concrete contents of a series of standards, these factors can be effectively controlled, so that the food that meets the food standard can prevent the food from contaminating toxic and harmful substances and ensure the hygienic quality of the food.

2. The basis for the state administration of the food industry. In order to ensure food quality, to macro-control the industrial structure and development direction of the food industry, and to standardize and stabilize the food market, the state must carry out effective management of food enterprises, such as the inspection of production facilities, sanitary conditions and product quality, etc. These inspections are based on relevant food standards.

3. The basis of scientific management of food enterprises. Only by means of test methods,

则、操作程序、工作方法、工艺规程等各类标准,才能统一生产和工作的程序和要求,保证每项工作的质量,使有关生产、经营、管理工作走上低耗高效的轨道,使企业获得最大经济效益和社会效益。

(4)促进交流合作,推动贸易。通过标准可以在企业间、地区间或国家间传播技术信息,促进科学技术的交流与合作,加速新技术、新成果的应用和推广,并推动国际贸易的健康发展。

3)食品标准制定的依据

(1)法律依据。《食品安全法》《标准化法》等法律及有关法规是制定食品标准的法律依据。

(2)科学技术依据。食品标准是科学技术研究和生产经验总结的产物。在标准制定的过程中,应尊重科学,尊重客观规律,保证标准的真实性;应合理使用已有的科研成果,善于总结和发现与标准有关的各种技术问题;应充分利用现代科学技术条件,促进标准具有较高的先进性。

(3)有关国际组织的规定。世界贸易组织(WTO)制定的《卫生和植物卫生措施协定》(SPS)、《贸易技术壁垒协定》(TBT)是食品贸易中必须遵守的两项协定。这两项协定都明确指出,国际食品法典委员会(CAC)的标

inspection rules, operating procedures, working methods, process rules and other standards, food enterprises can unify production and work procedures and requirements, ensure the quality of each work, make production, operation and management work on the low consumption and efficient track, so that enterprises can get maximum economic and social benefits.

4. Promote exchange and cooperation, promote trade. Through standards, we can disseminate technical information among enterprises, regions or countries, promote the exchange and cooperation of science and technology, accelerate the application and promotion of new technologies and achievements, then promote the healthy development of international trade.

C. The Basis for the Formulation of Food Standards

1. Legal basis. *Food Safety Law*, *Standardization Law* and other laws and regulations are the legal basis for the formulation of food standards.

2. Scientific and technological basis. Food standard is the product of scientific and technological research and production experience summary. In the process of setting standards, we should respect science, respect objective laws, guarantee the authenticity of standards, make rational use of existing scientific research achievements, and be good at summing up and discovering all kinds of technical problems related to standards. We should make full use of the conditions of modern science and technology and promote the progressiveness of the standard.

3. The provisions of international organizations. The SPS and TBT made by the WTO are two agreements that must be observed in the food trade. Both the SPS and the TBT agreement clearly point out that the SPS standards can be used as the basis for resolving international trade

准可作为解决国际贸易争端,协调各国食品卫生标准的依据。因此,每一个世界贸易组织的成员国都必须履行世界贸易组织有关食品标准制定和实施的各项协议和规定。

4)食品标准的制定程序

标准制定是指标准制定部门对需要制定标准的项目编制计划、组织草拟、审批编号、发布的活动。它是标准化工作的任务之一,也是标准化活动的起点。

中国国家标准制定程序划分为九个阶段,即预备阶段、立项阶段、起草阶段、征求意见阶段、审查阶段、批准阶段、发布出版阶段、复审阶段、废止阶段。

(1)预备阶段。该阶段的任务是提出新工作项目建议。对将要立项的新工作项目进行研究和论证,提出新工作项目建议,包括标准草案或标准大纲(如标准的范围、结构、相互关系等)。

每项技术标准的制定,都是按一定的标准化工作计划进行的。技术委员会根据需要,对将要立项的新工作项目进行研究及必要的论证,并在此基础上提出新工作项目建议,包括技术标准草案或技术标准的大纲。如拟起草的技术标准的名称和范围,制定该技术标准的依据、目的、意义及主要工作内容,国内外相应技术标准及有关科学技术成就的简要说明,工作步骤及计划进度,工作分工,制定过程中可能出现的问题和解决措施,经费预

disputes and coordinating national food hygiene standards. Therefore, every member country of the WTO must comply with the various agreements and regulations relating to the formulation and implementation of food standards in relation to WTO.

D. The Procedures for the Formulation of Food Standards

Standard-setting refers to the activities of the standard-setting department to draw up plans for the projects that need to be formulated, to organize the drafting, examination and approval of numbers, and to issue them. It is one of the standardization work and task, and also the starting point of standardization activities.

The procedures for the establishment of Chinese national standards are divided into nine stages: preparatory phase, proposal stage, drafting phase, committee stage, examination phase, approval stage, publication stage, review stage, abolition stage.

1. Preparatory phase. Phase tasks: propose the project proposal of new work. Research and demonstrate the new work items to be approved, and propose new work items, including the standard draft or standard outline (such as the scope, structure, interrelationship of the standard, etc.).

Each technical standard is formulated according to a certain standardized work plan. As necessary, the Technical Committee studies and justifies the new work items to be established, and on this basis, they propose new work items, including the technical standard draft and the outline of technical standards, such as the name and scope of the technical standard to be drafted, the basis, purpose, significance and main work content of the technical standard, the brief description of relevant technical standards at home and abroad and related scientific and technological

算等。

(2)立项阶段。该阶段的任务是提出新工作项目。对新工作项目建议进行审查、汇总、协调、确定,然后下达计划。

主管部门对有关单位提出的新工作项目建议进行审查、汇总、协调、确定,直至列入技术标准制定计划并下达给负责起草的单位。

(3)起草阶段。该阶段的任务是提出标准草案征求意见稿。组织标准起草工作直至完成标准草案征求意见稿。

负责起草的单位接到下达的计划项目后,即应组织有关专家成立起草工作组,通过调查研究,起草技术标准草案征求意见稿。

①调查研究。各类技术资料是起草技术标准的依据,是否充分掌握有关资料,直接影响技术标准的质量。因此,必须进行广泛的调查研究,这是制定好技术标准的关键环节。主要应收集的资料有试验验证资料、与生产制造有关的资料、国内外有关标准资料。

②起草征求意见稿。经过调查研究之后,根据标准化的对象和目的,按技术标准编写要求起草技术标准草案征求意见稿,同时起草编制说明。

achievements, the work steps and schedule, the division of labor, the possible problems and solutions in the process of formulation, the budget of funds, etc.

2. Proposal stage. Phase tasks: propose new work items. Review, aggregate, coordinate, determine and plan the proposals for new work items.

The competent department shall review, aggregate, coordinate and determine the proposals for new work items put forward by the relevant units until they are included in the technical standard-setting plan and issued to the responsible drafting unit.

3. Drafting phase. Phase tasks: propose the comments of draft standard. Organize the drafting of standards until the comments of draft standard are completed.

After the unit in charge of drafting has received the planned project, it should organize the relevant experts to set up a working group to draft the technical standard draft through investigation and study.

a. Investigation and study. All kinds of technical materials are the basis for drafting technical standards, whether to fully grasp the relevant information directly affects the quality of technical standards. Therefore, it is necessary to carry out extensive investigation and study, which is a key link in the formulation of technical standards. The main data to be collected are test verification data, data related to production and manufacturing as well as the relevant standard data at home and abroad.

b. Draft for advice. After investigation and study, according to the object and purpose of standardization, according to the technical standard to prepare the draft of technical standards for comments, while drafting the preparation of

(4)征求意见阶段。该阶段的任务是提出标准草案送审稿。对标准征求意见稿征求意见,根据返回意见完成意见汇总处理表和标准草案送审稿。

征求意见应广泛。还可以对一些主要问题组织专题讨论,直接听取意见。工作组对反馈意见要认真收集整理、分析研究、归并取舍,完成意见汇总处理,对征求意见稿及编制说明进行修改,完成技术标准草案送审稿。

(5)审查阶段。该阶段的任务是提出标准草案报批稿。对标准草案送审稿组织审查(可采取会审和函审),形成会议纪要(或函审结论)和标准草案报批稿。

(6)批准阶段。该阶段的任务是提供标准出版稿。主管部门对标准草案报批稿及材料进行审核;国家标准审查部对标准草案报批稿及材料进行审查;国务院标准化行政主管部门批准、发布。

(7)出版阶段。该阶段的任务是提供标准出版物。
技术标准出版稿统一由制定的出版机构负责印刷、出版和发行。

notes.

4. Committee stage. Phase tasks: propose the draft standard for examination. Solicit comments on standard for comment, complete the opinion summary processing table and the draft standard for examination according the return of the opinions.

Solicit opinions should be widely. Organize special discussions on some major issues and listen to the opinions directly. The working group should conscientiously collect, analyze, study, merge and choose from the feedback, complete the collection and processing of the opinions, revise the draft of the solicitation of opinions and the preparation of the instructions, and complete the draft of the technical standards to be submitted to the draft for review.

5. Examination phase. Phase tasks: propose the examination and approval of the draft standard. Organize and review (may be taken by means of joint hearing and examination by letter) the examination of the draft standard, and form the meeting summary (or letter to the conclusion) and the examination and approval of the draft standard.

6. Approval stage. Phase tasks: provide the standard published manuscripts. The competent department shall examine and verify the examination and approval of the draft standard and materials; the National Standards Review Department shall review the examination and approval of the draft standard and materials; and the competent administrative department for standardization under the State Council shall approve and issue the draft standards.

7. Publication stage. Phase tasks: provide the standard publications.

The publication of technical standards shall be uniformly printed, published and distributed by

(8)复审阶段。该阶段的任务是定期复审。对实施周期达5年的标准进行复审,以确定是否确认、修改、修订或废止。

(9)废止阶段。对复审后确定为无必要存在的标准,经主管部门审核同意后发布,予以废止。

对下列情况,制定国家标准可以采用快速程序。

①对等同采用、等效采用国际标准或国外先进标准的标准制、修订项目,可直接由立项阶段进入征求意见阶段,省略起草阶段。

②对现有国家标准的修订项目或中国其他各级标准的转化项目,可直接由立项阶段进入审查阶段,省略起草阶段和征求意见阶段。

5)食品标准的主要内容

食品标准主要有食品卫生标准、食品产品标准、食品检验标准、食品包装材料和容器标准、食品添加剂标准、食品标签通用标准、食品企业卫生规范、食品工业基础及相关标准等。

(1)食品卫生标准。食品卫生标准包括食品生产车间、设备、环境、人员等生产设施的卫生标准,食品原料、产品的卫生标准等。食品卫生标准内容包括环境感官指标、理化指标和微生物指标。

the publishing agency that has been formulated.

8. Review stage. Phase tasks: periodic reviews. Review the standards for the implementation of the cycle for more than 5 years to determine whether or not to confirm, modify, revise or abolish.

9. Abolition stage. The standards which are determined to be unnecessary after reexamination shall be issued after examination and approval by the competent authorities and shall be annulled.

Rapid procedures may be adopted for the establishment of national standards in the following cases.

a. For the standard system or revision project which is equivalent to adopting international standard or adopting foreign advanced standard, it can be directly entered into the stage of soliciting opinions from the stage of project establishment and omitting the stage of drafting.

b. The revision of existing national standards or the transformation of other standards at all levels in China may be carried out directly from the stage of project establishment to the stage of examination, and omit the drafting stage and the consultation stage.

E. Main Contents of Food Standards

Food standards include food hygiene standards, food product standards, food inspection standards, food packaging materials and containers standards, food additives standards, food labeling standards, food enterprise hygiene standards, food industry basis and related standards, etc.

1. Food hygiene standards. Include the hygienic standard food raw materials and hygienic standard for products of food producing departments, equipment, environment, personnel and other production facilities. The content of food hygiene standard includes the environmental sensory index, physical and chemical index and microbial index.

（2）食品产品标准。食品产品标准内容较多，一般包括范围、引用标准、相关定义、技术要求、检验方法、检验规则、标志包装、运输和储存等。其中技术要求是标准的核心部分，主要包括原辅材料要求、感官要求、理化指标、微生物指标等。

（3）食品检验标准。食品检验标准包括适用范围、引用标准、术语、原理、设备和材料、操作步骤、结果计算等内容。

（4）食品包装材料和容器标准。食品包装材料和容器标准的内容包括卫生要求和质量要求。

（5）其他食品标准。例如食品工业基础标准、质量管理、标志包装储运、食品机械设备等。

6）食品标准的分类

（1）根据标准适用的范围，我国的食品标准分为四级：国家标准、行业标准、地方标准和企业标准。从标准的法律级别上来讲，国家标准高于行业标准，行业标准高于地方标准，地方标准高于企业标准。但从标准的内容上来讲却不一定与级别一致，一般来讲，企业标准的某些技术指标应严于地方标准、行业标准和国家标准。

（2）根据标准的性质分类，通常把标准分为基础标准、技术标准、管理标准和工作标准四大类。

2. Food product standards. There are many contents, including scope, reference standards, relevant definitions, technical requirements, inspection methods, inspection rules, marking, packaging, transportation and storage. The technical requirements are the core part of the standard, including raw and auxiliary material requirements, sensory requirements, physical and chemical indicators, microbial indicators and so on.

3. Food inspection standards. Include the scope of application, quoted standards, terminology, principles, equipment and materials, operating procedures, calculation of results and so on.

4. Food packaging materials and containers standards. Its contents include hygiene requirements and quality requirements.

5. Other food standards. For example, the basic standards of food industry, quality management, marking packaging, storage and transportation, food machinery and equipment.

F. Classification of Food Standards

1. Scope of application according to the standard. China's food standards are divided into four levels: national standards, industry standards, local standards and enterprise standards. From the legal level of the standard, the national standard is higher than the industry standard, the industry standard is higher than the local standard, and the local standard is higher than the enterprise standard. However, the content of the standard is not necessarily consistent with the level. Generally speaking, some technical indicators of enterprise standards should be more stringent than local standards, industry standards and national standards.

2. Classification according to the nature of the standard. Generally, standards are divided into four categories: basic standards, technical standards, management standards and working standards.

①基础标准是在一定范围内作为其他标准的基础并普遍使用,是具有广泛指导意义的标准。例如术语、符号、代号、代码、计量与单位标准等都是目前广泛使用的综合性基础标准。

②技术标准是指对标准化领域中需要协调统一的技术事项所制定的标准。技术标准包括基础技术标准、产品标准、工艺标准、检测标准,以及安全、卫生、环保标准等。

③管理标准是指对标准化领域中需要协调统一的管理事项所制定的标准,主要规定人们在生产活动和社会生活中的组织结构、职责权限、过程方法、程序文件以及资源分配等事宜。它是合理组织国民经济,正确处理各种生产关系,正确实现合理分配,提高生产效率和效益的依据。管理标准包括管理基础标准、技术管理标准、经济管理标准、行政管理标准、生产经营管理标准等。

④工作标准是指对工作的责任、权利、范围、质量要求、程序、效果、检查方法、考核办法所制定的标准。工作标准一般包括部门工作标准和岗位(个人)工作标准。

(3)根据法律的约束性分类,国家标准和行业标准分为强制性标准和推荐性标准。

①强制性标准是国家通过法律的形式明确要求对标准所规定的技术内

a. The basic standard is the standard which is widely used as the basis of other standards in a certain range and has wide guiding significance. Such as terminology, symbols, marks, codes, measurement and unit standards are currently widely used comprehensive basic standards.

b. The technical standards are standards for technical matters that need to be harmonized in the area of standardization. Technical standards include basic technical standards, product standards, process standards, test standards and safety, hygiene, environmental standards and so on.

c. The management standards refer to standards for management matters that need to be harmonized in the area of standardization. It mainly provides for the organizational structure, responsibility and authority, process methods, procedural documents and resource allocation of people in production activities and social life. It is the basis that rationally organizes the national economy, correctly handle all kinds of production relations, correctly realize the rational distribution, and improve the efficiency and benefit of production. Management standards include basic management standards, technical management standards, economic management standards, administrative standards, production and operation management standards and so on.

d. The working standard refers to the standards set up for the responsibilities, rights, scope, quality requirements, procedures, effects, methods of inspection, and assessment methods of the work.

3. Classification according to the binding nature of the law. National standards and industry standards are divided into mandatory standards and recommended standards.

a. The mandatory standard is that the state clearly requires the implementation of the technical

容和要求必须执行,不允许以任何理由或方式加以违反、变更,这样的标准称之为强制性标准,包括强制性的国家标准、行业标准和地方标准。对违反强制性标准的,国家将依法追究当事人的法律责任。一般保障人民身体健康、人身财产安全的标准是强制性标准。

②推荐性标准是指国家鼓励自愿采用的具有指导作用而又不宜强制执行的标准,即标准所规定的技术内容和要求具有普遍的指导作用,允许使用单位结合自己的实际情况,灵活加以选用。虽然推荐性标准本身并不要求有关各方遵守该标准,但在一定的条件下,推荐性标准可以转化成强制性标准,具有强制性标准的作用。如以下几种情况:被行政法规、规章所引用;被合同、协议所引用;被使用者声明其产品符合某项标准。

食品卫生标准属于强制性标准,因为它是食品的基础性标准,关系到人体健康和安全。食品产品标准,一部分为强制性标准,也有一部分为推荐性标准。我国加入世界贸易组织后,更多地采用国际标准或国外先进标准,食品标准的约束性也会根据具体情况进行调整。

(4)根据标准化的对象和作用分类。

content and requirements stipulated by the standard in the form of the law, and that it is not allowed to violate or modify them for any reason or in any way. Such standards are called mandatory standards. Including mandatory national standards, industry standards and local standards. For violations of mandatory standards, the State will investigate the parties' legal responsibility according to law. In general, the standard of guaranteeing the health of the people and the safety of personal property is the mandatory standard.

b. Recommended standards refer to standards that the State encourages to adopt voluntarily but is not suitable for enforcement. That is, the technical content and requirements set out in the standards have a universal guiding role and allow the use of units to combine their actual situation, choose flexibly. Although the recommended standard itself does not require the parties concerned to comply with the standard, but under certain conditions, the recommended standard can be transformed into a mandatory standard, with the role of mandatory standards. As follows: cited by administrative laws and regulations; cited by contract, agreement; declared by users that its product meets a certain standard.

Food hygiene standard is a mandatory standard because it is the basic standard of food, which is related to human health and safety. Some of the food product standards are mandatory standards, but also some are recommended standards. After China's entry into WTO, more international standards or advanced foreign standards are adopted, and the binding nature of food standards are adjusted according to specific conditions.

4. Classification according to standardized objects and roles.

①产品标准。为保证产品的适用性,对产品必须达到的某些或全部特性要求所制定的标准,包括品种、规格、技术要求、试验方法、检验规则、包装、标志、运输和储存要求等。

②方法标准。以试验、检查、分析、抽样、统计、计算、测定、作业等各种方法为对象而制定的标准。

③安全标准。以保护人和物的安全为目的制定的标准。

④卫生标准。为保护人的健康,对食品、医药及其他方面的卫生要求制定的标准。

⑤环境保护标准。为保护环境和有利于生态平衡,对大气、水体、土壤、噪声、振动、电磁波等环境质量、污染管理、监测方法及其他事项制定的标准。

a. Product standards. In order to ensure the suitability of the product, the standards for some or all of the characteristics that the product must meet, including variety, specifications, technical requirements, test methods, inspection rules, packaging, marking, transportation and storage requirements, etc.

b. Method standard. The standard established by various methods such as test, inspection, analysis, sampling, statistics, calculation, measurement, operation, etc.

c. Safety standards. A standard designed to protect the safety of people and objects.

d. Hygiene standards. Standards for the protection of human health and hygiene requirements for food, medicine, and other aspects.

e. Environmental protection standards. Standards for the protection of the environment and for the ecological balance of environmental quality, pollution management, monitoring methods and other matters such as atmosphere, water, soil, noise, vibration, electromagnetic waves, etc.

7) 标准的代号和编号

(1) 国家标准代号及编号。国家标准代号由汉字"国标"拼音首字母"GB"构成,强制性国家标准代号为"GB",推荐性国家标准代号为"GB/T"。

国家标准编号由国家标准的代号、国家标准发布的顺序号和国家标准发布的年号构成。

①强制性国家标准编号(图2-2)。

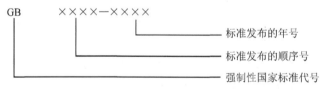

图2-2 强制性国家标准编号

例如:GB 2760—2014《食品安全国家标准 食品添加剂使用标准》。

②推荐性国家标准编号(图2-3)。

例如:GB/T 19001—2016《质量管理体系 要求》。

(2) 行业标准代号及编号。行业标准编号由行业标准代号、标准顺序号及年号组成。行业标准代号由国务院标准化行政主管部门规定,如轻工为QB,机械为JB,商业为SB。

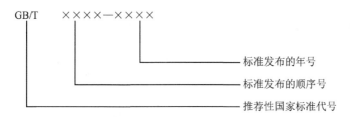

图 2-3 推荐性国家标准编号

①强制性行业标准编号(图 2-4)。

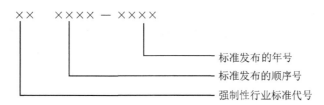

图 2-4 强制性行业标准编号

例如:QB 2394—2007《食品添加剂 乳酸链球菌素》。

②推荐性行业标准编号(图 2-5)。

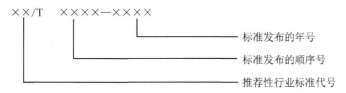

图 2-5 推荐性行业标准编号

例如:QB/T 4892—2015《冷冻调制食品检验规则》。

(3)地方标准代号及编号。地方标准代号由汉字"地标"拼音首字母"DB"加上省、自治区、直辖市行政区划代码前两位数字及斜线组成。

①强制性地方标准编号(图 2-6)。

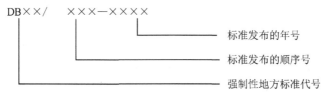

图 2-6 强制性地方标准编号

例如:DB12/356—2008《污水综合排放标准》,本标准适用于天津市辖区内的排污单位水污染物的排放管理、建设项目的环境影响评价、建设项目环境保护设施设计、竣工验收及其投产后的排放管理。

②推荐性地方标准编号(图 2-7)。

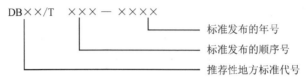

图 2-7 推荐性地方标准编号

例如:DB12/T 510—2014《地理标志产品 黄花山核桃》,本标准适用于天津市蓟州区孙各庄满族乡、下营镇两个乡镇现辖行政区域。

(4)企业标准代号及编号。企业标准代号由汉字"企"的拼音首字母"Q"加斜线和企业代号组成。企业代号由企业名称简称的四个汉语拼音第一个大写字母组成。企业标准编号由企业标准代号、顺序号、食品标准代号 S、年号组成(图 2-8)。

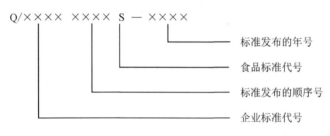

图 2-8 企业标准编号

例如:Q/NTLY 0001S—2010《速冻调理油豆腐》(南通隆源食品有限公司),南通市的企业标准。

8)食品国际标准简介

食品及相关产品标准化的国际组织有国际标准化组织(ISO)、联合国粮食和农业组织(FAO)、联合国世界卫生组织(WHO)、食品法典委员会(CAC)、国际谷类加工食品科学技术协会(ICC)、国际乳制品联合会(IDF)、国际葡萄与葡萄酒局(IWO)、国际分析化学家协会(AOAC)等。其中食品法典委员会和国际标准化组织的标准被广泛认同和采用。

(1)食品法典标准。食品法典委员会制定并向各成员国推荐的食品产品标准、农药残留限量、卫生与技术规范、准则和指南等,通称为食品法典。食品法典共由 13 卷构成,其主要内容有:卷 1A 为通用要求法典标准;卷 1B 为通用要求(食品卫生)法典标准;卷 2 为食品中农药残留法典标准;卷 3 为食品中兽药——最大残留限量法典标准;卷 4 为特殊饮食用途的食品法典标准;卷 5A 为速冻水果和蔬菜的加工处理法典标准;卷 5B 为热带新鲜水果和蔬菜法典标准;卷 6 为水果汁和相关制品法典标准;卷 7 为谷类、豆类、豆荚、相关产品、植物蛋白法典标准;卷 8 为食用油、脂肪及相关产品法典标准;卷 9 为鱼及水产品法典标准;卷 10 为肉及肉制品法典标准;卷 11 为糖、可可制品和巧克力及其他产品法典标准;卷 12 为乳及乳制品法典标准;卷 13 为分析方法与取样法典标准。

食品法典一般准则提倡成员国最大限度地采纳法典标准。法典的每一项标准本身对其成员国政府来讲并不具有自发的法律约束力,只有在成员国政府正式声明采纳之后才具有法律约束力。在食品贸易领域,一个国家只要采用了食品法典委员会的标准,就被认为是与世界贸易组织制定的《卫生和植物卫生措施协定》《贸易技术壁垒协定》的要求一致。

(2)国际标准化组织食品标准。国际标准化组织下设许多专门领域的技术委员会(TC),其中 TC34 为农产食品技术委员会。TC34 主要制定农产品各领域的产品分析方法标准。为避免重复,凡国际标准化组织制定的产品分析方法标准都被食品法典委员会直接采用。

国际标准化组织还发布了适用广泛的系列质量管理标准,其中已在食品行业普遍采用的是 ISO9000 体系。2005 年 9 月 1 日又颁布了 ISO22000 标准,该标准通过对食品链中任何组织在生产(经营)过程中可能出现的危害进行分析,确定关键控制点,将危害降低到消费者可以接受的水平。该标准是对各国现行的食品安全管理标准和法规的整合,是一个可以通用的国际标准。

2. 食品标准的检索

1)国内食品标准的检索

(1)检索工具。选择合适的检索工具,如《中华人民共和国国家标准和行业标准目录》《中华人民共和国国家标准目录》《标准化年鉴》《中国国家标准汇编》《中国标准化》《中国食品工业标准汇编》《食品卫生标准汇编》等书目检索工具,利用手工检索办法从中找到有关的食品标准。

(2)网站检索。登录国内的专业网站检索食品标准,主要有食品安全国家标准数据检索平台(http://bz.cfsa.net.cn/db)、国家市场监督管理总局(http://samr.saic.gov.cn)、国家标准化管理委员会(http://www.sac.gov.cn)、食品伙伴网(http://www.foodmate.net)、万方数据库(http://www.wanfangdata.com.cn)等。另外,中国标准出版社读者服务部、各省市自治区的标准化研究院均提供专门的标准查询检索服务,可以快速检索到需要的标准文献。

2)国外食品标准的检索

(1)检索工具。选择合适的检索工具,如《国际标准化组织标准目录》《世界卫生组织出版物目录》《美国国家标准目录》《法国国家标准目录》《英国标准年鉴》《日本工业标准目录》《日本工业标准年鉴》《德国技术规程目录》等书目检索工具,利用手工检索办法从中找到有关的食品标准。

(2)网站检索。登录国外的专业网站检索食品标准,主要有国际标准化组织(http://www.iso.org)、德国标准学会(http://www.din.de)、法国标准化协会(http://catafnor.afnor.fr)、日本工业标准调查会(http://www.jisc.go.jp)、美国国家标准系统网络(http://www.nssn.org)等。

2.3.2 实训任务

查找乳制品企业的相关标准。

实训组织:对学生进行分组,每个组参照"基础知识"的内容并利用网络资源,查找和列出涉及乳制品企业的标准(表 2-2)。

表 2-2　乳制品企业适用的标准清单

序号	标准	标准文件编号	实施日期

任务 4　食品安全标准

2.4.1　基础知识

现行食品标准覆盖了所有食品范围,基本涵盖了从原料到产品中涉及健康危害的各种卫生安全指标,包括食品产品生产加工过程中原料收购与验收、生产环境、设备设施、工艺条件、卫生管理、产品出厂前检验等各个环节的卫生要求。

The current food standards cover all kinds of food, basically covering a wide range of health and safety indicators, from raw materials to products, involving health hazards. Including raw material purchase and acceptance, production environment, equipment and facilities, technological conditions, hygiene management, pre-factory inspection and other hygiene requirements in the process of food production and processing.

《食品安全国家标准目录》(截至 2016 年 9 月)中将食品安全国家标准分为通用标准、食品产品标准、特殊膳食食品标准、食品添加剂质量规格标准、食品营养强化剂质量规格标准、食品相关产品标准、生产经营规范标准、理化检验方法标准、微生物检验方法标准、毒理学检验方法与规程标准和兽药残留检测方法标准。食品安全标准按照内容分类,可分为食品安全基础标准、生产规范、产品标准、检验检测方法等,与国际食品法典标准分类基本一致。

The National Food Safety Standard Catalog (as of September 2016) divides the national standards of food safety into general standards, food product standards, special dietary food standards, food additive quality specifications, food nutrition fortifier quality standard, food related product standard, production and operation standard, physical and chemical inspection method standard, microbial inspection method standard, methods and standards for toxicology test as well as the standard for detection of veterinary drug residues. Food safety standards can be classified into basic food safety standards, production standards, product standards, inspection methods and so on, which are basically consistent with the international classification of codex alimentarius standard.

1. 食品产品标准

食品产品标准一般包括范围、引用标准、相关定义、技术要求、检验方法、检验规则、标志包装、运输和储存等。其中技术要求是标准的核心部分,主要包括原辅材料要求、感官要求、理化指标、微生物指标等。

以 GB 7099—2015《食品安全国家标准 糕点面包》为例,该标准主要结构如下:

1 范围	3.2 感官要求
2 术语和定义	3.3 理化指标
2.1 糕点	3.4 污染物限量
2.2 面包	3.5 微生物限量
3 技术要求	3.6 食品添加剂和食品营养强化剂
3.1 原料要求	

2. 生产经营规范标准

生产经营规范标准一般包括食品生产车间、设备、环境、人员等生产设施的卫生标准,食品原料、产品的卫生标准等。

以 GB 8957—2016《食品安全国家标准 糕点、面包卫生规范》为例,该标准主要结构如下:

1 范围	7 食品原料、食品添加剂和食品相关产品
2 术语和定义	7.1 一般要求
2.1 冷加工间	7.2 食品原料
2.2 饼店(面包坊)	7.3 食品添加剂
3 选址及厂区环境	7.4 食品相关产品
3.1 选址	7.5 其他
3.2 厂区环境	8 生产过程的食品安全控制
4 厂房和车间	8.1 产品污染风险控制
4.1 设计和布局	8.2 生物污染的控制
4.2 建筑内部结构与材料	8.3 化学污染的控制
5 设施与设备	8.4 物理污染的控制
5.1 设施	8.5 包装
5.2 设备	9 检验
6 卫生管理	10 食品的贮存和运输
6.1 卫生管理制度	11 饼店(面包坊)的销售
6.2 厂房及设施卫生管理	12 产品召回管理
6.3 食品加工、经营人员健康管理与卫生要求	13 培训
	14 管理制度和人员
6.4 虫害控制	15 记录和文件管理
6.5 废弃物处理	附录 A 糕点、面包加工过程的微生物监控程序指南
6.6 工作服管理	

3. 检验方法标准

检验方法标准包括适用范围、引用标准、术语、原理、设备和材料、操作步骤、结果计算等内容。

以 GB 5009.12—2017《食品安全国家标准 食品中铅的测定》为例,该标准主要结构

如下:
1 范围
第一法 石墨炉原子吸收光谱法
2 原理
3 试剂和材料
3.1 试剂
3.2 试剂配制
3.3 标准品
3.4 标准溶液配制
4 仪器和设备
5 分析步骤
5.1 试样制备
5.2 试样前处理
5.3 测定
6 分析结果的表述
7 精密度
8 其他
第二法 电感耦合等离子体质谱法(见 GB 5009.268)
第三法 火焰原子吸收光谱法
9 原理
10 试剂和材料
10.1 试剂
10.2 试剂配制
10.3 标准品
10.4 标准溶液配制
11 仪器和设备
12 分析步骤
12.1 试样制备
12.2 试样前处理
12.3 测定
13 分析结果的表述
14 精密度
15 其他
第四法 二硫腙比色法
16 原理
17 试剂和材料
17.1 试剂
17.2 试剂配制
17.3 标准品
17.4 标准溶液配制
18 仪器和设备
19 分析步骤
19.1 试样制备
19.2 试样前处理
19.3 测定
20 分析结果的表述
21 精密度
22 其他
附录 A 微波消解升温程序
附录 B 石墨炉原子吸收光谱法仪器参考条件
附录 C 火焰原子吸收光谱法仪器参考条件

2.4.2 实训任务

为某月饼企业查询相关食品标准。

实训组织:对学生进行分组,每个组参照"基础知识"中的内容并利用网络资源为某月饼企业查询相关食品标准,然后汇报讲解。

任务5 食品安全企业标准的编写及备案

2.5.1 基础知识

对于没有食品安全国家标准和地方标准的食品,应当制定食品安全企

For foods that do not have national or local standards for food safety, enterprise standards for

业标准。企业制定严于国家标准或地方标准的食品安全企业标准，应当如实提交必要的依据和验证材料。除以上情形外，对已有食品安全国家标准或者地方标准的，或者国家另有相关规定的，不再备案相关的企业标准。

food safety shall be formulated. When an enterprise formulates enterprise standards for food safety that are stricter than national or local standards, it shall truthfully submit the necessary basis and verification materials. In addition to the above, where there is a national or local standard for food safety, or where the state has other relevant provisions, the relevant enterprise standards shall no longer be put on record.

1. 企业标准编制依据与原则

1）食品安全企业标准编制依据

编制食品安全企业标准应依据以下法律法规及标准：《中华人民共和国食品安全法》、《中华人民共和国标准化法》、《中华人民共和国标准化法实施条例》、GB/T 1.1《标准化工作导则》、国家强制性卫生标准、同类产品国家标准或行业标准或地方标准、企业标准化管理办法等。

2）制定食品安全企业标准的原则

（1）贯彻食品安全法，严格执行强制性国家标准、行业标准和地方标准。

（2）保证安全、卫生，充分考虑使用要求，保护消费者利益，保护环境。

（3）有利于企业技术进步，保证和提高产品质量，改善经营管理和增加社会经济效益。

（4）积极采用国际标准和国外先

Ⅰ. **Basis and Principles for the Preparation of Enterprise Standards**

A. Basis for the Formulation of Standards for Food Safety Enterprises

The standards for food safety enterprises shall be formulated in accordance with the following laws, regulations and standards: *The Food Safety Law of the People's Republic of China*, *The Standardization Law of the People's Republic of China*, *The Regulations on the Implementation of the Standardization Law of the People's Republic of China*, GB/T 1.1 *The Guidelines for Standardization Work*, the national mandatory health standards, the management methods of the national standard or industry standard or local standard and enterprise standardization of the same kind of products and so on.

B. Principles for Formulating Enterprise Standards for Food Safety

1. Implement the Food Safety Law and strictly enforce mandatory national standards, industry standards and local standards.

2. Ensure safety and hygiene, fully consider the requirements of use, protect the interests of consumers and protect the environment.

3. Conducive to the technological progress of enterprises, ensure and improve the quality of products, improve management and increase social and economic benefits.

4. Actively adopt national standards and foreign

进标准。

（5）有利于合理利用国家资源、能源，推广科学技术成果，有利于产品的通用互换，符合使用要求，技术先进，经济合理。

（6）有利于对外经济技术合作和对外贸易。

（7）企业内的企业标准之间应协调一致。

2. 企业标准基本内容及编制说明

1）企业标准基本内容

（1）封面。封面主要内容可分为上、中、下三部分。

封面上部内容包括标准的类别、标准的标志、标准的编号、备案号。

封面中部的内容包括标准的中文名称、标准对应的英文名称、与国际标准一致性程度的标志。

封面下部的内容包括标准的发布及实施日期、标准的发布部门或单位。

（2）前言。前言应依次包含下列信息：本标准由×××提出，本标准由×××批准，本标准由×××归口，本标准的起草单位，本标准的主要起草人，标准首次发布、历次修订或复审确认的年、月。

前言的特定部分也可给出关于标准的一些重要信息，包括标准本身的结构，标准与所采用的标准的差异，标准附录的性质以及与前一版本变化的说明等。

5. Conducive to the rational utilization of national resources and energy, the popularization of scientific and technological achievements, the universal exchange of products, in line with the requirements of use, advanced technology, and economic rationality.

6. Beneficial to foreign economic and technological cooperation and foreign trade.

7. The enterprise standards within the enterprise should be harmonized.

Ⅱ. **Basic Content and Compilation of Enterprise Standards**

A. Basic Content of Enterprise Standards

1. Cover. The main contents of the cover can be divided into three parts: upper, middle and lower.

The upper part of the cover includes the standard category, the standard logo, the standard number, the record number.

The middle part of the cover includes the Chinese name of the standard, the English name corresponding to the standard, and the symbol of the degree of consistency with the international standard.

The content of the lower part of the cover includes the publication and implementation date of the standard, the department or unit of the publication of the standard.

2. Preface. The preface shall contain the following information in turn: this standard is proposed by ×××, this standard is approved by ×××, this standard is adopted by ×××, the drafting unit of this standard, the main drafter of this standard, the year and month of the first publication, revision or review of the standard.

Some important information about the standard may also be given in the particular part of the preface, including the structure of the standard itself, the difference between the standard and the standard adopted, the nature of the appendix to

(3)范围。范围应明确表明标准的对象和所涉及的各个方面,指明标准的适用界限,必要时可说明不适用界限。

(4)规范性引用文件。引用的所有规范性文件一定要在标准中提及,没有提及的文件不应作为规范性引用文件。资料性引用文件、尚未发布过的文件或不能公开得到的文件,不能列入规范性引用文件中。

(5)术语和定义。采用国家或行业标准已规定的术语和定义,如GB/T 12140《糕点术语》、GB/T 15109《白酒工业术语》、SB/T 10295《调味品名词术语 综合》等。只有用于特定的含义或者可能引起歧义时,才有必要对术语进行定义,不能给食品品名、俗称、品牌名下定义。

(6)技术要求。技术要求的目的要明确,性能特性要量化,规定的性能特性和描述性特性要可证实,尽量引用现行相关标准。

①原料和添加剂要求。应对食品的主要原料、添加剂作出规定。食品原料和添加剂必须符合国家有关法律、法规和强制性标准的要求,确保人体健康和生命安全,不得使用违禁物质。

②感官要求。应从食品的色泽、

the standard and the description of the change from the previous version.

3. Range. The range should clearly indicate the object of the standard and the various aspects involved, indicate the limits of its application, and if necessary, its non-application should be indicated.

4. Normative references. All normative references cited must be referred to in the standard, and documents not mentioned should not be cited as normative references. Documents referenced by information, documents that have not yet been published, or documents that are not publicly available may not be included in the normative references.

5. Terms and definitions. Use terms and definitions already provided for in national or industry standards. For example, GB/T 12140 *Pastry Terms*, GB/T 15109 *White Spirit Industry Terms*, SB/T 10295 *Condiment Noun Terminology Synthesis*, etc. It is necessary to define the term only when it is used in a particular meaning or may cause ambiguity, and cannot define the food product name, common name or brand name.

6. Technical requirements. The purpose of the technical requirements should be clear, the performance characteristics should be quantified, the specified performance characteristics and descriptive characteristics should be verifiable, and the existing relevant standards should be used as far as possible.

a. Requirements for raw materials and additives. Provisions shall be made on the main raw materials and additives of food. Food raw materials and additives must meet the requirements of the relevant laws, regulations and mandatory standards of the state, ensure human health and life safety, and prohibit the use of prohibited substances.

b. Sensory requirements. The product should

组织状态、滋味与气味、质地等方面对产品提出要求。

③理化要求。应对食品的物理、化学以及污染物指标作出规定。物理指标包括净含量、固形物含量、比容、密度、粒度、杂质等。化学指标包括水分、灰分、酸度、总糖、营养素的含量，以及食品添加剂和营养强化剂的允许使用量等。污染物限量指标包括农药残留限量、有害金属和有害非金属限量、兽药残留限量等。

④微生物要求。应对食品的生物学特性和生物性污染作出规定，如活性酵母、乳酸菌、菌落总数、大肠菌群、致病菌、霉菌、生物毒素、寄生虫、虫卵等。对能定量表示的要求，应在标准中以最合理的方式规定其限值，或规定上下限，或只规定上限或下限。

⑤质量等级要求。根据质量要求能分级的食品，应作出合理分级。

（7）生产加工过程要求。生产加工过程应符合食品企业通用卫生规范。

（8）检验方法。

①一般应直接引用已发布的有关专业的标准试验方法的现行有效版本；需要制定的试验方法如与现行标准试验方法的原理、步骤基本相同，仅是个别操作步骤不同，应在引用现行标准的前提下只规定其不同部分，不宜重复制定；对于没有上级试验方法

be required from the color, organization, taste, smell and texture of the food.

c. Physical and chemical requirements. Provisions should be made for physical, chemical and pollutant indicators of food. Physical indexes include net content, solid content, specific volume, density, particle size, impurity and so on. Chemical indicators include the content of moisture, ash, acidity, total sugar, nutrient, the allowable use content of food additives and nutritional fortifier. The limits of pollutants include pesticide residue limit, harmful metal and harmful non-metal limit, the veterinary drug residue limit and so on.

d. Microbial requirements. Provisions should be made for biological characteristics and biological contamination of food. For example, active fermentation, lactobacillus, total numbers of colonies, coliform bacteria, pathogenic bacteria, mould, biotoxin, parasites, eggs, etc. The requirements that can be expressed quantitatively should be specified in the standard in the most reasonable manner, or the upper and lower limits, or set only the upper or lower limits.

e. Quality rating requirements. Food classified according to quality requirements shall be graded reasonably.

7. Requirements for the process of production and processing. The production and processing process should conform to the general hygiene standards of food enterprises.

8. Testing method.

a. In general, the current valid version of the published standard testing method for the profession concerned should be quoted directly; if the testing method that needs to be developed is basically the same as the principle and procedure of the current standard testing method, only a few operational steps are different, on the premise of

的，应明确试验原理、操作步骤和试验条件及所用的仪器设备等。

②"要求"章节中的每项要求，均应有相应的检验和试验方法。

(9)检验规则。

①抽样的主要内容应包括根据食品特点、抽样条件、抽样方法、抽样数量，易变质的产品应规定储存样品的容器及保管条件。标准中具体选择哪一种较为适合的抽样方案，应根据食品特点，参考GB/T 13393—2008《验收抽样检验导则》编制。

②检验规则的主要内容应包括检验分类、检验项目、组批规则、判定原则和复检规则。

(10)标志、包装、储存、运输。

①标志。标志是产品的"标识"，它包括标签、图形、文字和符号。产品标志应符合《中华人民共和国产品质量法》《中华人民共和国消费者权益保护法》《食品标识管理规定》等法律法规和强制性标准的规定，一般可直接引用 GB 7718—2011《食品安全国家标准 预包装食品标签通则》、GB 13432—2013《食品安全国家标准 预包装特殊膳食用食品标签》等。

quoting the existing standards, only different parts of the standards should be stipulated and should not be formulated repeatedly. For those without superior test methods, the test principles, operating procedures, test conditions and the instruments and equipment used should be clearly defined.

b. For each requirement in the section of "requirements", there shall be a corresponding test and testing method.

9. Testing rules.

a. The main contents of sampling should include according to the characteristics of food, sampling conditions, sampling methods, sampling quantity, perishable products should specify storage containers and storage conditions. According to the characteristics of food, we should refer to the GB/T 13393—2008 *Guidelines for Acceptance Sampling Inspection* to select a more suitable sampling scheme in the standard.

b. The main contents of the inspection rules should include inspection classification, inspection items, group and batch rules, judgment principles and reinspection rules.

10. Marking, packaging, storage, transportation.

a. Marking. The marking is the "logo" of a product. It includes labels, graphics, text, and symbols. Product marking shall comply with the following laws and regulations and the provisions of mandatory standards, such as the *Product Quality Law of the People's Republic of China*, *Law on the Protection of the Rights and Interests of Consumers in People's Republic of China*, *Regulations for the Management of Food Labeling*. Generally, the following criteria can be directly quoted: GB 7718—2011 *National Food Safety Standard—General Standard for the Labeling of Prepackaged Foods*, GB 13432—2013 *National Food Safety Standard—Labeling of Prepackaged*

②标签。食品标签应包括产品名称、配料表、营养素名称及含量、生产日期、保质期（安全使用期或失效日期）、生产者名称和地址、质量等级、净含量、执行标准号、许可证号、认证标志、警示说明或标志、食用方法、适用人群、功效成分、热量、有效的商品条码、商标、规格、数量等。应根据不同食品类别，按照《食品安全国家标准 预包装食品标签通则》的要求将上述内容列出来。

③包装。国家标准或行业标准中对包装环境、包装物、包装方法有规定的，应当引用现行的国家标准或行业标准；没有标准的，可以制定单独的标准，也可在一项产品标准中规定包装材料、包装形式、包装量以及对包装的试验等。食品包装材料要防止食品发生污染、损害。

④运输和储存。应根据产品的特点对储存场所、储存条件、储存方式、储存期限作出相应的规定。对运输要规定装卸方式、温度，以及运输过程中可能造成影响的其他因素。

（11）产品标准格式，符合GB/T 1.1—2009要求。

2) 编制说明

企业标准编制说明应当详细说明企业标准制定过程，以及与相关国家标准、地方标准、国际标准、国外标准的比较情况。标准比较适用下列原

Foods for Special Dietary Uses and so on.

b. Label. Food label shall include product name, ingredient list, nutrient name and content, production date, shelf life, producer name and address (safe life or expiry date), quality grade, net content, execution standard, registration number, license number, certification mark, warning note or mark, edible method, applicable crowd, effective component, heat, valid commodity bar code, trademark, specification, quantity, etc. The above contents shall be listed in accordance with the requirements of *National Food Safety Standard—General Standard for the Labeling of Prepackaged Foods* according to different food categories.

c. Packaging. Where there are provisions in national standards or industry standards on packaging environment, packaging and packaging methods, the existing national standards or industry standards shall be cited. If there are no standards, separate standards may be formulated. Packaging materials, packaging forms, packaging volumes and testing of packaging can also be specified in a product standard. Food packaging materials to prevent food pollution and damage.

d. Transportation and storage. According to the characteristics of the product, the storage place, storage conditions, storage methods, storage period shall be accordingly stipulated. The mode of loading and unloading, the temperature and other factors that may have an impact on the transport shall be specified.

11. Product standard format. Comply with the requirements of GB/T 1.1—2009.

B. Compilation Illustration

The enterprise standard compilation instructions shall specify the process of enterprise standard formulation and the comparison with relevant national standards, local standards,

则:有国家标准或者地方标准时,与国家标准或者地方标准比较;没有国家标准和地方标准时,与国际标准比较;没有国家标准、地方标准、国际标准时,与两个以上国家或者地区的标准比较。

(1)工作简况(目的和意义、工作过程等)。

(2)标准编制的原则和确定的标准主要内容(技术指标参数性能要求和试验方法等的说明)。

(3)主要试验(验证)分析、综述报告,技术经济和预期效果说明。

(4)采用国际标准及标准水平分析。

(5)与现有法律、法规、国家(行业、地方)标准的关系。

(6)重大分歧意见处理过程和依据。

(7)其他。

international standards and foreign standards. The following principles shall apply to the standards: when there is a national or local standard, it is compared with the national or local standard; when there is no national or local standard, it is compared with the international standard; and when there is no national, local or international standard, it is compared with the standards of more than two countries or regions.

1. Work profile (purpose and meaning, working process, etc.).

2. The principles of the standard preparation and the main contents of the standard (specification of technical parameters, performance requirements and test methods, etc.).

3. Main test (validation) analysis, summary report; technical economy and expected effect description.

4. Use international standards and standard level analysis.

5. Relationship with existing laws, regulations, national (industry, local) standards.

6. The process and basis for handling major differences of opinion.

7. Others.

3. 企业标准备案

食品生产企业依法制定、发布食品安全企业标准后,应当按照规定将企业标准向省级卫生计生行政部门备案,由省级卫生计生行政部门存档备查。食品企业对其制定的企业标准内容的真实性、合法性负责,并对备案后的企业标准的实施后果依法承担责任。备案的企业标准,在本企业内部适用。企业标准中凡不符合食品安全国家标准或地方标准的,一经发现,备案企业应当修订其企业标准。企业应当依据法律法规和食品安全标准要求,组织食品生产经营,确保食品安全。一般企业标准备案时需提交的材料有企业标准备案登记表、企业标准文本及电子版、企业标准编制说明和省级卫生行政部门规定的其他资料。

省级卫生行政部门收到企业标准备案材料时,应当对提交的材料是否齐全等进行核对,并根据下列情况分别作出处理:企业标准依法不需要备案的,应当即时告知当事人不需备案;提交的材料不齐全或者不符合规定要求的,应当立即或者在5个工作日内告知当事人补正;提交的材料齐全,符合规定要求的,受理其备案。省级卫生行政部门受理企业标准备案后,应当在受理之日起10个工作日内在备案登记表上标注备案号并加盖备案章。标注的备案号和加盖的备案章作为企业标准备案凭证。

天津市企业标准备案流程如图2-9所示。

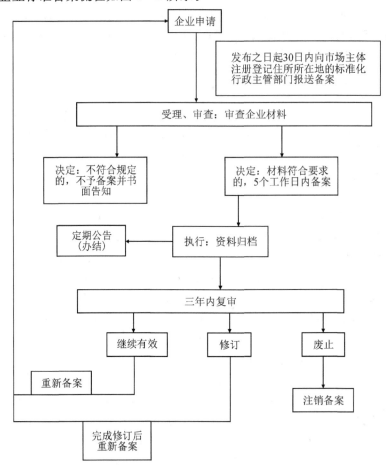

图2-9　天津市企业标准备案流程图

2.5.2　实训任务

为某月饼企业编写食品安全企业标准。

实训组织：对学生进行分组，每个组参照"基础知识"中的内容并利用网络资源，为某月饼生产企业编写食品安全企业标准（框架）。

思考题

1. 我国的食品法律法规主要有哪些？
2. 《食品安全法》的主要内容有哪些？
3. 简述《食品安全法》对食品行业的意义。
4. 食品标准的作用是什么？
5. 编写食品安全企业标准的基本内容有哪些？

拓展学习网站

1. 国家市场监督管理总局(http://samr.saic.gov.cn)
2. 食品安全标准与监测评估司(http://www.nhfpc.gov.cn/sps/new_index.shtml)
3. 天津市市场监督管理委员会(http://scjg.tj.gov.cn)

项目3 食品企业生产许可证申办

项目概述

本项目主要介绍了食品生产许可制度、食品生产许可证的申请和审查方法。

任务1 食品生产许可制度

3.1.1 基础知识

1. 食品生产许可制度

食品生产许可制度是工业产品许可证制度的一个组成部分,是为保证食品的质量安全,由国家主管食品生产领域质量监督工作的行政部门制定并实施的一项旨在控制食品生产加工企业生产条件的监控制度。

凡在中华人民共和国境内从事以销售为最终目的的食品生产加工活动的国有企业、集体企业、私营企业、三资企业,以及个体工商户、具有独立法人资格企业的分支机构和其他从事食品生产加工经营活动的每个独立生产场所,都必须申请食品生产许可证。食品生产许可实行一企一证原则,即同一个食品生产者从事食品生产活动,应当取得一个食品生产许可证。

Ⅰ. Food Production License System

The food production license system is an integral part of the licensing system for industrial products and is intended to ensure the quality and safety of food. It is a monitoring system designed and implemented by the national administrative department in charge of quality supervision in the field of food production aimed at controlling the production conditions of food production and processing enterprises.

All state-owned enterprises, collective enterprises, private enterprises and foreign-funded enterprises engaged in food production and processing activities with the ultimate aim of selling in the territory of the People's Republic of China, as well as individual industrial and commercial households, branches of enterprises with independent legal personality and every other independent production place engaged in food production and processing activities, they must apply for the Food Production License. The principle of one enterprise and one certificate shall be applied in the food production license, that is, if the same food producer engages in food production activities, he shall obtain a food production license.

没有取得食品生产许可证的企业不得生产食品,任何企业和个人不得销售无证食品。保健食品、特殊医学用途配方食品、婴幼儿配方食品的生产许可由省、自治区、直辖市食品药品监督管理部门负责。

An enterprise without a Food Production License shall not produce food, and no enterprise or individual may sell unlicensed food. The departments of food and drug supervision and administration of provinces, autonomous regions and municipalities directly under the Central Government shall be responsible for the production and licensing of health foods, formulated foods for special medical purposes and infant formula foods.

2. 食品生产许可证编号

从2015年10月起,我国开始启用新版食品生产许可证。2018年10月1日及以后生产的食品一律不得继续使用原包装和标签以及"QS"标志。食品包装袋上印制的"QS"标志(全国工业产品生产许可证),将被"SC"(食品生产许可证)替代。"SC"体现了食品生产企业在保证食品安全方面的主体地位,而监管部门则从单纯发证,变成了事前、事中、事后的持续监管。食品生产许可证编号一经确定便不再改变,以后申请许可延续及变更时,许可证书编号也不再改变。

1)编号结构

食品生产许可证编号应由"SC"("生产"的汉语拼音首字母)和14位阿拉伯数字组成。编号14个数字从左至右依次为:3位食品类别编码、2位省(自治区、直辖市)代码、2位市(地)代码、2位县(区)代码、4位顺序码、1位校验码(图3-1)。

图3-1 食品生产许可证编号

2)食品、食品添加剂类别编码

食品、食品添加剂类别编码用3位数字标识。

第1位数字代表食品、食品添加剂生产许可识别码,阿拉伯数字"1"代表食品,阿拉伯数字"2"代表食品添加剂。

第2、3位数字代表食品、食品添加剂类别编号。其中,食品类别编号按照《食品生产许可管理办法》第十一条所列食品类别顺序依次标识,即"01"代表粮食加工品,"02"代表食用油、油脂及其制品,"03"代表调味品……"27"代表保健食品,"28"代表特殊医学用途配方食品,"29"代表婴幼儿配方食品,"30"代表特殊膳食食品,"31"代表其他食品。食品添加剂类别编号标识为:"01"代表食品添加剂,"02"代表食品用香精,"03"代表复配食品添加剂。如表3-1所示。

表 3-1 食品、食品添加剂类别编码

类别编码	类别名称	类别编码	类别名称	类别编码	类别名称
101	粮食加工品	113	糖果制品	125	豆制品
102	食用油、油脂及其制品	114	茶叶及相关制品	126	蜂产品
103	调味品	115	酒类	127	保健食品
104	肉制品	116	蔬菜制品	128	特殊医学用途配方食品
105	乳制品	117	水果制品	129	婴幼儿配方食品
106	饮料	118	炒货食品及坚果制品	130	特殊膳食食品
107	方便食品	119	蛋制品	131	其他食品
108	饼干	120	可可及焙烤咖啡产品		
109	罐头	121	食糖	201	食品添加剂
110	冷冻饮品	122	水产制品	202	食品用香精
111	速冻食品	123	淀粉及淀粉制品	203	复配食品添加剂
112	薯类和膨化食品	124	糕点		

3) 省级行政区划代码

省级行政区划代码按《中华人民共和国行政区划代码》(GB/T 2260—2007)执行,按照该标准中表1"省、自治区、直辖市、特别行政区代码表"中的"数字码"的前两位数字取值,2位数字。

4) 市级行政区划代码

市级行政区划代码按《中华人民共和国行政区划代码》(GB/T 2260—2007)执行,按照该标准中表2至表32"各省、自治区、直辖市代码表"中各地市的"数字码"中间两位数字取值,2位数字。

5) 县级行政区划代码

县级行政区划代码按《中华人民共和国行政区划代码》(GB/T 2260—2007)执行,按照该标准中表2至表32"各省、自治区、直辖市代码表"中各区县的"数字码"后两位数字取值,2位数字。

6) 顺序码

许可机关按照准予许可事项的先后顺序,依次编写许可证的流水号码,一个顺序码只能对应一个生产许可证,且不得出现空号。

7) 校验码

用于检验本体码的正确性,采用 GB/T 17710—1999 中规定的"MOD11,10"校验算法,1位数字。

8) 食品生产许可证编号的赋码和使用

食品生产许可证编号应按照以下原则进行赋码和使用。

(1) 属地性。食品生产许可证编号坚持"属地编码"原则,第4位至第9位数字组合表示获证生产者的具体生产地址所在地县级行政区划代码,涉及两个及以上县级行政区划生产地址的,第8、9位代码可任选一个生产地址所在县级行政区划代码加以标识。

(2) 唯一性。食品生产许可证编号在全国范围内是唯一的,任何一个从事食品、食品添加剂生产活动的生产者只能拥有一个许可证编号,任何一个许可证编号只能赋给一个生产者。

(3) 不变性。生产者在从事食品、食品添加剂生产活动存续期间,许可证编号保持不变。

(4) 永久性。食品生产许可证注销后,该许可证编号不再赋给其他生产者。

3. 食品生产许可证证书

食品生产许可证发证日期为许可决定作出的日期,有效期为5年。食品生产许可证分为正本、副本。正本、副本具有同等法律效力。食品生产者应当妥善保管食品生产许可证,不得伪造、涂改、倒卖、出租、出借、转让。食品生产者应当在生产场所的显著位置悬挂或者摆放食品生产许可证正本。

食品生产许可证应当载明生产者名称、社会信用代码(个体生产者为身份证号码)、法定代表人(负责人)、住所、生产地址、食品类别、许可证编号、有效期、日常监督管理机构、日常监督管理人员、投诉举报电话、发证机关、签发人、发证日期和二维码。日常监督管理人员为负责对食品生产活动进行日常监督管理的工作人员。日常监督管理人员发生变化的,可以通过签章的方式在许可证上变更。副本还应当载明食品明细和外设仓库(包括自有和租赁)的具体地址。生产保健食品、特殊医学用途配方食品、婴幼儿配方食品的,还应当载明产品注册批准文号或者备案登记号;接受委托生产保健食品的,还应当载明委托企业名称及住所等相关信息。

食品生产许可证有效期内,现有工艺设备布局和工艺流程、主要生产设备设施、食品类别等事项发生变化,需要变更食品生产许可证载明的许可事项的,食品生产者应当在变化后10个工作日内向原发证的食品药品监督管理部门提出变更申请。生产场所迁出原发证的食品药品监督管理部门管辖范围的,应当重新申请食品生产许可证。食品生产许可证副本载明的同一食品类别内的事项、外设仓库地址发生变化的,食品生产者应当在变化后10个工作日内向原发证的食品药品监督管理部门报告。

3.1.2　实训任务

了解、熟悉食品生产许可证编号。

实训组织:对学生进行分组,每个组参照"基础知识",选择一个附近的超市调研一类产品包装上的 SC 编号,解析产品 SC 编号。

任务2　食品生产许可的申请

3.2.1　基础知识

1. 食品生产许可的申请条件

1)申请人

申请食品生产许可,应当先行取得营业执照等合法主体资格。企业法人、合伙企业、个人独资企业、个体工商户等,以营业执照载明的主体作为申请人。申请人应当如实向食品药品监督管理部门提交有关材料和反映真实情况,对申请材料的真实性负责,并在申请书等材料上签名或者盖章。

Ⅰ. Application Conditions for Food Production License

A. Applicant

An application for a food production license shall first obtain the legal subject qualification such as a business license. Business entity, partnership business, individual ownership, privately or individually-owned business shall use the subject specified in the business license as the applicant. The applicant shall truthfully submit the relevant materials to the food and drug administration department and reflect the true situation, be responsible for the authenticity of the application materials, and sign or seal the

2)食品类别

申请食品生产许可,应当按照以下食品类别提出:粮食加工品,食用油、油脂及其制品,调味品,肉制品,乳制品,饮料,方便食品,饼干,罐头,冷冻饮品,速冻食品,薯类和膨化食品,糖果制品,茶叶及相关制品,酒类,蔬菜制品,水果制品,炒货食品及坚果制品,蛋制品,可可及焙烤咖啡产品,食糖,水产制品,淀粉及淀粉制品,糕点,豆制品,蜂产品,保健食品,特殊医学用途配方食品,婴幼儿配方食品,特殊膳食食品,其他食品等。

3)企业条件

取得食品生产许可,应当符合食品安全标准,并符合下列要求:

(1)具有与生产的食品品种、数量相适应的食品原料处理和食品加工、包装、储存等场所,保持该场所环境整洁,并与有毒、有害场所以及其他污染源保持规定的距离;

(2)具有与申请生产许可的食品品种、数量相适应的生产设备或者设施,有相应的消毒、更衣、盥洗、采光、照明、通风、防腐、防尘、防蝇、防鼠、防虫、洗涤以及处理废水、存放垃圾和废弃物的设备或者设施;

(3)有专职或兼职的食品安全管理人员和保障食品安全的规章制度;

(4)具有合理的设备布局、工艺流

materials such as the application form.

B. Food Categories

An application for a food production license shall be made in accordance with the following categories of food: grain processing products, edible oils, oils and fats and their products, condiments, meat products, dairy products, beverages, convenience foods, biscuits, canned food, frozen drinks and frozen foods, potato and puffed food, confectionery products, tea and related products, alcohol, vegetable products, fruit products, roasted foods and nuts products, egg products, cocoa and roasted coffee products, sugar, aquatic products, starch and starch products, pastries, bean products, beehives, health foods, special medical formula foods, infant formula foods, special dietary foods, other foods and so on.

C. Conditions of the Enterprise

In obtaining a food production license, the food safety standards shall be met and the following requirements shall be met:

1. The place of food raw material handling and food processing, packing and storage suitable for the variety and quantity of the production of food shall be kept clean and tidy, and keep the prescribed distance from toxic, harmful, and other sources of pollution;

2. Production equipment or facilities with the application of production licenses which adapt to the varieties and quantities of food, there is a corresponding disinfection, washing, dressing, daylighting, lighting, ventilation, anticorrosion, dustproof, fly control, rodent control, insect control, washing and treatment of waste water, storage of garbage and waste, etc.;

3. Have full-time or part-time food safety administrators and rules and regulations to ensure food safety;

4. It has reasonable equipment layout and

程,防止待加工食品与直接入口食品、原料与成品交叉污染,避免食品接触有毒物、不洁物;

(5)法律、法规规定的其他条件。

2. 食品生产许可申请的受理

县级以上地方食品药品监督管理部门对申请人提出的食品生产许可申请,应当根据下列情况分别作出处理。

(1)申请事项依法不需要取得食品生产许可的,应当即时告知申请人不受理。

(2)申请事项依法不属于食品药品监督管理部门职权范围的,应当即时作出不予受理的决定,并告知申请人向有关行政机关申请。

(3)申请材料存在可以当场更正的错误的,应当允许申请人当场更正,由申请人在更正处签名或者盖章,注明更正日期。

(4)申请材料不齐全或者不符合法定形式的,应当当场或者在 5 个工作日内一次告知申请人需要补正的全部内容。当场告知的,应当将申请材料退回申请人;在 5 个工作日内告知的,应当收取申请材料并出具收到申请材料的凭据。逾期不告知的,自收到申请材料之日起即为受理。

technological process to prevent cross-contamination between processed food and directly imported food, raw materials and finished products, and to avoid contact with toxic and unclean food;

5. Other conditions prescribed by laws and regulations.

Ⅱ. Acceptance of Application of Food Production License

The food and drug supervision and administration departments at or above the county level shall apply for the application of food production licenses to the applicant, which shall be dealt with according to the following conditions.

1. Where the application items do not require a food production license according to law, the applicant shall be informed immediately that the application is not accepted.

2. Where the application is not within the competence of the food and drug administration department according to law, a decision of inadmissibility shall be made immediately and the applicant shall be informed of the application to the relevant administrative mechanism.

3. Where there is an error in the application material that can be corrected on the spot, the applicant shall be allowed to correct it on the spot, and the applicant shall sign or seal the correction with the date of correction.

4. If the application materials are not complete or do not conform to the legal form, the applicant shall be informed on the spot or within five working days of all the contents that need to be corrected. If the applicant is informed on the spot, the application material shall be returned to the applicant; if the application is notified within five working days, the application material shall be received and the certificate of receipt of the application material shall be issued. Where no information is given within the time limit, the

（5）申请材料齐全、符合法定形式，或者申请人按照要求提交全部补正材料的，应当受理食品生产许可申请。

县级以上地方食品药品监督管理部门对申请人提出的申请决定予以受理的，应当出具受理通知书；决定不予受理的，应当出具不予受理通知书，说明不予受理的理由，并告知申请人依法享有申请行政复议或者提起行政诉讼的权利。

application shall be accepted as of the date of receipt of the application materials.

5. If the application materials are complete, in accordance with the statutory form, or if the applicant submits all the supplementary and correction materials in accordance with the requirements, the application for food production license shall be accepted.

Where a food and drug administration department at or above the county level accepts an application decision submitted by the applicant, it shall issue an acceptance notice; if it decides not to accept the application, it shall issue an inadmissibility notice and explain the reasons for rejection, and inform the applicant of the right to apply for administrative reconsideration or to bring an administrative suit in accordance with the law.

食品生产许可证核发、变更（委托下放）流程如图3-2所示。

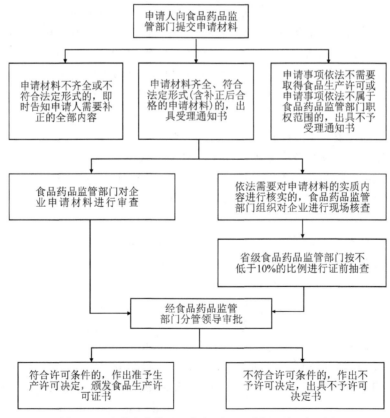

图3-2 食品生产许可证核发、变更（委托下放）流程图

3. 食品生产许可的申请材料

1)申请食品生产许可

申请食品生产许可,应当向申请人所在地县级以上地方食品药品监督管理部门提交下列材料:

(1)食品生产许可申请书;

(2)营业执照复印件(营业执照载明的经营范围应当覆盖其申请的食品许可类别);

(3)食品生产加工场所及其周围环境平面图、各功能区间布局平面图、工艺设备布局图和食品生产工艺流程图;

(4)食品生产主要设备、设施清单;

(5)进货查验记录、生产过程控制、出厂检验记录、食品安全自查、从业人员健康管理、不安全食品召回、食品安全事故处置等保证食品安全的规章制度;

(6)申请保健食品、特殊医学用途配方食品、婴幼儿配方食品的生产许可,还应当提交与所生产食品相适应的生产质量管理体系文件以及相关注册和备案文件。

申请人委托他人办理食品生产许可申请的,代理人应当提交授权委托书以及代理人的身份证明文件。

2)申请食品添加剂生产许可

申请食品添加剂生产许可,应当

Ⅲ. Application Materials for Food Production License

A. Application for Food Production License

In applying for a food production license, the following materials shall be submitted to the local food and drug regulatory authorities at or above the county level where the applicant is located:

1. An application for a food production license;

2. A photocopy of the business license (the business scope specified in the business license shall cover the category of food license it applies for);

3. The plan of the food production and processing place and its surrounding environment, the layout plan of each function interval, the layout plan of the process equipment and the flow chart of the food production process;

4. List of major equipment and facilities for food production;

5. Rules and regulations for ensuring food safety, such as inspection records of incoming goods, control of production processes, inspection record of foodstuffs exiting factory, self-inspection of food safety, health management of employees, recall of unsafe food, disposal of food safety accidents, etc.;

6. When applying for the production license of health food, special medical formula food and infant formula food, it shall also be submitted the production quality management system documents and relevant registration and filing documents in accordance with the produced food.

Where the applicant entrusts another person to apply for a food production license, the agent shall submit the power of attorney and the identity document of the agent.

B. Application for Food Additive Production License

An application for a food additive production

具备与所生产食品添加剂品种相适应的场所、生产设备或者设施、食品安全管理人员、专业技术人员和管理制度，应当向申请人所在地县级以上地方食品药品监督管理部门提交下列材料：

(1)食品添加剂生产许可申请书；

(2)营业执照复印件；

(3)食品添加剂生产加工场所及其周围环境平面图和生产加工各功能区间布局平面图；

(4)食品添加剂生产主要设备、设施清单及布局图；

(5)食品添加剂安全自查、进货查验记录、出厂检验记录等保证食品添加剂安全的规章制度。

license shall have a place, production equipment or facility, food safety administration personnel, professional and technical personnel and management system suitable for the variety of food additives produced. The following materials shall be submitted to the local food and drug regulatory authorities at or above the county level where the applicant is located：

1. An application for the production license of food additives；

2. The photo copy of business license；

3. A plan for the production and processing of food additives and their surrounding environment, and a layout plan for each functional section of production and processing；

4. The list and layout of the main equipment and facilities for the production of food additives；

5. Food additive safety self-inspection, incoming inspection records, factory inspection records and other rules and regulations to ensure the safety of food additives.

3.2.2　实训任务

填写食品生产许可申请书。

实训组织：对学生进行分组，每个组参照"基础知识"的内容并利用网络资源，填写食品生产许可申请书。

食品生产许可申请书

　　　　　□ 食品　　　　□ 食品添加剂
　　　　□ 首次　　　□ 变更　　　□延续
　　　　　　申请人名称：×××
　　　　　　　（签字或盖章）
　　　　申请日期：××××年××月××日

声　明

按照《中华人民共和国食品安全法》及《食品生产许可管理办法》要求，本申请人提出食品生产许可申请。所填写申请书及其他申请材料内容真实、有效（复印件与原件相符）。

特此声明。

<table>
<tr><td colspan="4">一、申请人基本情况</td></tr>
<tr><td>申请人名称</td><td colspan="3"></td></tr>
<tr><td>法定代表人
（负责人）</td><td colspan="3"></td></tr>
<tr><td>食品生产许可证
编号</td><td colspan="3">（变更、延续申请时填写）</td></tr>
<tr><td>营业执照注册号</td><td colspan="3"></td></tr>
<tr><td>社会信用代码
（身份证号码）</td><td colspan="3"></td></tr>
<tr><td>住　　所</td><td colspan="3"></td></tr>
<tr><td>生产地址</td><td colspan="3"></td></tr>
<tr><td>联 系 人</td><td></td><td>联系电话</td><td></td></tr>
<tr><td>传 真</td><td></td><td>电子邮件</td><td></td></tr>
<tr><td>变更事项</td><td colspan="3">（变更、延续申请时填写）</td></tr>
<tr><td>备 注</td><td colspan="3"></td></tr>
</table>

<table>
<tr><td colspan="5">二、食品安全管理及专业技术人员</td></tr>
<tr><td>序号</td><td>姓名</td><td>身份证号</td><td>职务</td><td>文化程度、专业</td></tr>
<tr><td></td><td></td><td></td><td></td><td></td></tr>
<tr><td></td><td></td><td></td><td></td><td></td></tr>
<tr><td></td><td></td><td></td><td></td><td></td></tr>
</table>

<table>
<tr><td colspan="6">三、产品信息表</td></tr>
<tr><td>序号</td><td>食品、食品添加剂类别</td><td>类别编号</td><td>类别名称</td><td>品种明细</td><td>执行标准</td></tr>
<tr><td></td><td></td><td></td><td></td><td></td><td></td></tr>
<tr><td></td><td></td><td></td><td></td><td></td><td></td></tr>
<tr><td></td><td></td><td></td><td></td><td></td><td></td></tr>
</table>

注：具体填写参照食品分类目录表。

<table>
<tr><td colspan="3">四、食品生产加工场所信息表</td></tr>
<tr><td>序号</td><td>生产场所、工艺、工序名称</td><td>生产场所、工艺、工序所在地</td></tr>
<tr><td></td><td></td><td></td></tr>
<tr><td></td><td></td><td></td></tr>
<tr><td></td><td></td><td></td></tr>
</table>

注：本表所报工序必须覆盖审查细则规定的各工艺要求。

食品生产许可其他申请材料清单

根据《食品生产许可管理办法》,申请食品生产许可,申请人还需提交的材料如下:

(1)营业执照复印件;

(2)食品生产加工场所及其周围环境平面图;

(3)食品生产加工场所各功能区间布局平面图;

(4)工艺设备布局图;

(5)食品生产工艺流程图;

(6)食品生产主要设备、设施清单;

(7)保证食品安全的规章制度清单及文本;

(8)试制食品检验合格报告;

(9)保健食品、特殊医学用途配方食品、婴幼儿配方食品与所生产食品相适应的生产质量管理体系文件以及相关注册和备案文件;

(10)其他材料。

注:上述材料中第(8)项材料在现场核查时提供。

食品生产主要设备、设施清单

设备、设施			
序号	名称	规格/型号/生产厂家	数量
检验仪器			
序号	检验仪器名称	精度等级/生产厂家	数量

食品安全管理制度清单

序号	管理制度名称	文本编号
1	进货查验记录管理制度	
2	生产过程控制管理制度	
3	出厂检验记录管理制度	
4	食品安全自查管理制度	
5	从业人员健康管理制度	
6	不安全食品召回管理制度	
7	食品安全事故处置管理制度	
8	其他制度	

任务3 食品生产许可的审查

3.3.1 基础知识

2015年8月31日,国家食品药品监督管理总局发布了《食品生产许可管理办法》(国家食品药品监督管理总局令第16号),以下简称《许可办法》,并于2015年10月1日起施行。《许可办法》中明确规定了国家食品药品监督管理总局负责制定食品生产许可审查通则和细则,作为《许可办法》的配套技术文件,用以指导食品生产许可审查工作。

1. 《食品生产许可审查通则》

为加强食品生产许可管理,规范食品生产许可审查工作,依据《中华人民共和国食品安全法》及其实施条例、《食品生产许可管理办法》等有关法律、法规、规章和食品安全国家标准,国家食品药品监督管理总局组织制定了《食品生产许可审查通则》(以下简称《通则》),自2016年10月1日起施行。

1)《通则》的适用范围

《通则》适用于食品药品监管部门对申请人的食品(含保健食品、特殊医学用途配方食品、婴幼儿配方食品)、食品添加剂生产许可申请以及许可的

On August 31, 2015, the State Food and Drug Administration issued the *Regulations on the Management of Food Production License* (the sixteenth order of the State Food and Drug Administration), hereinafter referred to as the *Licensing Method*, and it came into effect on October 1, 2015. The *Licensing Method* clearly stipulates that the State Food and Drug Administration is responsible for formulating the general rules and detailed rules for the examination of food production license, as a supporting technical document of the *Licensing Method*, which is used to guide the examination of food production license.

Ⅰ. *General Principles for the Examination of Food Production License*

In order to strengthen the administration of food production license and standardize the examination of food production licenses, in accordance with the relevant laws and regulations, rules and national standards for food safety of the *Food Safety Law of the People's Republic of China* and its implementing regulations, as well as the *Measures for the Administration of Food Production License*, the State Food and Drug Administration has organized and formulated the *General Principles for the Examination of Food Production License* (hereinafter referred to as the *General Principles*), which shall come into effect on October 1, 2016.

A. Scope of Application of the *General Principles*

The *General Principles* are applicable to the examination of the food and drug supervision and administration departments, such as the application for the applicant's food (including

变更、延续等审查工作,包括申请材料审查和现场核查。《通则》应结合相关审查细则开展审查。地方特色食品依据生产许可审查细则开展审查的,审查细则应符合《许可办法》的规定。保健食品、特殊医学用途配方食品和婴幼儿配方食品,以及另有法律、法规、规章规定的,应从其规定。本通则不适用于食品生产加工小作坊,其审查依照各省、自治区、直辖市的相关规定执行。

2)材料审查

食品药品监督管理部门应当对申请人提交的申请材料进行审查。

(1)审查部门应当对申请人提交的申请材料的完整性、规范性进行审查。

(2)审查部门应当对申请人提交的申请材料的种类、数量、内容、填写方式以及复印材料与原件的符合性等方面进行审查。

(3)申请材料均须由申请人的法定代表人或负责人签名,并加盖申请人公章。复印件应当由申请人注明"与原件一致",并加盖申请人公章。

(4)食品生产许可申请书应当使

health food, formula foods for special medical purposes, formula foods for infants and young children) and food additive production license, as well as the modification and renewal of the permit, including the examination of the application materials and on-site verification. The *General Principles* shall be reviewed in conjunction with the relevant review rules. If a local characteristic food is examined in accordance with the rules for examination of production licenses, the rules shall conform to the provisions of the *Licensing Method*. Health foods, formulated foods for special medical purposes, formula foods for infants and young children, as well as other laws, regulations and rules, shall be subject to such provisions. These principles are not applicable to small workshops for food production and processing, and their examination shall be carried out in accordance with the relevant provisions of the provinces, autonomous regions and municipalities directly under the Central Government.

B. Material Review

The food and drug supervision and administration departments shall examine the application materials submitted by the applicants.

1. The examination department shall examine the completeness and standardization of the application materials submitted by the applicant.

2. The examination department shall examine the types, quantity, contents, filling methods and conformity of the photocopied materials with the original documents submitted by the applicant.

3. The application materials must be signed by the applicant's legal representative or responsible person and stamped with the official seal of the applicant. The photocopy shall be marked "consistent with the original" by the applicant and affixed with the official seal of the applicant.

4. The application for food production license

用钢笔、签字笔填写或打印,字迹应当清晰、工整,修改处应当签名并加盖申请人公章。申请书中各项内容填写完整、规范、准确。

(5)申请人名称、法定代表人或负责人、社会信用代码或营业执照注册号、住所等填写内容应当与营业执照一致,所申请生产许可的食品类别应当在营业执照载明的经营范围内,且营业执照在有效期限内。

(6)申证产品的类别编号、类别名称及品种明细应当按照食品生产许可分类目录填写。

(7)申请材料中的食品安全管理制度设置应当完整。

(8)申请人应当配备食品安全管理人员及专业技术人员,并定期进行培训和考核。

(9)申请人及从事食品生产管理工作的食品安全管理人员应当未受到从业禁止。

(10)食品生产加工场所及其周围环境平面图、食品生产加工场所各功能区间布局平面图、工艺设备布局图、食品生产工艺流程图等图表清晰,生产场所、主要设备设施布局合理,工艺流程符合审查细则和所执行标准规定的要求。

(11)食品生产加工场所及其周围环境平面图、食品生产加工场所各功

shall be filled in or printed with a pen or roller pen, the handwriting shall be clear and neat, and the place of amendment shall sign and affix the official seal of the applicant. The contents of the application shall be complete, standardized and accurate.

5. The name of the applicant, the legal representative or person in charge, the social credit code or business license registration number and domicile shall be the same as the business license, and the category of food for which the production license is applied for shall be within the scope of business specified in the business license, and the business license is within the valid period.

6. The category number, category name and various details of the products applying for the license shall be completed in accordance with the classification list of the food production license.

7. The provisions of the food safety administration system in the application materials shall be complete.

8. The applicant shall be provided with food safety management personnel and professional and technical personnel, and shall conduct regular training and assessment.

9. The applicant and the food safety administration personnel engaged in food production administration shall not be prohibited from practicing.

10. The plans of the food production and processing sites and their surrounding environment, the layout plans of the functional sections of the food production and processing sites, the layout plans of the process equipment, the flow charts of the food production processes, etc. shall be clear, the layout of production sites and main equipment facilities is reasonable, and the process flow is in accordance with the requirements of the detailed rules for examination and the standards to be implemented.

11. The plans for food production and processing sites and their surrounding environment,

能区间布局平面图、工艺设备布局图应当按比例标注。

许可机关发现申请人存在隐瞒有关情况或者提供虚假申请材料的,应当及时依法处理。申请材料经审查,按规定不需要现场核查的,应当按规定程序由许可机关作出许可决定。许可机关决定需要现场核查的,应当组织现场核查。

3)现场核查

(1)需要现场核查的情况。食品药品监督管理部门需要对申请材料的实质内容进行核实的,应当进行现场核查。下列情形,应当组织现场核查。

①申请生产许可的,应当组织现场核查。

②申请变更的,申请人声明其生产场所发生变迁,或者现有工艺设备布局和工艺流程、主要生产设备设施、食品类别等事项发生变化的,应当对变化情况组织现场核查;其他生产条件发生变化,可能影响食品安全的,也应当就变化情况组织现场核查。

③申请延续的,申请人声明生产条件发生变化,可能影响食品安全的,应当组织对变化情况进行现场核查。

④申请变更、延续的,审查部门决

the layout plans for each functional section of food production and processing sites, and layout plans for process equipment shall be marked proportionately.

Where the licensing authority discovers that the applicant has concealed relevant information or provides false application materials, it shall promptly handle the case in accordance with the law. If the application materials are examined and there is no need for on-site verification according to the regulations, the licensing authority shall make a decision on the application in accordance with the prescribed procedures. Where the licensing authority decides that on-site verification is required, it shall organize on-site verification.

C. Site Verification

1. Situation requiring site verification. Where the food and drug supervision and administration departments needs to verify the substance of the application materials, it shall carry out site verification. In the following cases, the site verification shall be organized.

a. Where a production license is applied for, the site inspection shall be organized.

b. In the case of an application for change, the applicant shall declare that the place of production has changed, or the layout and technological process of the existing process equipment, the main production equipment and facilities, the food category has changed, and the change situation shall be organized for the site verification; where other production conditions change and may affect food safety, the site verification shall also be organized on the basis of the changes.

c. Where the applicant requires the continuation and declares that the conditions of production have changed, which may affect the safety of food, the applicant shall organize the site verification of the change.

d. Where an application for change or renewal

定需要对申请材料内容、食品类别、与相关审查细则及执行标准要求相符情况进行核实的,应当组织现场核查。

⑤申请人的生产场所迁出原发证的食品药品监督管理部门管辖范围的,应当重新申请食品生产许可,迁入地许可机关应当依照本通则的规定组织申请材料审查和现场核查。

⑥申请人食品安全信用信息记录载明监督抽检不合格、监督检查不符合、发生过食品安全事故,以及其他保障食品安全方面存在隐患的,应当组织现场核查。

⑦法律、法规和规章规定需要实施现场核查的其他情形。

（2）现场核查范围。现场核查范围主要包括生产场所、设备设施、设备布局和工艺流程、人员管理、管理制度及其执行情况,以及按规定需要查验试制产品检验合格报告。

在生产场所方面,核查申请人提交的材料是否与现场一致,其生产场所周边和厂区环境、布局和各功能区划分、厂房及生产车间相关材质等是否符合有关规定和要求。申请人在生产场所外建立或者租用外设仓库的,应当承诺符合"食品、食品添加剂生产许可现场核查评分记录表"中关于库房的

is applied for, the examination department decides that it is necessary to verify the content of the application materials, the category of food, the conformity with the relevant examination rules and the requirements of the implementation standards, it shall organize the site verification.

e. Where the applicant's place of production moves out of the jurisdiction of the food and drug administration department originally issued, the applicant shall apply for a new food production license, and the licensing authority of the place of entry shall organize the application material examination and the site verification in accordance with the provisions of the general rules.

f. The applicant's food safety credit information record indicates that the supervision and inspection is not up to standard, the supervision and inspection is not in conformity, the food safety accident has occurred, and there are hidden dangers in other aspects of ensuring food safety.

g. Other circumstances requiring the site verification as provided for by laws, regulations and rules.

2. Site verification scope. The site verification scope mainly includes the production place, the equipment facility, the equipment layout and the technological process, the personnel management, the management system and the implementation situation, as well as the inspection and the qualified report of the trial-production product according to the stipulation.

With regard to the production site, they should check whether the materials submitted by the applicant are consistent with the site, whether the surrounding and factory environment, layout and the division of each functional area, the relevant materials of the plant and the production workshop, and so on meet the relevant regulations and requirements. Where an applicant establishes

要求,并提供相关影像资料。必要时,核查组可以对外设仓库实施现场核查。

在设备设施方面,核查申请人提交的生产设备设施清单是否与现场一致,生产设备设施材质、性能等是否符合规定并满足生产需要;申请人自行对原辅料及出厂产品进行检验的,是否具备审查细则规定的检验设备设施,性能和精度是否满足检验需要。在设备布局和工艺流程方面,核查申请人提交的设备布局图和工艺流程图是否与现场一致,设备布局、工艺流程是否符合规定要求,并能防止交叉污染。

实施复配食品添加剂现场核查时,核查组应当依据有关规定,根据复配食品添加剂品种特点,核查复配食品添加剂配方组成、有害物质及致病菌是否符合食品安全国家标准。

在人员管理方面,核查申请人是否配备申请材料所列明的食品安全管理人员及专业技术人员;是否建立生产相关岗位的培训及从业人员健康管理制度;从事接触直接入口食品工作的食品生产人员是否取得健康证明。

or rents an external warehouse outside the place of production, they shall undertake to meet the requirements of the warehouse on "Food and Food Additive Production License Site Verification Scoring Record Form" and provide the relevant image materials. When necessary, the verification team may set up an external warehouse for the site verification.

With regard to equipment and facilities, they should verify whether the list of production facilities submitted by the applicant is consistent with the site, whether the material and performance of the equipment and facilities meet the requirements and meet the needs of production. If the applicant carries out the inspection of the raw and auxiliary materials and the ex-factory products on his own, does he have the inspection facilities as stipulated in the detailed rules of the examination, and whether the performance and precision meet the inspection needs. In terms of equipment layout and process flow, check whether the equipment layout and process flow chart submitted by the applicant are consistent with the site, whether the equipment layout and process flow meet the prescribed requirements, and can prevent cross contamination.

When carrying out the site verification of compound food additives, the verification team shall verify whether the formula composition, harmful substances and pathogenic bacteria of the compound food additives conform to the national standards for food safety in accordance with relevant regulations and the various characteristics of the compound food additives.

In terms of personnel management, they should check whether the applicant is equipped with food safety management personnel and professional and technical personnel listed in the application materials; whether to establish a training system for production related posts and health management system

在管理制度方面,核查申请人的进货查验记录、生产过程控制、出厂检验记录、食品安全自查、不安全食品召回、不合格品管理、食品安全事故处置及审查细则规定的其他保证食品安全的管理制度是否齐全,内容是否符合法律法规等相关规定。

在试制产品检验合格报告方面,现场核查时,核查组可以根据食品生产工艺流程等要求,按申请人生产食品所执行的食品安全标准和产品标准核查试制食品检验合格报告。

实施食品添加剂生产许可现场核查时,可以根据食品添加剂品种,按申请人生产食品添加剂所执行的食品安全标准核查试制食品添加剂检验合格报告。

试制产品检验合格报告可以由申请人自行检验,或者委托有资质的食品检验机构出具。试制产品检验报告的具体要求按审查细则的有关规定执行。

审查细则对现场核查相关内容进行细化或者有补充要求的,应当一并核查,并在"食品、食品添加剂生产许可现场核查评分记录表"中记录。申

for practitioners; whether the food producer who is engaged in the work of directly importing food has obtained a health certificate.

With regard to the management system, they should check whether the applicant's purchase inspection record, production process control, factory inspection records, food safety self-examination, unsafe food recall, unqualified product management, food safety accident disposal and review of other food safety management system are complete, whether the content is in accordance with the laws and regulations and other relevant provisions.

With regard to the report on the quality of the trial-produce products, when the site verification is carried out, according to the requirements of food production and process, the verification team can verify the test-produced food inspection report in accordance with the food safety standards and product standards implemented by the applicant's production of food.

When carrying out the the site verification of the production license for food additives, it may verify the qualified report on the testing of trial-produce food additives according to the variety of food additives and the food safety standards implemented by the applicant for the production of food additives.

The trial-produce production test qualified report may be inspected by the applicant himself or may be issued by a qualified food inspection agency. The specific requirements of the trial-produce production test qualified report shall be carried out in accordance with the relevant provisions of the detailed rules.

Where the relevant contents of the site verification are refined by the rules of examination or if there are supplementary requirements, they shall be checked together and recorded in the "Food and

请变更及延续的,申请人声明其生产条件发生变化的,审查部门应当依照本通则的规定就申请人声明的生产条件变化情况组织现场核查。

(3)现场核查结论。现场核查按照"食品、食品添加剂生产许可现场核查评分记录表"的项目得分进行判定。核查项目单项得分无0分项且总得分率≥85%的,该食品类别及品种明细判定为通过现场核查;核查项目单项得分有0分项或者总得分率＜85%的,该食品类别及品种明细判定为未通过现场核查。

因申请人下列原因导致现场核查无法正常开展的,核查组应当如实报告审查部门,本次核查按照未通过现场核查作出结论:①不配合实施现场核查的;②现场核查时生产设备设施不能正常运行的;③存在隐瞒有关情况或提供虚假申请材料的;④其他因申请人主观原因导致现场核查无法正常开展的。

因不可抗力原因,或者供电、供水等客观原因导致现场核查无法正常开展的,申请人应当向许可机关书面提出许可中止申请。中止时间应当不超过10个工作日,中止时间不计入食品

Food Additive Production License Site Verification Scoring Record Form". If the applicant changes or continues, if the applicant declares that its production conditions are changed, the examining department shall organize the site verification according to the provisions of the general rules on the change of the production conditions declared by the applicant.

3. Site verification conclusion. Site verification shall be determined in accordance with the item score of the "Food and Food Additive Production License Site Verification Scoring Record Form". If the single item score of the verification project does not have 0 points and the total score rate is equal to or above 85%, the category and various details of the food shall be judged as having passed the site verification; if the single item score of the verification item is 0 or the total score rate is less than 85%, the category and various details of the food shall be classified as not having passed the site verification.

If the applicant is unable to carry out the site verification normally due to the following reasons, the verification team shall report to the examination department truthfully. In this verification, according to the conclusion of failure to pass the site verification: a. not cooperate with the implementation of the site verification; b. the production equipment and facilities fail to function normally during the site verification; c. the existence of concealment of relevant information or the provision of false application materials; d. other cases caused by the applicant's subjective reasons can not be carried out the site verification normally.

Due to force majeure or objective reasons such as power supply and water supply, the site verification can not be carried out normally, the applicant shall submit a written application to the licensing authority for the suspension of the

生产许可审批时限。因申请人涉嫌食品安全违法且被食品药品监督管理部门立案调查的,许可机关应当中止生产许可程序,中止时间不计入食品生产许可审批时限。

(4)现场核查文件。

①"现场核查首末次会议签到表"。参加核查组首、末次会议人员应当在"现场核查首末次会议签到表"上签到。

②"食品、食品添加剂生产许可现场核查评分记录表"。核查组实施现场核查时,应当依据"食品、食品添加剂生产许可现场核查评分记录表"中所列核查项目,采取核查现场、查阅文件、核对材料及询问相关人员等方法实施现场核查。

③"食品、食品添加剂生产许可现场核查报告"。核查组对核查情况和申请人的反馈意见进行会商后,应当根据不同食品类别的现场核查情况分别进行评分判定,并汇总评分结果,形成核查结论,填写"食品、食品添加剂生产许可现场核查报告"。

④"食品、食品添加剂生产许可核查材料清单"。核查组应当自接受现场核查任务之日起10个工作日内完

application. The suspension time shall not exceed 10 working days, and the suspension time shall not be included in the time limit for the examination and approval of food production license. Where the applicant is suspected of violating the law of food safety and is placed on file for investigation by the food and drug administration department, the licensing authority shall suspend the production license procedure, and the suspension time shall not be included in the time limit for the examination and approval of the food production license.

4. Site verification document.

a. "Check-in Form for the First and Last Meeting of the Site Verification". The personnel attending the first and last meeting of the verification team shall sign on the "Check-in Form for the First and Last Meeting of the Site Verification".

b. "Food and Food Additive Production License Site Verification Scoring Record Form". When the verification team carries out the site verification, it shall carry out the site verification by means of verifying the site, consulting documents, checking materials and inquiring relevant personnel, according to the inspection items listed in the "Food and Food Additive Production License Site Verification Scoring Record Form".

c. "Site Verification Report on the Production License of Food and Food Additives". After the verification team has discussed the verification situation and the applicant's feedback, it shall make a grading judgment separately, and summarize the results of the score, so as to form a verification conclusion and fill in the "Site Verification Report on the Production License of Food and Food Additives", according to the different food types of the site inspection.

d. "List of Materials for the Verification of Food and Food Additives Production License". The verification team should complete the site

成现场核查,并将"食品、食品添加剂生产许可核查材料清单"所列的许可相关材料上报审查部门。

verification within 10 working days from the date of accepting the site verification task, and submit the related license materials in the "List of Materials for the Verification of Food and Food Additives Production License" to the censorship department.

现场核查文件如下。

附件 1　现场核查首末次会议签到表

申请人名称							
核查组	组长						
	组员						
	观察员						
首次会议	会议时间	年　月　日　时　分至　时　分					
	会议地点						
参加会议的申请人及有关人员签名							
签名	职务	签名	职务		签名		职务
末次会议	会议时间	年　月　日　时　分至　时　分					
	会议地点						
参加会议的申请人及有关人员签名							
签名	职务	签名	职务		签名		职务
备注							

附件 2　食品、食品添加剂生产许可现场核查评分记录表

申请人名称:×××
食品、食品添加剂类别及类别名称:
生产场所地址:
核查日期:××××年××月××日

	姓名(签名)	单位	职务	核查分工	核查员证书编号
核查组成员			组长		
			组员		
			组员		

使用说明

1. 本记录表依据《中华人民共和国食品安全法》《食品生产许可管理办法》等法律法规、部门规章以及相关食品安全国家标准的要求制定。

2. 本记录表应当结合相关食品生产许可审查细则要求使用。

3. 本记录表包括生产场所(24分)、设备设施(33分)、设备布局和工艺流程(9分)、人员管理(9分)、管理制度(24分)以及试制产品检验合格报告(1分)等六部分,共34个核查项目。

4. 核查组应当按照核查项目规定的"核查内容""评分标准"进行核查与评分,并将发现的问题具体详实地记录在"核查记录"栏目中。

5. 现场核查结论判定原则:核查项目单项得分无0分且总得分率≥85%的,该食品类别及品种明细判定为通过现场核查。

当出现以下两种情况之一时,该食品类别及品种明细判定为未通过现场核查:

(1)有一项及以上核查项目得0分的;

(2)核查项目总得分率＜85%的。

6. 当某个核查项目不适用时,不参与评分,并在"核查记录"栏目中说明不适用的原因。

一、生产场所(共24分)

序号	核查项目	核查内容	评分标准		核查得分	核查记录
1.1	厂区要求	1. 保持生产场所环境整洁,周围无虫害大量孳生的潜在场所,无有害废弃物以及粉尘、有害气体、放射性物质和其他扩散性污染源。各类污染源难以避开时应当有必要的防范措施,能有效清除污染源造成的影响	符合规定要求	3		
			有污染源防范措施,但个别防范措施效果不明显	1		
			无污染源防范措施,或者污染源防范措施无明显效果	0		
		2. 厂区布局合理,各功能区划分明显。生活区与生产区保持适当距离或分隔,防止交叉污染	符合规定要求	3		
			厂区布局基本合理,生活区与生产区相距较近或分隔不彻底	1		
			厂区布局不合理,或者生活区与生产区紧邻且未分隔,或者存在交叉污染	0		
		3. 厂区道路应当采用硬质材料铺设,厂区无扬尘或积水现象。厂区绿化应当与生产车间保持适当距离,植被应当定期维护,防止虫害孳生	符合规定要求	3		
			厂区环境略有不足	1		
			厂区环境不符合规定要求	0		

续表

序号	核查项目	核查内容	评分标准		核查得分	核查记录
1.2	厂房和车间	1. 应当具有与生产的产品品种、数量相适应的厂房和车间,并根据生产工艺及清洁程度的要求合理布局和划分作业区,避免交叉污染;厂房内设置的检验室应当与生产区域分隔	符合规定要求	3		
			个别作业区布局和划分不太合理	1		
			厂房面积与空间不满足生产需求,或者各作业区布局和划分不合理,或者检验室未与生产区域分隔	0		
		2. 车间保持清洁,顶棚、墙壁和地面应当采用无毒、无味、防渗透、防霉、不易破损脱落的材料建造,易于清洁;顶棚在结构上不利于冷凝水垂直滴落,裸露食品上方的管路应当有防止灰尘散落及水滴掉落的措施;门窗应当闭合严密,不透水、不变形,并有防止虫害侵入的措施	符合规定要求	3		
			车间清洁程度以及顶棚、墙壁、地面和门窗或者相关防护措施略有不足	1		
			严重不符合规定要求	0		
1.3	库房要求	1. 库房整洁,地面平整,易于维护、清洁,防止虫害侵入和藏匿。必要时库房应当设置相适应的温度、湿度控制等设施	符合规定要求	3		
			库房整洁程度或者相关设施略有不足	1		
			严重不符合规定要求	0		
		2. 原辅料、半成品、成品等物料应当依据性质的不同分设库房或分区存放。清洁剂、消毒剂、杀虫剂、润滑剂、燃料等物料应当与原辅料、半成品、成品等物料分隔放置。库房内的物料应当与墙壁、地面保持适当距离,并明确标识,防止交叉污染	符合规定要求	3		
			物料存放或标识略有不足	1		
			原辅料、半成品、成品等与清洁剂、消毒剂、杀虫剂、润滑剂、燃料等物料未分隔存放;物料无标识或标识混乱	0		
		3. 有外设仓库的,应当承诺外设仓库符合1.3.1、1.3.2条款的要求,并提供相关影像资料	符合规定要求	3		
			承诺材料或影像资料略不完整	1		
			未提交承诺材料或影像资料,或者影像资料存在严重不足	0		

二、设备设施(共 33 分)

序号	核查项目	核查内容	评分标准		核查得分	核查记录
2.1	生产设备	1. 应当配备与生产的产品品种、数量相适应的生产设备,设备的性能和精度应当满足生产加工的要求	符合规定要求	3		
			个别设备的性能和精度略有不足	1		
			生产设备不满足生产加工要求	0		
		2. 生产设备清洁卫生,直接接触食品的设备、工器具材质应当无毒、无味、抗腐蚀、不易脱落、表面光滑、无吸收性,易于清洁保养和消毒	符合规定要求	3		
			设备清洁卫生程度或者设备材质略有不足	1		
			严重不符合规定要求	0		
2.2	供排水设施	1. 食品加工用水的水质应当符合 GB 5749 的规定,有特殊要求的应当符合相应规定。食品加工用水与其他不与食品接触的用水应当以完全分离的管路输送,避免交叉污染,各管路系统应当明确标识以便区分	符合规定要求	3		
			供水管路标识略有不足	1		
			食品加工用水的水质不符合规定要求,或者供水管路无标识或标识混乱,或者供水管路存在交叉污染	0		
		2. 室内排水应当由清洁程度高的区域流向清洁程度低的区域,且有防止逆流的措施。排水系统出入口设计合理,并有防止污染和虫害侵入的措施	符合规定要求	3		
			相关防护措施略有不足	1		
			室内排水流向不符合要求,或者相关防护措施严重不足	0		
2.3	清洁消毒设施	应当配备相应的食品、工器具和设备的清洁设施,必要时配备相应的消毒设施。清洁、消毒方式应当避免对食品造成交叉污染,使用的洗涤剂、消毒剂应当符合相关规定要求	符合规定要求	3		
			清洁消毒设施略有不足	1		
			清洁消毒设施严重不足,或者清洁消毒的方式、用品不符合规定要求	0		
2.4	废弃物存放设施	应当配备设计合理、防止渗漏、易于清洁的存放废弃物的专用设施。车间内存放废弃物的设施和容器应当标识清晰,不得与盛装原辅料、半成品、成品的容器混用	符合规定要求	3		
			废弃物存放设施及标识略有不足	1		
			废弃物存放设施设计不合理,或者与盛装原辅料、半成品、成品的容器混用	0		

续表

序号	核查项目	核查内容	评分标准		核查得分	核查记录
2.5	个人卫生设施	生产场所或车间入口处应当设置更衣室,更衣室应当保证工作服与个人服装及其他物品分开放置;车间入口及车间内必要处,应当按需设置换鞋(穿戴鞋套)设施或鞋靴消毒设施;清洁作业区入口应当设置与生产加工人员数量相匹配的非手动式洗手、干手和消毒设施;卫生间不得与生产、包装或贮存等区域直接连通	符合规定要求	3		
			个人卫生设施略有不足	1		
			个人卫生设施严重不符合规范要求,或者卫生间与生产、包装、贮存等区域直接连通	0		
2.6	通风设施	应当配备适宜的通风、排气设施,避免空气从清洁程度要求低的作业区域流向清洁程度要求高的作业区域;合理设置进气口位置,必要时应当安装空气过滤净化或除尘设施。通风设施应当易于清洁、维修或更换,并能防止虫害侵入	符合规定要求	3		
			通风设施略有不足	1		
			通风设施严重不足,或者不能满足必要的空气过滤净化、除尘、防止虫害侵入的需求	0		
2.7	照明设施	厂房内应当有充足的自然采光或人工照明,光泽和亮度应能满足生产和操作需要,光源应能使物料呈现真实的颜色。在暴露食品和原辅料正上方的照明设施应当使用安全型或有防护措施的照明设施;如需要,还应当配备应急照明设施	符合规定要求	3		
			照明设施或者防护措施略有不足	1		
			照明设施或者防护措施严重不足	0		
2.8	温控设施	应当根据生产的需要,配备适宜的加热、冷却、冷冻以及用于监测温度和控制室温的设施	符合规定要求	3		
			温控设施略有不足	1		
			温控设施严重不足	0		
2.9	检验设备设施	自行检验的,应当具备与所检项目相适应的检验室和检验设备。检验室应当布局合理,检验设备的数量、性能、精度应当满足相应的检验需求	符合规定要求	3		
			检验室布局略不合理,或者检验设备性能略有不足	1		
			检验室布局不合理,或者检验设备数量、性能、精度不能满足检验需求	0		

三、设备布局和工艺流程(共 9 分)

序号	核查项目	核查内容	评分标准		核查得分	核查记录
3.1	设备布局	生产设备应当按照工艺流程有序排列,合理布局,便于清洁、消毒和维护,避免交叉污染	符合规定要求	3		
			个别设备布局不合理	1		
			设备布局存在交叉污染	0		
3.2	工艺流程	1. 应当具备合理的生产工艺流程,防止生产过程中造成交叉污染。工艺流程应当与产品执行标准相适应。执行企业标准的,应当依法备案	符合规定要求	3		
			个别工艺流程略有交叉,或者略不符合产品执行标准的规定	1		
			工艺流程存在交叉污染,或者不符合产品执行标准的规定,或者企业标准未依法备案	0		
		2. 应当制定所需的产品配方、工艺规程、作业指导书等工艺文件,明确生产过程中的食品安全关键环节。复配食品添加剂的产品配方、有害物质、致病性微生物等的控制要求应当符合食品安全标准的规定	符合规定要求	3		
			工艺文件略有不足	1		
			工艺文件严重不足,或者复配食品添加剂的相关控制要求不符合食品安全标准的规定	0		

四、人员管理(共 9 分)

序号	核查项目	核查内容	评分标准		核查得分	核查记录
4.1	人员要求	应当配备食品安全管理人员和食品安全专业技术人员,明确其职责。人员要求应当符合有关规定	符合规定要求	3		
			人员职责不太明确	1		
			相关人员配备不足,或者人员要求不符合规定	0		
4.2	人员培训	应当制订职工培训计划,开展食品安全知识及卫生培训。食品安全管理人员上岗前应当经过培训,并考核合格	符合规定要求	3		
			培训计划及计划执行略有不足	1		
			无培训计划,或者已上岗的相关人员未经培训或考核不合格	0		
4.3	人员健康管理制度	应当建立从业人员健康管理制度,明确患有国务院卫生行政部门规定的有碍食品安全疾病的或有明显皮肤损伤未愈合的人员,不得从事接触直接入口食品的工作。从事接触直接入口食品工作的食品生产人员应当每年进行健康检查,取得健康证明后方可上岗工作	符合规定要求	3		
			制度内容略有缺陷,或者个别人员未能提供健康证明	1		
			无制度,或者人员健康管理严重不足	0		

五、管理制度(共 24 分)

序号	核查项目	核查内容	评分标准		核查得分	核查记录
5.1	进货查验记录制度	应当建立进货查验记录制度,并规定采购原辅料时,应当查验供货者的许可证和产品合格证明,记录采购的原辅料名称、规格、数量、生产日期或者生产批号、保质期、进货日期,以及供货者名称、地址、联系方式等信息,保存相关记录和凭证	符合规定要求	3		
			制度内容略有不足	1		
			无制度,或者制度内容严重不足	0		
5.2	生产过程控制制度	应当建立生产过程控制制度,明确原料控制(如领料、投料等)、生产关键环节控制(如生产工序、设备管理、贮存、包装等)、检验控制(如原料检验、半成品检验、成品出厂检验等),以及运输和交付控制的相关要求	符合规定要求	3		
			个别制度内容略有不足	1		
			无制度,或者制度内容严重不足	0		
5.3	出厂检验记录制度	应当建立出厂检验记录制度,并规定食品出厂时,应当查验出厂食品的检验合格证和安全状况,记录食品的名称、规格、数量、生产日期或者生产批号、保质期、检验合格证号、销售日期,以及购货者名称、地址、联系方式等信息,保存相关记录和凭证	符合规定要求	3		
			制度内容略有不足	1		
			无制度,或者制度内容严重不足	0		
5.4	不安全食品召回制度及不合格品管理	1. 应当建立不安全食品召回制度,并规定停止生产、召回和处置不安全食品的相关要求,记录召回和通知情况	符合规定要求	3		
			制度内容有不足	1		
			无制度,或者制度内容严重不足	0		
		2. 应当规定生产过程中发现的原辅料、半成品、成品中不合格品的管理要求和处置措施	符合规定要求	3		
			管理要求和处置措施略有不足	1		
			无相关规定,或者管理要求和处置措施严重不足	0		
5.5	食品安全自查制度	应当建立食品安全自查制度,规定对食品安全状况定期进行检查评价,并根据评价结果采取相应的处理措施	符合规定要求	3		
			制度内容略有不足	1		
			无制度,或者制度内容严重不足	0		

续表

序号	核查项目	核查内容	评分标准		核查得分	核查记录
5.6	食品安全事故处置方案	应当建立食品安全事故处置方案，并规定食品安全事故处置措施及向相关食品安全监管部门和卫生行政部门报告的要求	符合规定要求	3		
			方案内容略有不足	1		
			无方案,或者方案内容严重不足	0		
5.7	其他制度	应当按照相关法律法规、食品安全标准以及审查细则规定，建立其他保障食品安全的管理制度	符合规定要求	3		
			个别制度内容略有不足	1		
			无制度,或者制度内容严重不足	0		

六、试制产品检验合格报告（共1分）

序号	核查项目	核查内容	评分标准		核查得分	核查记录
6.1	试制产品检验合格报告	应当提交符合审查细则有关要求的试制产品检验合格报告	符合规定要求	1		
			非食品安全标准规定的检验项目不全	0.5		
			无检验合格报告,或者食品安全标准规定的检验项目不全	0		

附件3　食品、食品添加剂生产许可现场核查报告

根据《食品生产许可审查通则》及生产许可审查细则,核查组于　　年　　月　　日至　　年　　月　　日对　　　　（申请人名称）进行了现场核查,结果如下。

一、现场核查结论

（一）现场核查正常开展,经综合评价,本次现场核查的结论是：

序号	食品、食品添加剂类别	类别名称	品种明细	执行标准及标准编号	核查结论
1					
2					

（二）因申请人的下列原因导致现场核查无法正常开展,本次现场核查的结论判定为未通过现场核查：

　　□ 不配合实施现场核查；
　　□ 现场核查时生产设备设施不能正常运行；
　　□ 存在隐瞒有关情况或提供虚假申请材料；
　　□ 因申请人的其他主观原因。

(三)因下列原因导致现场核查无法正常开展,中止现场核查:
□ 因不可抗力原因,或其他客观原因导致现场核查无法正常开展的;
□ 因申请人涉嫌食品安全违法且被食品药品监督管理部门立案调查的。

核查组组长签名:　　　　　　　　　　申请人意见:
核查组组员签名:
观察员签名:　　　　　　　　　　　　申请人签名(盖章):
　　　　年　　月　　日　　　　　　　　　　　年　　月　　日

二、食品、食品添加剂生产许可现场核查得分及存在的问题
食品、食品添加剂类别及类别名称:

核查项目	实际得分
生产场所	(分)
设备设施	(分)
设备布局和工艺流程	(分)
人员管理	(分)
管理制度	(分)
试制产品检验合格报告	(分)
总分:　(分);得分率:　%;单项得分为0分的共　项	
现场核查发现的问题	
核查项目序号	问题描述

核查组组长签名:　　　　　　　　　　申请人意见:
核查组组员签名:
观察员签名:　　　　　　　　　　　　申请人签名(盖章):
　　　　年　　月　　日　　　　　　　　　　　年　　月　　日

注:1. 申请人申请多个食品、食品添加剂类别的,应当按照类别分别填写本页。

2."现场核查发现的问题"应当详细描述申请人扣分情况;核查结论为"通过"的食品类别,如有整改项目,应当在报告中注明;对于核查结论为"未通过"的食品类别,应当注明否决项目;对于无法正常开展现场核查的,其具体原因应当注明。

附件4　食品、食品添加剂生产许可核查材料清单

1. 食品生产许可申请书;
2. 营业执照复印件;
3. 食品生产加工场所及其周围环境平面图;
4. 食品生产加工场所各功能区间布局平面图;
5. 工艺设备布局图;
6. 食品生产工艺流程图;

7. 食品生产主要设备、设施清单；

8. 食品安全管理制度清单；

9. 食品、食品添加剂生产许可现场核查通知书；

10. 现场核查首末次会议签到表；

11. 食品、食品添加剂生产许可现场核查评分记录表；

12. 食品、食品添加剂生产许可现场核查报告；

13. 许可机关要求提交的其他材料。

2. 食品生产许可审查细则

以《饼干生产许可证审查细则》为例。

依据：2006年2月22日《关于加强对食品中铝残留量检验的通知》（国质检食监函〔2006〕108号）；2005年9月26日《关于发布食品生产许可证审查细则修改单的通知》（国质检监函〔2005〕776号）之《饼干生产许可证审查细则》修改单（第1号）；2005年1月17日《关于印发小麦粉等15类食品生产许可证审查细则（修订）的通知》（国质检监〔2005〕15号）。

1）发证产品范围及申证单元

实施食品生产许可证管理的饼干产品包括以小麦粉、糖、油脂等为主要原料，加入疏松剂和其他辅料，按照一定工艺加工制成的各种饼干，如酥性饼干、韧性饼干、发酵饼干、薄脆饼干、曲奇饼干、夹心饼干、威化饼干、蛋圆饼干、蛋卷、粘花饼干、水泡饼干。饼干的申证单元为1个。

在生产许可证上应当注明获证产品名称即饼干。饼干生产许可证有效期为3年，其产品类别编号为0801。

2）基本生产流程及关键控制环节

(1)基本生产流程：配粉和面→成型→烘烤→包装。

(2)关键控制环节：配粉、烤制、灭菌。

(3)容易出现的质量安全问题：

①食品添加剂超范围和超量使用；

②残留物质变质、霉变等；

③水分和微生物超标。

3）必备的生产资源

(1)生产场所。饼干生产企业除必备的生产环境外，还应当有与企业生产相适应的原辅料库、生产车间、成品库。如生产线不是连续的，必须有冷却车间。生产发酵产品的企业还必须具备发酵间。

(2)必备的生产设备：机械式配粉设备如和面机；成型设备；烤炉；机械式包装机。

企业必备的生产设备中成型设备应与企业生产的品种相符合。主要成型设备有：

①酥性饼干：辊印成型机等。

②韧性饼干：叠层机、辊切成型机等。

③发酵饼干：叠层机、辊印成型机等。

④薄脆饼干：辊印或辊切成型机等。

⑤曲奇饼干：叠层机、辊印或辊切成型机等。

⑥威化饼干:制浆设施、叠层机、切割机等。
⑦蛋圆饼干:制浆设施、辊印成型机等。
⑧蛋卷:制浆设施、浇注设备、烘烤卷制成型机等。
⑨粘花饼干:辊印或辊切成型机等。
⑩水泡饼干:辊印或辊切成型机等。
⑪压缩饼干:辊印或辊切成型机、压缩机等。
生产夹心类产品的应具备夹心设备。

4）产品相关标准

GB 7100—2003《饼干 卫生标准》；QB/T 1253—2005《饼干通用技术条件》；QB/T 1254—2005《饼干试验方法》；QB/T 1433.1—2005《饼干 酥性饼干》；QB/T 1433.2—2005《饼干 韧性饼干》；QB/T 1433.3—2005《饼干 发酵饼干》；QB/T 1433.4—2005《饼干 压缩饼干》；QB/T 1433.5—2005《饼干 曲奇饼干》；QB/T 1433.6—2005《饼干 夹心饼干》；QB/T 1433.7—2005《饼干 威化饼干》；QB/T 1433.8—2005《饼干 蛋圆饼干》；QB/T 1433.9—2005《饼干 蛋卷和煎饼》；QB/T 1433.10—2005《饼干 装饰饼干》；QB/T 1433.11—2005《饼干 水泡饼干》；备案有效的企业标准。

5）原辅材料的有关要求

企业生产饼干的原辅材料必须符合国家标准、行业标准和有关规定。采购纳入生产许可证管理的原辅材料时,应当选择获得生产许可证企业生产的产品。对夹心类产品的心料等如有外购情况的,应制定进货验收制度并实施。

6）必备的出厂检验设备

分析天平(0.1 mg);干燥箱;灭菌锅;无菌室或超净工作台;微生物培养箱;生物显微镜。

7）检验项目

饼干的发证检验、监督检验和出厂检验项目按表3-2中列出的检验项目进行。出厂检验项目中注有"*"标记的,企业每年应当进行2次检验。

表3-2 饼干质量检验项目表

序号	检验项目	发证	监督	出厂	备注
1	感官	√	√	√	
2	净含量	√	√	√	
3	水分	√	√	√	
4	碱度	√	√		酥性、韧性(可可型除外)、压缩、曲奇(可可型除外)、威化(可可型除外)、蛋圆、水泡饼干、蛋卷及煎饼(不发酵产品)检此项目
5	酸度	√	√		发酵饼干、蛋卷及煎饼(发酵产品)检此项目
6	脂肪	√	√		曲奇饼干检此项目
7	酸价	√	√	*	
8	pH值	√	√	*	可可韧性、可可曲奇、可可威化饼干检此项目
9	松密度	√	√	*	压缩饼干检此项目

续表 3-2

序号	检验项目	发证	监督	出厂	备注
10	总砷	√	√	*	
11	铅	√	√	*	
12	铝	√			
13	过氧化值	√	√	*	
14	食品添加剂:甜蜜素、糖精钠	√	√	*	
15	菌落总数	√	√	√	
16	大肠菌群	√	√	√	
17	致病菌	√	√	*	
18	霉菌计数	√	√	*	
19	标签	√	√		

注：(1)夹心饼干的检验项目按照饼干单片相应品种的要求检验。

(2)装饰饼干的检验项目按照饼干单片相应品种的要求检验。

8) 抽样方法

根据企业申请发证产品的品种，随机抽取 1 种产品进行检验。如果企业生产夹心类产品，应抽取夹心类产品。

在企业的成品库内随机抽取发证检验样品。所抽样样品须为同一批次保质期内的产品，以同班次、同规格的产品为抽样基数，抽样基数不少于 200 袋（盒），随机抽样不少于 4 kg 且不少于 8 个最小包装的样品。样品分成 2 份，1 份检验，1 份备查。

样品确认无误后，由抽样人员与被抽样单位有关人员在抽样单上签字、盖章，当场封存样品，并加贴封条。封条上应当有抽样人员签名、抽样单位盖章及封样日期。如果抽取的样品为夹心类产品，在抽样单上应注明。

3.3.2 实训任务

实训组织：对学生进行分组，参照"基础知识"中的内容，每个组模拟对企业进行食品生产许可审查，并完成审查报告的编写。

思考题

1. 企业取得食品生产许可，应当符合哪些要求？
2. 申请食品生产许可，应当向申请人所在地县级以上地方食品药品监督管理部门提交哪些材料？
3. 材料审查的基本内容是什么？
4. 现场核查的范围是什么？
5. 哪些情形应当组织现场核查？
6. 食品生产许可证编号有几位数字，各代表什么含义？

拓展学习网站

1. 国家市场监督管理总局(http://samr.saic.gov.cn)
2. 天津市市场监督管理委员会(http://scjg.tj.gov.cn)

项目4　食品质量安全管理常用工具应用方法

项目概述

某乳制品企业2013年生产的巴氏乳出现了质量安全问题,请利用质量控制(QC)常用工具指导该企业进行质量安全管理。

项目导入案例

企业名称:天津乳业有限公司

公司产品:200 mL巴氏塑料袋包装乳、300 mL巴氏涂塑复合纸袋包装乳

企业人数:8人

企业设计生产能力:日处理50 t鲜奶

巴氏乳工艺流程:原料乳的验收★→预处理→标准化★→均质→巴氏杀菌★→冷却→灌装★→检验→冷藏★(标注★为关键控制点)

巴氏乳工艺步骤:

(1)原料乳的验收:按《食品安全国家标准生乳》和企业制定的《原料奶验收标准》进行验收。

(2)预处理:原料乳的预处理包括脱气、过滤和净化。

(3)标准化:利用在线标准化设备使含脂率>3.1%,蛋白质>2.9%,非脂乳固体>8.1%。

(4)均质:均质压力16.7~20.6 MPa。

(5)巴氏杀菌:72~75 ℃,保持15~20 s。

(6)冷却:冷却至4~5 ℃。

(7)灌装:杀菌冷却后立即灌装。

(8)检验:成品检验合格。

(9)冷藏:产品温度10 ℃以下,6 ℃以下避光贮藏运输。

任务1　食品质量安全数据及随机变量

4.1.1　基础知识

1. 质量数据的性质

数据可分为两大类,即计量值数据和计数值数据。

1)计量值数据

计量值数据是可以连续取值的数据,通常是使用量具、仪器进行测量而取得的,如长度、温度、重量、时间、压力、化学成分等。对于长度,在1~2 mm

Ⅰ. Nature of Quality Data

Data can be divided into two major categories, namely, variable data and count value data.

A. Variable Data

Variable data is the data which can be measured continuously, and is usually obtained by using measuring tools and instruments to continue the measurement. Such as length, temperature,

就可以连续测出 1.1 mm、1.2 mm、1.3 mm等数值；而在 1.1~1.2 mm 还可以进一步连续测出 1.11 mm、1.12 mm、1.13 mm 等数值。

2）计数值数据

计数值数据是不能连续取值，而只能以个数计算的数据。这类数据一般不用量具仪器进行测量就可以"数"出来，它具有离散性，如不合格品数、罐头瓶数、发酵罐数等。

计数值数据还可以细分为计件值数据和计点值数据。计件值数据是指按件计数的数据，如不合格品件数等；计点值数据是指按点计数的数据，如菌落斑点数、单位缺陷数等。

计量值数据与计数值数据的划分并非绝对。如细菌的直径大小，用测量仪检查时所得到的质量特性值的数据是计量值数据；而用计数方法检查时，得到的就是以个数表示产品质量的计数值数据。

计数值为离散性数据，虽以整数值来表示，但它不是划分计数值数据与计量值数据的尺度。计量值是具有连续性的数据，往往表现为非整数，但也不能由此得出只要是非整数值就一定是计量值数据的结论。例如，大麦吸水率为 67.5%，是一个非整数，但此数据并非是测量仪取得的结果，也不具备连续性质，而是通过计算大麦吸水率＝[（大麦吸水后质量－原大麦质量)/原大麦质量×100%]得到的，它

weight, time, pressure, chemical composition, etc. If the length is between 1 mm and 2 mm, the values of 1.1 mm, 1.2 mm, 1.3 mm can be measured continuously, and the values of 1.11 mm, 1.12 mm, 1.13 mm can be further measured continuously between 1.1 mm and 1.2 mm.

B. Count Value Data

Count value data is data that cannot be taken continuously, but can only be calculated in number. This type of data is generally measured without a measuring instrument, it can be "numbered" out, it is discrete. Such as the number of unqualified products, the number of cans and the number of fermentors, etc.

Count value data can also be subdivided into piecework value data and point value data. Piecework value data refers to the data counted by the parts, such as the number of unqualified products; the point value data refers to the date counted by the point, such as the number of colony spots, the number of defects per unit, etc.

The division between the variable data and count value data is not absolute. For example, the diameter of the bacteria, the data of the quality characteristics obtained from the measuring instrument is the variable data, while the counting method is used to check and the count value data you got that represents the quality of the product.

The count value is discrete data, which is expressed as integer value, but it is not the scale of dividing the count value data and the variable data. The variable is a continuous data, which is often expressed as a non-integer value, but it can not be concluded that as long as it is a non-integer value, it must be the variable data. For example, the water absorption of barley is 67.5%, which is a non-integer, but the data is not obtained by the measuring instrument, nor is it of a continuous nature. It is obtained by calculating the water

是计数值的相对数性质的数据。

对于上述相对数,判断其是计数值数据还是计量值数据,通常依照下述原则,即依照分子的数据性质来确定。如分子数据性质是计数值,则其分数值为计数值;如分子数据性质为计量值,则其分数值为计量值。

2. 质量数据的收集方法

1)收集数据的目的

(1)掌握和了解生产现状,如调查食品质量特性值的波动,推断生产状态。

(2)分析质量问题,找出产生问题的原因,以便找到问题的症结所在。

(3)对加工工艺进行分析、调查,判断其是否稳定,以便采取措施。

(4)调节、调整生产,如测量pH值,然后使之达到规定的标准状态。

(5)对一批加工食品的质量进行评价和验收。

2)收集数据的方法

运用现代科学方法,开展质量管理,需要认真收集数据。在收集数据时,应当如实记录,根据不同的数据选用合适的收集方法。在质量管理中,主要通过"抽样法"或"试验法"获得数据。

(1)抽样法。收集数据一般采用

absorption rate of barley = [(mass of barley after water absorption − raw barley mass)/raw barley mass × 100%], which is the counting value of the property of relative numbers.

For the relative number mentioned above, it is judged whether it is the count value data or the variable data, usually according to the following principles: that is determined according to the nature of the molecular data. If the property of the molecular data is a counting value, its fractional value is a counting value; if the molecular data property is a variable, the fractional value is a variable.

Ⅱ. **Methods of Collecting Quality Data**

A. Purpose of Data Collection

1. Master and understand the current production situation. Such as investigating the fluctuation of food quality characteristics and infer the production status.

2. Analyze the quality problem. Find out the cause of the problem so as to find the crux of the problem.

3. The processing technology is analyzed and investigated to determine whether it is stable or not, so as to take measures.

4. Regulate and adjust production. Such as the measurement of pH value, and then make it to the specified standard state.

5. The quality of a batch of processed food was evaluated and accepted.

B. Methods of Data Collection

The use of modern scientific methods and quality management requires careful data collection. When collecting data, it should be recorded truthfully, and the appropriate collection methods should be selected according to different data. In quality management, data are obtained mainly through "sampling" or "testing".

1. Sampling. A sampling method is used to

的是抽样方法,即先从一批产品(总体)中抽取一定数量的样品,然后经过测量或判断,作出质量检验结果的数据记录。

收集的数据应能客观地反映被调查对象的真实情况。因此对抽样总的要求是随机抽取,即不挑不拣,使一批产品里每一件产品都有均等的机会被抽到。

(2)试验法。试验法是用来设计试验方案、分析试验结果的一种科学方法,它是数理统计学的一个重要分支。这种方法能在考察范围内以最少的试验次数和最合理的试验条件,取得最佳的试验结果,并根据试验所获得的数据,对产品或某一质量指标进行估计。

3. 产品质量的波动

在生产过程中,尽管所用的设备是高精度的,操作是很谨慎的,但产品质量还会有波动。因此,反映产品质量的数据也相应地表现出波动,即表现为数据之间的参差不齐。例如,同一批次乳制品蛋白含量不完全相同等。总之,我们所收集到的数据,都具有这样一个基本特征,即它们毫不例外都具有分散性。数据的分散性乃产品质量本身的差异所致,是由生产过程中条件变化和各种误差造成的。即使条件相同、原料均匀、操作谨慎,生产出来的产品质量数据也是不会相同的,但这仅仅是数据特征的一个方面。另一方面,如果我们收集数据的方法恰当,数据又足够多,经过仔细观察或适当整理,将会发现它们都在一定范围内围绕着一个中心值分散。越靠近中心值,数据出现的机会越多;而离中心值越远,出现的机会就越少。

collect data, that is, a certain number of samples are taken from a batch of products (population), and then measured or judged to record the results of quality inspection.

The collected data should objectively reflect the true situation of the subjects surveyed. Therefore, the total requirement of sampling is random sampling. That is, not picking. So that each product in a batch of products have an equal opportunity to be drawn.

2. Testing. Testing is a scientific method used to design test scheme and analyze test results. It is an important branch of mathematical statistics. This method can obtain the best test results with the fewest test times and the most reasonable test conditions within the scope of investigation, and estimate the product or a certain quality index according to the data obtained from the test.

Ⅲ. Fluctuations in Product Quality

In the production process, although the equipment used is high precision and the operation is very careful, the quality of the product also have fluctuation. Therefore, the data that reflect the product quality also show the fluctuation correspondingly, its performance is the difference between the data. For example, the same batch of dairy protein content is not exactly the same. In short, the data we have collected all have a basic characteristic: they are all decentralized without exception. The dispersibility of the data is caused by the difference of the product quality itself, which is caused by the change of the conditions and various errors in the process of production. Even if the conditions are the same, the raw materials are uniform, and the operation is careful, the product quality data produced will not be the same, but this is only one aspect of the data characteristics. On the other hand, if we collect the data in the right way, and if we have enough data, after

从统计学角度来看,可以把产品质量波动分为正常波动和异常波动两类。

1)正常波动

正常波动是由偶然因素或随机因素(随机原因)引起的产品质量波动。这些偶然因素(随机因素)在生产过程中大量存在,对产品质量经常发生影响,但其所造成的质量特性值波动往往较小。如原材料的成分和性能上的微小差异、机器设备的轻微振动、温湿度的微小变化,以及操作方面、测量方面、检测仪器的微小差异等。对这些波动随机因素的消除,在技术上难以达到,在经济上代价又很大,因此一般情况下这些波动在生产过程中是允许存在的,所以称为正常波动。公差就是承认这种波动的产物。把仅有正常波动的生产过程称为过程处于统计控制状态,简称为受控状态或稳定状态。

特点:①影响因素多;②造成的波动范围小;③无方向性(逐件不同);④作用时间长;⑤对产品质量的影响小;⑥完全消除偶然因素的影响,在技术上有困难或在经济上不允许。所以由随机因素引起的产品质量的随机波动是不可避免的。

careful observation or proper collation, we will find that they are scattered around a central value within a certain range, the closer to the central value, the more chances it will appear; the further away from the center value, the less chance it will appear.

From the statistical point of view, the product quality fluctuation can be divided into two categories: normal fluctuation and abnormal fluctuation.

A. Normal Fluctuation

Normal fluctuation is a product quality fluctuation caused by accidental or random factors (random causes). These accidental factors(random factors) exist in large quantities in the process of production, which often affect the quality of the product, but the fluctuation of the quality characteristic value is often small. For example, there are slight differences in composition and performance of raw materials, slight vibration of machine and equipment, slight change of temperature and humidity, operation aspect, measurement aspect and tiny difference of detection instruments. The elimination of random factors of these fluctuations is difficult to achieve in a technical and costly in economy. Therefore, these fluctuations are generally allowed to exist in the production process, so they are called normal fluctuations. The public errand is the product of acknowledging this fluctuation. The production process with only normal fluctuations is called the process in the state of statistical control, which is referred to as the controlled state or the stable state.

Characteristics: a. a lot of influence factors; b. resulting in a small fluctuation range; c. no directivity (each is different); d. long acting time; e. little impact on product quality; f. completely eliminates the influence of accidental factors, which are technically difficult or economically unacceptable. Thus, random fluctuation of product

2) 异常波动

异常波动是由异常因素或系统因素(系统原因)引起的产品质量波动。这些系统因素在生产过程中并不大量存在,对产品质量不经常发生影响,一旦存在,对产品质量的影响就比较显著。如原材料不符合规定要求,机器设备带病运转,操作者违反操作规程,测量工具的系统误差等。由于这些因素引起的质量波动大小和作用方向一般具有周期性和倾向性,因此异常波动比较容易查明,容易预防和消除;又由于异常波动对质量特性的影响较大,一般来说生产过程中是不允许其存在的。把有异常波动的生产过程称为过程处于非统计控制状态,简称为失控状态或不稳定状态。

特点:①影响因素相对较少;②造成的波动范围大;③往往具有单向性、周期性;④作用时间短;⑤对产品质量的影响较大;⑥异常因素易于消除或减弱,在技术上不难识别和消除,在经济上往往也是允许的。所以由异常因素造成的产品质量波动在生产过程中是不允许存在的,只要发现产品质量有异常波动,就应尽快找出其异常因素加以消除,并采取措施使之不再出现。

质量管理的一项重要工作,就是

quality caused by random factors is inevitable.

B. Abnormal Fluctuation

Abnormal fluctuation is the fluctuation of product quality caused by abnormal factors or system factors (system causes). These system factors do not exist in a large number in the production process, and do not often affect the quality of products. Once they exist, the impact on product quality is more significant. Such as raw materials do not meet the requirements, machinery and equipment with the operation of malfunction, operators violate the operation rules, measurement tools system errors. As the magnitude and direction of mass fluctuations caused by these factors are generally cyclical and tendentious, anomalous fluctuations are relatively easy to identify, easy to prevent and eliminate, and because abnormal fluctuations have a greater impact on quality characteristics, generally speaking, it is not allowed to exist in the production process. The production process with abnormal fluctuations is referred to as the process in non-statistical control state, which is referred to as runaway state or unstable state.

Characteristics: a. the influencing factors are relatively few; b. the fluctuation range is large; c. it often with unidirectional and periodicity; d. the action time is short; e. it has a great influence on the quality of the product; f. abnormal factors are easy to eliminate or weaken, are not difficult to identify and eliminate technically, and are often economically permissible. So the fluctuation of product quality caused by abnormal factors is not allowed in the process of production, as long as there are abnormal fluctuations in product quality, we should find out the abnormal factors as soon as possible, eliminate them, and take measures to make them disappear.

An important task of quality management is to

要找出产品质量波动规律,把正常波动控制在合理范围内,消除系统原因引起的异常波动。

find out the fluctuation law of product quality, so as to control the normal fluctuation within a reasonable range and eliminate the abnormal fluctuation caused by the system reasons.

从微观角度看,引起产品质量波动的原因主要来自六个方面,即5M1E——工序六大因素(Man, Machine, Material, Method, Measurement, Environment)。

From the microcosmic point of view, the causes of product quality fluctuation come from six main factors, namely, 5M1E—six major factors of process (Man, Machine, Material, Method, Measurement, Environment).

正常波动与异常波动的对比如表4-1所示。

表4-1 正常波动与异常波动的对比

项目	正常波动	异常波动
引起质量波动的原因(因素)	一般原因/普通原因/偶然原因/随机原因/偶然因素或随机(性)因素	异常原因/可查明原因/系统原因/特殊原因/异常因素或系统(性)因素
识别性	不易识别	可识别或不难识别
属性	过程所固有的	非过程所固有
影响因素的多少	影响因素多	影响因素相对较少
造成的波动范围大小	造成的波动范围小	造成的波动范围大
方向性/周期性	无方向性(逐件不同)	往往具有单向性/周期性
作用时间长短	一直起作用(时间长)	在一定时间内对生产过程起作用
对产品质量的影响大小	对产品质量的影响微小	对产品质量的影响较大
能否消除	完全消除偶然因素的影响,在技术上有困难或在经济上不允许(不值得)	异常因素易于消除或减弱,在技术上不难识别、测量,且采取措施不难消除,在经济上也往往是允许的,是必须消除的
解决途径	需要管理、配置资源,以改进过程和系统,如更换高精度的加工设备、模具,改变现有的加工工艺,这需要高层决策	对5M1E进行调整,现场班组长甚至操作者都有权利和能力,故称为局部措施
能否避免/可否允许存在	由随机因素引起的产品质量的随机波动是不可避免的	由异常因素造成的产品质量波动在生产过程中是不允许存在的,只要发现产品质量有异常波动,就应尽快找出其异常因素加以消除,并采取措施使之不再出现
质量特性值分布状态	由偶然因素造成的质量特性值分布状态不随时间的变化而变化	由异常因素造成的质量特性值分布状态随时间的变化可能发生各种变化

通过以上的分析可以得出这样的结论,造成产品不合格的根本原因就是波动。

4. 乳制品企业产品质量的分布规律

食品工业中搜集到的数据大多为正态分布,乳制品企业产品质量分布也不例外。表 4-2 为某乳品企业收集的 100 次原料奶的蛋白质含量数据。

表 4-2　某乳品企业收集的原料奶蛋白质含量数据(单位:g)

3.77	3.95	3.73	3.11	3.77	3.30	3.82	3.36	3.55	3.70
3.55	3.54	3.32	4.03	3.78	2.94	4.55	3.78	4.35	3.86
4.33	3.82	3.83	3.84	3.99	3.75	3.54	4.10	3.83	4.10
4.53	3.70	3.21	4.38	3.59	3.19	4.15	4.17	2.99	3.61
4.34	3.38	3.76	4.17	3.80	3.94	3.91	3.93	3.55	3.58
3.89	3.64	3.77	3.44	3.96	3.95	3.67	3.50	3.84	4.74
3.60	4.28	4.24	4.09	3.14	3.56	3.53	3.99	3.71	3.72
3.82	3.66	3.90	3.49	4.50	3.89	3.82	3.57	3.61	3.38
3.70	3.80	3.41	3.76	3.44	4.18	3.67	3.89	3.98	3.85
4.13	3.95	3.81	3.59	3.80	3.81	3.72	3.24	3.44	3.89

为找出这些数据的统计规律,将它们分组、统计、作直方图,如图 4-1 所示,直方图的高度与该组数据出现的频数成正比。

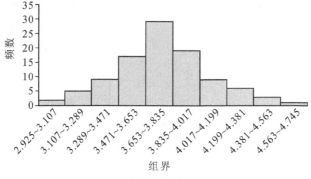

图 4-1　直方图

4.1.2　实训任务

收集乳品企业某项产品的质量数据,并对其进行分类。

实训提升技能点:加深对质量数据波动的认识。

专业技能点:①质量数据;②数据收集方法。

职业素养技能点:①调研能力;②沟通能力。

实训组织:对学生进行分组,每个组参照"基础知识"中的内容并利用网络资源,收集附近乳品企业某项产品的质量数据,对其进行分类,在班级进行汇报。

实训提交文件:收集的质量数据。

实训评价:由主讲教师进行评价。

学生姓名	数据收集的完整性 （20分）	数据描述的正确性 （20分）	回答质疑的准确性 （10分）	质量数据 （50分）

思考题

1. 简述数据的来源和分类。
2. 如何收集数据？
3. 乳制品的检测指标都有哪些？
4. 参照《数据说话：基于统计技术的质量改进》和《质量管理学》等相关资料学习5M1E——工序六大因素。

任务2 分层法的应用

4.2.1 基础知识

1. 分层法的概念及应用

分层法也叫分类法或分组法，是分析影响质量（或其他问题）原因的一种方法。它把所搜集到的质量数据依照使用目的，按其性质、来源和影响因素等进行分类，把性质相同、在同一生产条件下收集到的质量特性数据归在一组，把划分的组叫作"层"，通过数据分层，把错综复杂的影响质量的因素分析清楚，以便采取措施加以解决。

数据分层与收集数据的目的性紧密相联，目的不同，分层的方法和粗细程度也不同。另外，还与人们对生产情况掌握的程度有关，如果对生产过程的了解甚少，分层就比较困难。所以，分层要结合生产实际情况进行。分层法经常同质量管理中的其他方法一起使用，可将数据分层之后再进行加工，整理成分层排列图、分层直方

Ⅰ. The Concept and Application of Stratification Method

The stratification method is also called the classification method or the grouping method, which is a method to analyze the causes of the quality (or other questions). It classifies the quality data collected according to its purpose, according to its properties, sources and influencing factors, and classifies the quality characteristics data that are collected under the same production condition and have the same quality, and classifies the group as "layer". Through data stratification, the intricate factors affecting quality are analyzed clearly in order to take measures to solve them.

Data stratification is closely related to the purpose of collecting data. The purpose of data stratification is different, and the method and thickness of stratification are also different. In addition, it is also related to the degree of mastery of the production situation, which is more difficult to stratify if knowing a little about production process. Therefore, stratification should be combined with the actual production situation.

图、分层控制图和分层散布图等。

2. 常用的分层法

(1)按不同的时间分,如按不同的班次、不同的日期进行分类。

(2)按操作人员分,如按男女工、不同工龄、不同技术等级分类。

(3)按使用设备分,如按设备型号、新旧设备分类。

(4)按操作方法分,如按切削用量、温度、压力等分类。

(5)按原材料分,如按供料单位、进料时间、批次等分类。

(6)按不同检验手段、测量者、测量位置、仪器、取样方式等分类。

(7)其他分类,按不同的工艺、使用条件、气候条件等进行分类。

3. 应用案例

某酸奶生产企业某年上半年生产的酸奶发生菌落总数超标事件50次。为了找出原因,明确责任,进行改进,防止事件再次发生,可以对数据进行如下分层。

(1)按发生菌落总数超标的时间分层(图4-2)。

The stratification method is often used with other methods in quality management. It can be processed after data stratification and then processed into layering arrangement diagrams, hierarchical histogram, stratified control chart and layering scatter diagram, etc.

II. Commonly Used Stratification Method

1. Classification according to different time, such as according to different shifts, different dates.

2. Classification according to operators, such as new workers, male workers, female workers, different years of service, different technical grade.

3. Classification according to the use of equipment, such as according to the type of equipment, new and old equipment.

4. Classification according to operation methods, such as cutting parameters, temperature, pressure and so on.

5. Classification according to raw materials, such as feed units, feed time, batches, etc.

6. Classification according to different inspection means, surveyors, measuring positions, instruments, sampling methods and so on.

7. Other classification, according to different technology, use conditions, climate conditions and so on.

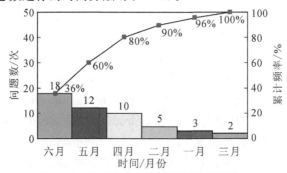

图4-2 按发生菌落总数超标的时间分层

(2) 按操作人员分层(图 4-3)。

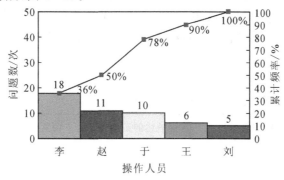

图 4-3 按操作人员分层

(3) 按原料来源基地分层(图 4-4)。

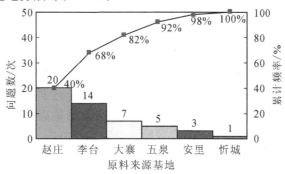

图 4-4 按原料来源基地分层

通过这三种分层可以看出,分层时标志的选择十分重要。标志选择不当就不能达到"把不同质的问题划分清楚"的目的。所以分层标志的选择应使层内数据尽可能均匀,层与层之间数据差异明显。

按发生菌落总数超标的时间分层时,各月差异不明显,而六月份差错稍多,可能是受天气温度过高的影响;按操作人员分层时,李某及赵某操作时出现菌落总数超标事件所占比重较大,应作为重点问题来解决;从按原料来源基地分层的情况来看,赵庄和李台两个奶源基地的原料造成菌落总数超标事件所占比重较大。经过分层就可以有针对性地分析原因,找出解决问题的办法。

分层法必须根据所研究问题的目的加以灵活运用。实践证明,分层法是分析处理质量问题成败的关键,使用时必须具有一定的经验和技巧才能分好层。

4.2.2 实训任务 1

酸奶企业质量安全数据的收集。

实训提升技能点:收集、分析食品质量安全数据的能力。

专业技能点:食品质量安全数据的收集。

职业素养技能点:①调研能力;②沟通能力。

实训组织：对学生进行分组，每个组参照"基础知识"中的内容，选择一个酸奶企业，对其生产过程中的质量安全数据进行收集，并自行设计调研表格。每组将调研结果与"基础知识"中的内容进行比较。调研结束后，每组在班级汇报调研结论和调研实训提升了自己哪些能力。

实训提交文件：调研报告。

实训评价：由酸奶企业或主讲教师进行评价。

学生姓名	数据收集的完整性 （20分）	数据描述的正确性 （20分）	回答质疑的准确性 （10分）	调研报告 （50分）

4.2.3　实训任务2

"分层法"在酸奶中大肠菌群来源分析中的应用。

实训提升技能点："分层法"的灵活应用能力。

专业技能点：①酸奶大肠菌群来源；②分层法。

职业素养技能点：①分析和解决问题的能力；②团队沟通协作能力。

实训组织：每个组针对收集到的酸奶生产中的质量安全数据，参照"基础知识"中的内容并利用网络资源，用"分层法"进行分析，按其性质、来源和影响因素等进行归类，并作出相应的图。

实训提交文件：分层排列图。

实训评价：由酸奶企业或主讲教师进行评价。

学生姓名	数据分类的正确性 （20分）	数据描述的正确性 （20分）	回答质疑的准确性 （10分）	分层排列图 （50分）

思考题

1. 分层法的意义是什么？
2. 分层法的应用范围和使用步骤是什么？
3. 参照《数据说话：基于统计技术的质量改进》和《质量管理学》等相关资料学习分层直方图、分层控制图和分层散布图的制作。

任务3　调查表法的应用

4.3.1　基础知识

1. 调查表的概念、意义和作用

调查表(data-collection form)，也叫检查表、核对表或统计分析表，是收集和积累数据的一种形式。调查表便于按统一的方式收集数据并进行分析，用于系统地收集数据，以获取对事实的明确认识，并可用于粗略的分析。调查表既适用于数字数据的收集和分析，也适用于非数字数据的收集和分析。调查表的格式多种多样，常见的有缺陷位置调查表、不良项目调查表、质量分布调查表和矩阵调查表。可根据检查目的的不同，使用不同的调查表。

调查表用来系统地收集资料和积累数据，在质量管理活动中，特别是在质量控制小组活动、质量分析和改进活动中得到广泛的应用。其意义和作用表现在以下几方面：

(1) 为质量管理和质量改进提供第一手资料；

(2) 为初步统计技术分析提供依据；

(3) 与生产过程同步完成，起到记录和检测作用；

(4) 调查表收集的资料着重于质量改进和质量控制应用；

Ⅰ. The Concept, Significance and Role of the Data-Collection Form

The data-collection form is also called a check form, checklists or statistical analysis form, which is the form of collecting and accumulating data. The data-collection form facilitates the collection and analysis of data in a uniform manner and is used for systematic data collection to obtain a clear picture of the facts and for rough analysis. The data-collection form applies not only to the collection and analysis of digital data, but also to the collection and analysis of non-digital data. The format of the data-collection form is varied, but the common ones are the defect location data-collection form, bad item data-collection form, quality distribution data-collection form and matrix data-collection form. Different data-collection forms may be used depending on the purpose of the inspection.

The data-collection form, which is used to collect information and accumulate data systematically and is widely used in quality management activities, especially in QC group activities, quality analysis and improvement activities. Its significance and function are manifested in the following aspects:

(1) Provide first-hand information for quality management and quality improvement;

(2) Provide the basis for the preliminary statistical technical analysis;

(3) Synchronize with the production process, play a recording and testing role;

(4) The information collected from the data-collection form focuses on quality improvement and QC application;

(5)调查表要求系统完成数据的积累,有利于技术档案的完善。

2. 调查表的应用步骤和注意事项

1)应用步骤

(1)明确收集资料的目的。目的必须明确,即要明确"为什么要调查"。

(2)明确为达到目的所需收集的资料。要达到已确立的目的而解决某项质量问题,则需以一定的数据为基础。那么,首先必须识别和明确为达到目的所需要的数据是什么、有哪些,即必须确定调查表的种类以及调查的项目。调查项目不要过于繁琐。

(3)确定资料的分析方法和负责人。收集的数据类型及其内容决定了"用怎样的统计工具"和"怎样进行分析",因此需要的数据确定后,应确定由谁以及如何分析这些数据。

(4)根据目的不同,设计用于记录资料的调查表格式。其内容应包括调查表的题目、调查对象和项目、调查方法、调查日期和期间、调查人、调查场所、调查结果的整理(合计、平均数、比例等的计算和考查)。

(5)未完成前应对收集和记录的部分资料进行预先检查,目的是审查表格设计的合理性。可在小范围内试用已设计好的调查表,收集和填写某些数据以初步测试调查表的有效性和可行性。

(5) The data-collection form requires the system to complete the accumulation of data, which is conducive to the improvement of technical files.

II. Application Steps and Precautions of the Data-Collection Form

A. Application Steps

1. Clear purpose of data collection. The purpose must be clear, that is, to clarify "why to investigate".

2. Identify the information needed to achieve the purpose. In order to achieve the established goal of solving a quality problem, it is necessary to base it on certain data. Then, first of all, it is necessary to identify and clarify what data is needed and what is needed to achieve the purpose, that is, to determine the type of data-collection form and the items of the survey. Don't make the investigation too cumbersome.

3. Determine the analytical method and the person responsible for the data. The type of data collected and its content determine "what statistical tools to use" and "how to analyze", so that the data needed should be determined by who and how to analyze the data.

4. The design of a data-collection form format for recording information depends on the purpose, and its contents should include the title of the data-collection form, the objects and items of the survey, the method of investigation, the date and period of the survey, the investigator, the place of investigation, the collation of the survey results (the calculation and examination of the total, average, proportion, etc.).

5. Prior to completion, part of the information collected and recorded should be pre-checked in order to review the reasonableness of the design of the form. The designed data-collection form can be tested on a small scale and selected data collected and filled in to initially test the validity and

(6)对于一些重要的调查表,初步完成后,如有必要,应评审和修改调查表格式。组织有关的具有丰富实际经验的各类人员对调查表进行全面的评估和审查,以使其在以后的使用中更加有效地发挥作用。

(7)正式使用调查表。针对调查表的对象和项目,仔细观察事实,将观察到的结果如实地填入调查表。

2)注意事项

(1)调查表一般在现场同步完成,由生产班组或者现场技术人员填写,不可事后补填,更不可提前杜撰。

(2)调查表要求的数据必须准确记录,可不作为绩效考核依据。

(3)调查表应在应用过程中不断修订完善,成为成熟的生产记录。

(4)提倡应用计算机汇总数据,并利用调查表展开阶段性统计分析,提出质量改进意见和质量改进策划。

feasibility of the data-collection form.

6. For some important data-collection forms, after initial completion, if necessary, the format of the questionnaire should be reviewed and revised. The data-collection form is comprehensively assessed and reviewed by a wide range of organization-related personnel with extensive practical experience in order to enable it to play a more effective role in future use.

7. Formal use of data-collection form. With regard to the objects and items of the data-collection form, carefully observe the facts and fill in the results truthfully.

B. Matters Needing Attention

1. The data-collection form is generally completed synchronously on the spot, filled in by production teams or technicians on the spot, not filled in after the event, and not fabricated in advance.

2. The data required by the data-collection form must be accurately recorded and may not be used as a basis for performance appraisal.

3. The data-collection form should be continuously revised and perfected in the process of application and become a mature production record.

4. Promote the use of computer data collection, and use the data-collection form to carry out stage statistical analysis, put forward quality improvement advice and quality improvement planning.

3. 调查表的分类

1)缺陷位置调查表

调查产品各不同部位的缺陷情况,可将其发生缺陷的位置标记在产品示意图或展开图上,将不同的缺陷采用不同的符号或颜色在图中标出以示区别。这种调查表会使缺陷的位置、性质、数量、程度等一目了然。

某汽车厂生产的汽车喷漆不良检查如表4-3所示。

表 4-3 汽车喷漆缺陷位置调查表

项 目	内 容
型号	××
工序	喷漆
检查目的	喷漆缺陷
检查部位	外表
检查件数	30 台
检查者	×××
检查时间	××××年×月×日

汽车喷漆有保护机身和使外观美观的作用,是比较重要的一道工序,常因种种原因造成喷漆缺陷,如色斑(●)、流漆(▲)和尘埃(◎),不同种类和不同位置的缺陷(图 4-5)常常是由不同的原因导致的。检查结果如表 4-4 所示。

图 4-5 汽车喷漆缺陷位置

表 4-4 汽车喷漆缺陷检查结果

部位	缺陷/件		
	色斑	流漆	尘埃
车顶	1	2	7
右侧面	1	2	1
前面	1	5	2
左侧面	1	4	2
后面	3	8	2
合计	7	21	14

从检查结果可知,色斑和流漆主要集中在后面,而尘埃主要集中在车顶。按以上线索深入调查分析,就找到了导致缺陷的原因。

2）不良项目调查表

不良项目调查表也称不合格项目调查表，是调查不良产品（如不合格品、有缺陷产品、废品）具体情况的调查表。它列出可能发生不良产品的具体项目，用一定的标记符号记录各不良项目的发生，并计算出相应的总的发生次数及其百分比。当调查结束后，就能立即知道任何一个项目的不良情况和不良情况发生的程度。不同的不良项目往往是由不同的原因造成的，着手于那些发生频率高的不良项目，分析导致其发生的原因，就可能找到改善质量的重要突破口。发生频率高或有增加倾向的不良项目都是进一步分析的重要线索。

某乳品企业在某月利乐枕包装巴氏奶抽样检验中，外观不合格项目调查记录如表4-5所示。

表 4-5 利乐枕包装巴氏奶外观不合格项目调查表

调查者：×××　　　　地点：包装车间　　　　日期：××××年×月

批次	产品规格	批量/箱	抽样数/袋	不合格品数/袋	不合格品率/%	封口不严	封口不平	标签模糊	标签擦伤	胀袋	批号模糊
1	利乐枕	100	50	1	2			1	1		
2	利乐枕	100	50	0	0						
3	利乐枕	100	50	2	4			2	1		
4	利乐枕	100	50	0	0						
⋮											
250	利乐枕	100	50	1	2	1		1			
合计		25000	12500	175	1.4	5	10	75	65	10	10

从表4-5可知，检查了25000箱共12500袋利乐枕巴氏奶，平均不合格品率为1.4%。不合格数因不良项目的不同而不同，标签模糊问题最突出，为75袋，占全部外观不合格数的42.9%；其次为标签擦伤的质量问题，为65袋，占全部外观不合格数的37.1%。应根据这些结论突出针对性的改进措施。

3）质量分布调查表

质量分布调查表也称工序分布调查表，是对计量值数据进行现场调查的有效工具。质量分布调查表是根据以往取得的资料，将某一质量特性项目的数据分布范围分成若干区间而制成的表格，用以记录和统计每一质量特性数据落在某一区间的频数。在能测量产品的尺寸、重量、纯度之类的计量值数据的工序中，运用质量分布调查表技术可以掌握这些工序的产品质量状况；有时也适用于服务过程的质量控制和检测，前提是表征此服务过程的参数为计量值（比如时间）。

某乳品企业在某月利乐枕包装巴氏奶抽样检验中，产品净含量调查记录如表4-6所示。

表 4-6 产品净含量实测值分布调查表

产品名称：利乐枕巴氏奶　　生产线：A　　调查者：×××　　日期：2013 年 12 月 12 日

净含量/mL	频数							小计
	5	10	15	20	25	30	35	
495.5~500.5								
500.5~505.5	/							1
505.5~510.5	//							2
510.5~515.5	/////	///						8
515.5~520.5	/////	/////						10
520.5~525.5	/////	/////	/////	/////	/			21
525.5~530.5	/////	/////	/////	/////	/////	////		29
530.5~535.5	/////	/////	/////					15
535.5~540.5	/////	///						8
540.5~545.5	////							4
545.5~550.5	//							2
550.5~555.5								
合计								100

从表格形式看，质量分布调查表与直方图的频数分布表相似。所不同的是，质量分布调查表的区间范围是根据以往资料进行划分，然后制成表格，以供现场调查记录数据；而频数分布表则是先收集数据，再适当划分区间，然后制成图表，以供分析现场质量分布状况之用。

4）矩阵调查表

矩阵调查表是一种多因素调查表，它要求把产生问题的对应因素分别排成行和列，在其交叉处标出调查到的各种缺陷、问题以及数量。

矩阵调查表在实际应用中要求能正确对项目进行分层，而且项目概念要明确，使分类数据易于归纳。按不同的标志进行分层，可以制作出各种不同形式的矩阵调查表，这对了解不良现象的具体原因十分方便。

经常采用如下集中分层标志：

(1)时间：白天、夜晚、上午、下午、月、季节等；

(2)人：作业者、男女、新老、熟练程度、班、组等；

(3)设备：机器、夹具、刀具、新旧、型号、用途等；

(4)材料：成分、尺寸、批号、厂家等；

(5)方法：作业方法、温度、压力、速度、操作条件等。

某酸奶生产企业某周对 1♯ 和 2♯ 两条生产线生产的酸奶产品菌落总数超标原因进行调查。现以生产线及相应的事件作为分层标志，调查菌落总数超标的原因所在。通过将实际调查的结果填入调查表，制成矩阵调查表，如表 4-7 所示。

表 4-7　某酸奶生产企业菌落总数矩阵调查表

产品名称：酸奶　　调查者：×××　　日期：2013 年 12 月 12 日

超标数量	73										化验时间	
化验员	××										场所	化验室

生产线	3月1日		3月2日		3月3日		3月4日		3月5日		3月6日	
	上午	下午	上午	下午	上午	下午	上午	下午	上午	下午	上午	下午
1#生产线	//	////	//	//	//	/	/// //	//	//	//	//	//
2#生产线	/	//	//	//	/	/	//	//		//	//	//

制成矩阵调查表后，经观察发现：1#生产线发生菌落总数超标的现象较多；3 月 4 日上午的产品菌落总数超标现象较严重。针对菌落总数超标这一现象，有关人员和部门根据菌落总数矩阵调查表揭示的这些信息，从根本上找到了导致菌落总数长期超标的原因：对 1#生产线每班次生产完后的清洗不彻底；生产操作人员为新进员工；3 月 4 日上午原料奶本身有较大的问题。

4.3.2　实训任务

"调查表法"在酸奶中菌落总数超标原因分析中的应用。

实训提升技能点："调查表法"在食品企业中的灵活应用能力。

专业技能点：①酸奶菌落总数来源；②调查表法。

职业素养技能点：①分析和解决问题的能力；②团队沟通协作能力。

实训组织：每个组针对收集到的酸奶生产中的质量安全数据，参照"基础知识"中的内容并利用网络资源，用"调查表法"进行分析，按其性质、来源和影响因素等进行归类，并作出相应的图表。

实训提交文件：调查表。

实训评价：由酸奶企业或主讲教师进行评价。

学生姓名	数据分类的正确性（20 分）	数据描述的正确性（20 分）	回答质疑的准确性（10 分）	调查表（50 分）

思考题

1. 调查表法的意义是什么？

2. 调查表法的应用范围和使用步骤是什么？

3. 参照《数据说话：基于统计技术的质量改进》和《质量管理学》等相关资料学习缺陷位置调查表、不良项目调查表、质量分布调查表、矩阵调查表等。

任务4　排列图法的应用

4.4.1　基础知识

1. 排列图的概念和结构

排列图也叫帕累托图，是找出影响产品质量的主要问题的一种有效方法。其形式如图4-6所示。

Ⅰ. Concept and Structure of the Permutations

The permutation is also called Pareto diagram, which is an effective way to find out the main problems that affect the quality of the product. The form is shown in Figure 4-6.

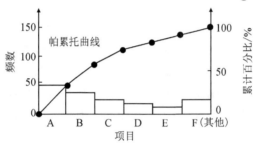

图4-6　排列图

排列图最早是意大利经济学家帕累托（Pareto）用来分析社会财富分布状况而得名的。他发现少数人占有大量财富，即所谓"关键的少数和次要的多数"的关系。后来，美国质量管理学家朱兰（J. M. Juran）把它的原理应用于质量管理，作为改善质量活动中寻找影响质量的主要因素的一种工具，它可以使质量管理者明确从哪里入手解决质量问题才能取得最好的效果。

1）概念

排列图是根据"关键的少数和次要的多数"的原理，将数据分项目排列，以直观的方法来表明质量问题的主次及关键所在的一种方法，是针对

The permutation was originally named after the Italian economist Pareto, who used it to analyze the distribution of social wealth. He found that minorities possessed a great deal of wealth, that is, the so-called relationship of the "critical minority and secondary majority". Later, J. M. Juran, a quality management scientist in the United States, applied its principles to quality management as a tool for improving quality activities in search of the main factors affecting quality, it can make quality managers clear where to solve quality problems to achieve the best results.

A. Concepts

According to the principle of the "critical minority and secondary majority", the permutation is a method to arrange the data subitems as the cause, and to show the primary, secondary and

各种问题按原因或状况分类,把数据从大到小排列而作出的累计柱状图。

2)结构

排列图的结构由两个纵坐标、一个横坐标、n 个柱形条和一条曲线组成。左边的纵坐标表示频数(件数、金额、时间等),右边的纵坐标表示频率(以百分比表示)。有时为了方便,也可把两个纵坐标都画在左边。横坐标表示影响质量的各个因素,按影响程度的大小从左至右排列。柱形条的高度表示某个因素影响的大小。曲线表示各影响因素大小的累计百分数,这条曲线称帕累托曲线(排列线)。

排列图在质量管理中的作用主要是用来抓质量的关键性问题。现场质量管理往往有各种各样的问题,应从何下手、如何抓住关键呢?一般说来,任何事物都遵循"少数关键,多数次要"的客观规律。例如,大多数废品由少数人员造成,大部分设备故障停顿时间由少数故障造成,大部分销售额由少数用户占有等。排列图正是能反映出这种规律的质量管理工具。

2. 排列图的作图步骤

1)确定评价问题的尺度(纵坐标)

排列图主要用来比较各问题(或一个问题的各原因)的重要程度。评价各问题的重要性,必须有一个客观

key position of the quality problem by intuitionistic method, which is the cumulative histogram of data from large to small in order to classify a variety of problems by reason or situation.

B. Structure

The structure of the permutation is composed of two vertical coordinates, one horizontal coordinate with cylindrical bars and one curve. The left represents the frequency (number of pieces, amount, time, etc.), and the right represents the rate (expressed as a percentage). Sometimes, for convenience, you can draw both vertical coordinates on the left. The horizontal coordinates represent the factors affecting the quality, arranged according to the degree of influence from left to right, the height of the column represents the size of a certain factor, and the curve represents the cumulative percentage of the size of each influence factor. This curve is called the Pareto curve (permutation line).

The function of permutation in quality management is mainly used to grasp the key problem of quality. Field quality management often has a variety of problems, how to start from, how to grasp the key? Generally speaking, everything follows the objective law of the " critical minority and secondary majority". For example, most of the waste products are caused by a small number of personnel, most of the equipment breakdown time is caused by a few failures, the majority of sales are held by a small number of users, and so on. The permutation is the quality management tool that can reflect this kind of law.

II. Mapping Steps of Permutations

A. Determination of the Scale of the Evaluation Problem (Vertical Coordinates)

The arrangement chart is mainly used to compare the importance of each problem (or all the reasons for a problem). There must be an

尺度。确定评价问题的尺度,即决定作图时的纵坐标的标度内容。

一般的纵坐标可取:①金额(包括把不合格品换算成损失金额);②不合格品件数;③不合格品率;④时间(包括工时);⑤其他。

2)确定分类项目(横坐标)

一个大的问题包括哪些小问题,或是一个问题与哪些因素有关,在作图时必须明确。分类项目表示在横坐标上,项目的多少决定横轴的长短。

一般可按不合格品项目、缺陷项目、作业班组、车间、设备、不同产品、不同工序、工作人员和作业时间等进行分类。

3)按分类项目搜集数据

笼统的数据是无法作图的。作图时必须按分类项目搜集数据。搜集数据的期间无原则性的规定,应随所要分析的问题而异。例如,可按日、周、旬、月、季、年等搜集数据。划分作图期间的目的是便于比较效果。

4)统计各个项目在该期间的记录数据,并按频数大小顺序排列

首先统计每个项目的发生频数,它决定直方图的高低。然后根据需要统计各项频数所占的百分比(频率)。最后,可按频数(频率)的大小顺序排列,并计算累计百分比,作排列图

objective yardstick for evaluating the importance of each issue. Determining the scale of the evaluation problem, that is the scale of the vertical coordinates of a drawing.

The normal vertical coordinates should be described: a. the amount (including the conversion of unqualified products into the amount of loss); b. the number of unqualified products; c. the rate of unqualified products; d. time (including working hours); e. others.

B. Determination of the Classification Items (Horizontal Coordinates)

In mapping, it is important to identify what small problems are included in a large problem, or what factors are involved in a problem. In horizontal coordinates, classification items indicate how much of the item determines the length of the horizontal axis.

Generally, they can be classified by unqualified items, defective items, job teams, workshops, equipment, different products, different processes, staff and work time, and so on.

C. Data Collection by Classification Items

General data is impossible to map. When drawing, data must be collected according to classification items. There is no rule of principle during the collection of data, but should vary according to the question to be analyzed, for example, by day, week, a period of ten days, month, season, year, etc. The purpose of dividing the mapping period is to facilitate comparison.

D. Make Statistics of the Recorded Data for Each Item During the Period, and Arrange in Order of Frequency

First, the occurrence frequency of each item is counted, which determines the level of histogram. The percentage (rate) of each frequency is then counted as needed. Finally, you can arrange them in order of frequency (rate) size, calculate the

用表。

5)画排列图中的直方图

可利用 Excel 进行画图,纵横坐标轴的标度要适当,纵轴表示评价尺度,横轴表示分类项目。在横轴上,按给出的频数大小顺序,把分类项目从左到右排列。"其他"一项不论其数值大小,务必排在最后一项。在纵轴上,以各项之频数为直方图高度,以横轴项目为底宽,一一画出对应的直方图。图宽应相同,每个直方柱之间不留间隙。如果需要分开,它们之间的间隔也要相同。

6)画排列线

为了观察各项累计占总体的百分比,可按右边纵坐标轴的标度画出排列线(又称帕累托线)。排列线的起点,可画在直方柱的中间、顶端中间或顶端右边的线上,其他各折点可按比例标注,并在折点处标上累计百分比。

7)在排列图上标注有关事项和标题

搜集数据的期间(何时至何时)、条件(检查方法、检查员等)、检查个数总数等,必须详细记载,在质量管理中这些信息都非常重要。

3. 绘制排列图的注意事项

绘制排列图时应注意以下事项。

cumulative percentage, and draw a table for the permutation diagram.

E. Draw the Histogram in the Permutation

Excel can be used for drawing, the scale of the vertical and horizontal coordinate axis should be appropriate, the vertical axis represents the evaluation scale, and the horizontal axis represents the classification items. On the horizontal axis, arrange the classification items from left to right in the order of the given frequency. The "other" item, regardless of its value, must be ranked last. On the vertical axis, take the frequency of each item as the histogram height, take the horizontal axis item as the base width, draw the corresponding histogram one by one. The width of the diagram should be the same, and there should be no gaps between each straight side, and if they need to be separated, the spacing between them should be the same.

F. Draw the Permutation Lines

In order to observe the cumulative percentage of the total, the permutation line (also known as the Pareto line) can be drawn on the scale of the right vertical axis. The starting point of the permutation line can be drawn in the middle of the straight column, the middle of the top, or the line on the right of the top, the other folding points can be marked proportionately, and the cumulative percentage is marked at the break point.

G. Mark the Related Items and Headings on the Permutation

The period during which data are collected (when to when), conditions (inspection methods, inspectors, etc.), total number of inspections, and so on must be documented in detail, and this information is important in quality management.

Ⅲ. Matters Need Attention in Drawing the Permutations

The following considerations should be taken

(1) 一般来说，主要原因是一两个，至多不超过三个，就是说它们所占的频率必须高于50%（如果分类项目少，则应高于70%或80%）；否则就失去了找主要问题的意义，要考虑重新进行分类。

(2) 纵坐标可以用"件数"或"金额""时间"等来表示，原则是以更好地找到"主要原因"为准。

(3) 不重要的项目很多时，为了避免横坐标过长，通常合并列入"其他"栏内，并置于最末一项。对于一些较小的问题，如果不容易分类，也可将其归入"其他"项里。如"其他"项的频数太多时，需要考虑重新分类。

(4) 为作排列图而取数据时，应考虑采用不同的原因、状况和条件对数据进行分类，如按时间、设备、工序、人员等分类，以取得更有效的信息。

4. 排列图的观察分析

利用ABC法分析确定重点项目。一般地讲，取图中前面的1~3项作为改善的重点就行了。若再精确些可采用ABC法确定重点项目。ABC分析法是把问题项目按其重要程度分为三级。

具体做法是把构成排列曲线的累计百分数分为三个等级：0~80%为A类，是累计百分数在80%以上的因素，它是影响质量的主要因素，作为解决的重点；累计百分数在80%~90%的为B类，是次要因素；累计百分数在90%~100%的为C类，在这一区间的因素是一般因素。

into account in drawing the permutations.

1. In general, the main reason is one or two, not more than three, that is, they must account for more than 50% of the frequency (if the number of classified items is small, it should be higher than 70% or more than 80%); otherwise, it will lose the sense of finding the main problem, and it is necessary to consider the reclassification.

2. The vertical coordinates can be expressed by "number of pieces" or "amount of money", "time", etc. The principle is to better find the "main cause".

3. When unimportant items are numerous, in order to avoid excessive horizontal coordinates, they are usually merged into the "other" column and placed in the last item. For smaller questions, if not easily classified, they could also be classified as "other". If "other" items have too many frequencies, you need to consider reclassifying them.

4. When taking data in order to make an permutation, different reasons, status and conditions should be taken into account to classify the data, such as time, equipment, process, personnel, etc., so as to obtain more effective information.

IV. Observation and Analysis of the Permutation

Use ABC analysis method to determine key items, generally speaking, take the 1/3 items in front of the map as the focus of improvement. If more accurate, the ABC method can be used to determine key projects. ABC analysis is to divide the problem item into three levels according to its importance.

The specific practice is to divide the cumulative percentage of the composition of the permutation curve into three levels. The 0~80% is regarded as category A, which is the factor with accumulative percentage above 80%. It is the main factor that affects the quality, and it is the key to solve the problem. The accumulative percentage of 80%~90% is category B, which is a secondary factor,

除了对排列图作 ABC 分析外,还可以通过排列图的变化对生产、管理情况进行分析:

(1)在不同时间绘制的排列图,项目的顺序有了改变,但总的不合格品数仍没有改变时,可认为生产过程是不稳定的;

(2)排列图的各分类项目都同样减小时,则认为管理效果是好的;

(3)如果改善后的排列图,其最高项和次高项一同减少,但顺序没变,则两个项目是相关的。

and the cumulative percentage of 90%～100% is category C, and the factor in this range is a general factor.

In addition to the ABC analysis of the permutations, we can also analyze the production and management through the changes in the permutations:

1. The production process can be considered unstable when the order of the items has changed but the total number of nonconforming products has not changed in drawing permutation diagrams at different times;

2. The classification items of the permutation are reduced equally, and the management effect is considered to be good;

3. If the improved arrangement reduces both the highest and the second highest items, but in the same order, the two items are related.

5. 应用案例

对某乳制品企业试生产的一批复合塑料袋装 UHT 灭菌乳 320 件产品的质量问题进行统计,并按问题项目作出统计表(表 4-8),最后作出排列图并进行分析。

表 4-8　某乳品企业某批次复合塑料袋装 UHT 灭菌乳产品质量问题统计表

问题项目	颜色褐变	脂肪上浮	蛋白凝固	坏包	异味包	其他
问题数/包	42	7	69	10	23	5

作图步骤如下。

(1)按排列图的作图要求将缺陷项目进行重新排列(表 4-9)。

表 4-9　排列图数据表

问题项目	蛋白凝固	颜色褐变	异味包	坏包	脂肪上浮	其他	总计
问题数/包	69	42	23	10	7	5	156
频率/%	44.2	26.9	14.7	6.4	4.5	3.2	100
累计频率/%	44.2	71.2	85.9	92.3	96.8	100	

(2)计算各排列项目所占百分比(频率)。
(3)计算各排列项目所占累计百分比(累计频率)。
(4)用 Excel 进行直方图的制作。选择项目、问题数、累计频率生成柱状图(图 4-7)。

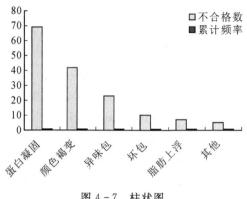

图 4-7 柱状图

(5)在图上选择累计频率图形,点击鼠标右键,选择更改图标类型,以此选择带标记的折线图。将累计频率的柱状图变为折线图(图 4-8)。

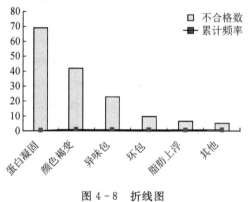

图 4-8 折线图

(6)更改成折线图后,选中折线图,点击鼠标右键,选择"设置数据系列格式",随后选择"次坐标轴",得到图 4-9。

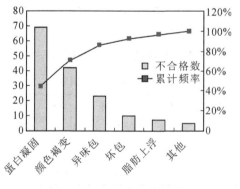

图 4-9 修改后的折线图

(7)选择累计频率的坐标轴,点击右键选择"设置坐标轴格式",将最大值改为 1,最小值改为 0,其他可按照需求选择默认设置。选择数量坐标轴,点击右键选择"设置坐标轴格式",将最大值设置为大于或等于累计问题数的最大值,最小值和间隔可按照需要选择(图 4-10)。

项目 4　食品质量安全管理常用工具应用方法　117

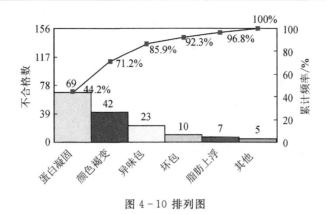

图 4-10　排列图

(8)根据各排列项目所占累计百分比画出排列图中的排列线。

从图 4-10 中可以看出,蛋白凝固、颜色褐变、异味包三项问题累计百分比占 85.9%,为 A 类因素,是要解决的主要问题。

4.4.2　实训任务

某乳品企业统计了 2013 年 2—7 月巴氏杀菌乳质量缺陷情况,如表 4-10 所示,请作出排列图。

表 4-10　巴氏杀菌乳质量缺陷情况统计表

质量缺陷	2013 年 2—7 月		
	产量/t	缺陷次数	(缺陷/产量)/%
胀包	1080	32	2.96
变味	1080	21	1.94
杂质	1080	18	1.67
脂肪上浮	1080	23	2.13
其他	1080	17	1.57

实训提升技能点:"排列图法"在食品质量中的应用。

专业技能点:用排列图分析巴氏乳质量缺陷。

职业素养技能点:①分析和解决问题的能力;②团队沟通协作能力。

实训组织:对学生进行分组,每个组参照"基础知识"中的内容,将上述乳品企业统计的巴氏杀菌乳质量缺陷统计情况作成排列图,并进行分析。

实训提交文件:排列图。

实训评价:由乳制品企业或主讲教师进行评价。

学生姓名	排列图制作的正确性 (20 分)	数据分类的正确性 (20 分)	回答质疑的准确性 (10 分)	排列图 (50 分)

思考题

1. 排列图的应用范围是什么？
2. 排列图的制作步骤有哪些？
3. 排列图作图中应注意什么问题？
4. 排列图还可以在哪些方面进行应用？
5. 参照《数据说话：基于统计技术的质量改进》和《质量管理学》等相关资料学习排列图在食品质量中的应用。

任务5　因果图法的应用

4.5.1　基础知识

因果图由日本质量管理专家石川馨（Kaoru Ishikawa）最早提出，于1953年首先开始在日本川崎制铁所中应用。由于其非常实用有效，在日本的企业中得到了广泛的应用，很快又被世界上许多国家采用，成为现代工业质量改进的基本工具。因果图也称"石川图"。

1. 因果图的概念和结构

任何一项质量问题的发生或存在都是有原因的，而且经常是多种复杂因素平行或交错地共同作用所致。要有效地解决质量问题，首先要从不遗漏地找出这些原因入手，而且要从粗到细地追究到最原始的因素。因果图正是解决这一问题的有效工具。

因果图又叫特性因素图，因其形状颇像树枝和鱼刺，也被称为树枝图或鱼刺图。它是把对某项质量特性具有影响的各种主要因素加以归类和分解，并在图上用箭头表示其间关系的一种工具。由于它使用起来简便有效，在质量管理活动中应用广泛。

The causality diagram was first proposed by Kaoru Ishikawa, a Japanese quality management expert. It was first applied in the Kawasaki Iron Institute in Japan in 1953. Because it is very practical and effective, it has been widely used in Japanese enterprises. Soon adopted by many countries in the world, it has become a basic tool for improving the quality of modern industry. The causality diagram is also called "Ishikawa Diagram".

I. The Concept and Structure of Causality Diagrams

Any quality problem occurs or exists for a reason, and is often caused by multiple complex factors in parallel or interlaced. In order to solve the quality problem effectively, we first must find out these reasons never missed, and from rough to sewing to the most primitive factors, the causality diagram is an effective tool to solve this problem.

Causality diagram is also called characteristic factor graph, because its shape is similar to branches and fish spines, also known as branch chart or fishbone diagrams. It is a tool that classifies and decomposes the main factors that have an effect on a quality characteristic and can indicate the relationship between them with arrows on the graph. Because it is simple and effective to

因果图是由以下几部分组成的(图 4-11)。

The causality diagram consists of the following parts (see Figure 4-11).

use, it is widely used in quality management activities.

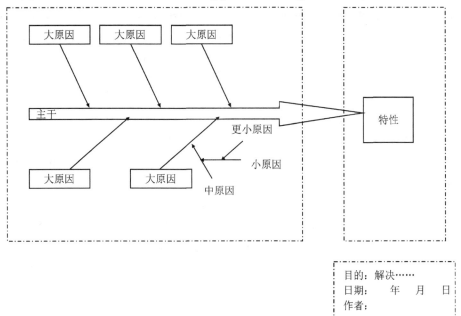

图 4-11　因果图

(1) 特性,即生产过程或工作过程中出现的结果,一般指与质量有关的特性,如产量、不合格率、缺陷数、事故件数、成本等与工作质量有关的特性。因果图中所提出的特性,是指要通过管理工作和技术措施予以解决并能够解决的问题。

(2) 原因,即对质量特性产生影响的主要因素,一般是导致质量特性发生分散的几个主要来源。原因通常又分为大原因、中原因、小原因等。

(3) 枝干,是表示特性(结果)与原因间关系,或原因与原因间关系的各种箭头。

1. Characteristics, that is, the production process or the results of the work process, generally refer to quality related characteristics, such as output, failure rate, number of defects, number of accidents, costs and other characteristics related to the quality of work. The characteristics set forth in the causality diagram refer to problems that are to be addressed and can be solved through management and technical measures.

2. The reason, that is, the main factor influencing the quality characteristics, which is generally the main source of the dispersion of the quality characteristics. The causes are usually divided into major causes, medium causes, small causes and so on.

3. A branch is a variety of arrowheads that indicate the relationship between characteristic (result) and cause or between cause and cause.

2. 因果图的作图步骤

(1)确认质量特性(结果)。质量特性是准备改善和控制的对象,应当通过有效的调查研究加以确认,也可以通过画排列图确认。

(2)画出特性(结果)与主干。

(3)选取影响特性的大原因。先找出影响质量特性的大原因,再进一步找出影响质量特性的中原因、小原因,最后画出中枝、小枝和细枝等。注意所分析的各层次原因之间的关系必须是因果关系,分析原因直到能采取措施为止。

(4)检查各项主要因素和细分因素是否有遗漏。

(5)对特别重要的原因要附以标记,用明显的记号将其框起来。特别重要的原因,即对质量特性影响较大的因素,可通过排列图来确定。

(6)记载必要的有关事项,如因果图的标题、制图者、时间及其他备查事项。

3. 绘制因果图的注意事项

绘制因果图时,应注意以下事项。

(1)主干线箭头指向的结果(要解决的问题)只能是一个,即分析的问题只能是一个。

(2)因果图中的原因是可以归类的,类与类之间的原因不发生联系,要注意避免归类不当和因果倒置的错误。

Ⅱ. Mapping Steps for Causality Diagrams

1. Confirm quality characteristics (results). Quality characteristics are the objects to be improved and controlled. It should be confirmed by effective investigation and study, or by drawing an permutation diagram.

2. Draw the characteristics (results) and the trunk.

3. Select the major reasons that affect the characteristics. First, find out the major reasons that affect the quality characteristics, and then further find out the medium cause and small reason of the quality characteristic, and draw the middle branch, small branch and twig, etc. It is important to note that the relationship between the various causes of the analysis must be a causal relationship and to analyze the causes until it is possible to take measures.

4. Check whether the major factors and subdivision factors are missing.

5. Attach a mark to the cause of particular importance and frame it with a clear mark. Especially important reasons, that is, the factors that have a great influence on the quality characteristics, can be determined by the permutation diagram.

6. Record the necessary relevant items, such as the title, the charting person, the time and other items for reference of the causality diagram.

Ⅲ. Considerations for Drawing Causality Diagrams

In drawing causality diagrams, the following considerations should be taken into account.

1. The main arrow points to the result (the problem to be solved) can only be one, that is, the analysis of the problem can only be one.

2. The causes in the causality diagram can be classified, and there is no relation between the causes of the class and the class, we should pay attention to avoid the improper classification and

(3)在分析原因时,要设法找到主要原因,注意大原因不一定都是主要原因。为了找出主要原因,可作进一步调查、验证。

(4)要广泛而充分地汇集各方面的意见,包括技术人员、生产人员、检验人员,以至辅助人员。因为各种问题的涉及面很广,各种可能因素不是少数人能考虑周全的。另外要特别重视有实际经验的现场人员的意见。

the error of cause and effect inversion.

3. In the analysis of the reasons, we should try to find the main reasons, pay attention to the major reasons are not necessarily the main reasons. In order to find out the main reasons, we can make further investigation and verification.

4. It is necessary to gather opinions from all aspects widely and fully, including technicians, producers, inspectors and even auxiliary personnel. Because of the wide range of problems involved, various possible factors are not fully considered by a small number of people. In addition, special attention should be paid to the opinions of field personnel with practical experience.

4. 应用案例

某乳品企业对"巴氏杀菌乳大肠菌群超标"进行原因分析,他们首先收集质量数据,请有关人员共同讨论分析巴氏杀菌乳大肠菌群超标的原因。

与会人员踊跃发言,先从大的方面找原因,问题主要来自以下几个方面:人员、机器、材料、环境等。画出果图,把这些大原因放在主干线两侧的大原因箭线的尾端。

而后大家又针对每个大原因进一步突出了许多具体的原因,经过进一步讨论分析和验证,把具体原因分别标在相应的位置上,因果图也就画好了。

然后进行表决,确定了五个主要原因。大家认为造成巴氏杀菌乳大肠菌群超标的主要原因是工艺卫生与个人卫生差、杀菌时的温度过低、杀菌时间不够、生产用水大肠菌群较高、室内卫生及孳生微生物源。把这五项主要原因标上标记(★),最终画出因果图(图4-12)。

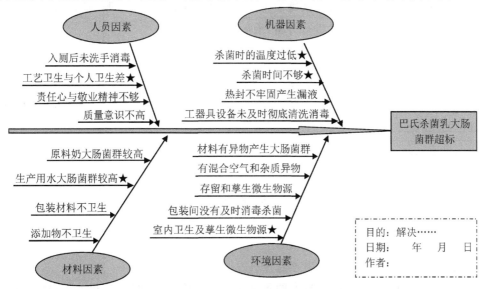

图4-12 巴氏杀菌乳大肠菌群超标因果图

记录必要的有关事项,如参加讨论的人员、绘制日期、绘制者等。

对主要原因制订对策表,落实改进措施。

4.5.2 实训任务 1

收集啤酒生产中常出现的质量缺陷信息。

实训提升技能点:收集、分析食品质量安全数据的能力。

专业技能点:食品质量安全数据的收集。

职业素养技能点:①调研能力;②沟通能力。

实训组织:对学生进行分组,每个组参照"基础知识"中的内容,选择附近一个啤酒生产企业,对其生产过程中遇到的质量缺陷信息进行收集,并自行设计调研表格。调研结束后,每组在班级中汇报调研结论和调研实训提升了自己哪些能力。

实训提交文件:调研报告。

实训评价:由啤酒企业或主讲教师进行评价。

学生姓名	数据收集的完整性 (20分)	数据描述的正确性 (20分)	回答质疑的准确性 (10分)	调研报告 (50分)

4.5.3 实训任务 2

"因果图法"在啤酒企业食品质量安全管理中的应用。

实训提升技能点:"因果图法"在食品企业中的灵活应用能力。

专业技能点:①啤酒生产中常见的质量问题;②因果图法。

职业素养技能点:①分析和解决问题的能力;②团队沟通协作能力。

实训组织:每个组针对收集到的啤酒生产中的质量缺陷信息,参照"基础知识"中的内容并利用网络资源,用"因果图法"进行分析,最后作出因果图。

实训提交文件:啤酒质量缺陷因果图。

实训评价:由啤酒企业或主讲教师进行评价。

学生姓名	因果图制作的正确性 (20分)	数据分类的正确性 (20分)	回答质疑的准确性 (10分)	因果图 (50分)

思考题

1. 因果图的构成有哪些?
2. 因果图的制作步骤及注意事项是什么?
3. 参照《数据说话:基于统计技术的质量改进》和《质量管理学》等相关资料学习因果图在食品质量中的应用。

任务6 对策表法的应用

4.6.1 基础知识

1. 对策表的概念

对策表也称措施表或措施计划表,是针对存在的质量问题制定解决对策的质量管理工具。利用排列图找到主要的质量问题(即主要矛盾),但问题并未迎刃而解,再通过因果图找到产生主要问题的主要原因,问题还是依然存在。为彻底解决问题,就应求助于对策表了。

对策表是一种矩阵式的表格,其中包括序号、问题(或原因)、对策(或措施)、执行人、检查人(或负责人)、期限、备注等栏目。基本格式如表4-11所示。

Ⅰ. The Concept of Countermeasure Table

The countermeasure table is also called the measure table or the measure plan table. It is a quality management tool to solve the existing quality problems. Using the permutation diagram to find the main quality problem (that is, the main contradiction), but the problem has not been solved easily, and then through the causality diagram to find the main cause of the main problem, the problem still exists. In order to solve the problem once and for all, we should ask the countermeasure table for help.

The countermeasure table is a matrix table. These include the serial number, problem (or cause), countermeasure (or measure), executor, inspector (or person in charge), time limit, remark and so on. The basic format is shown in Table 4-11.

表4-11 对策表

序号	质量问题	对策	执行人	检查人	期限	备注
(1)	(2)	(3)	(4)	(5)	(6)	(7)
1						
2						
3						
4						

2. 对策表的制作及注意事项

对策表各栏目的设置,可在基本格式的基础上根据实际需要进行增删或变换。如在第(1)栏与第(3)栏之间增设"目标"一栏,在第(3)栏与第(4)栏之间增加"地点"一栏,在第(6)栏之后增加"检查记录"一栏等。对于栏目

Ⅱ. Making and Noticing Matters of the Countermeasures Table

Each column of the countermeasure table can be added, deleted or changed according to the actual needs on the basis of the basic format. Add the column "objective" between columns (1) and (3), add the column "location" between columns (3) and (4), add the column "inspection record"

名称,第(2)栏也可改为"问题现状";当对策表与排列图、因果图构成"两图一表"联用时,第(2)栏应改为"主要原因"或"主要因素"等。

制定对策表的程序是:首先根据需要设计表格,填好表头名称;然后在讨论制定对策(或措施)后,逐一将有关内容填入表内。

填写对策表各栏目的具体内容时,应注意前后相对应。如第(2)栏填写一条问题(或原因)之后,可以与其他一条或几条对策(或措施)相对应。对策(措施)要尽量具体明确,有可操作性。

after column (6), etc. As to column name, column (2) could also be changed to "status of the problem"; when the countermeasures table was used in conjunction with the permutation diagram and causality diagram, column (2) should be replaced by "main cause" or "major factor", etc.

The procedure of developing the countermeasure table is: first, design the form according to the need, fill in the name of the table head; then, after discussing the development of countermeasures (or measures), fill in the related contents to table one by one.

When filling out the specific contents of each column of the countermeasure table, we should pay attention to the correspondence between the front and back. If a question (or cause) has been filled in column (2), it may correspond to one or more other countermeasures (or measures). Countermeasures (measures) should be as specific as possible and operational.

3. 应用案例

(1)某乳品企业对"巴氏杀菌乳大肠菌群超标"进行原因分析,经质量控制小组分析找出造成此问题的主要原因为:

①工艺卫生与个人卫生差;
②杀菌时的温度过低;
③杀菌时间不够;
④生产用水大肠菌群较高;
⑤室内卫生及孳生微生物源。

(2)召开质量控制小组会议,针对造成质量问题的主要原因制定对策,并对每一项对策进行分工,明确完成期限等。

(3)绘制"对策表"(表4-12),将有关内容填入表内。

(4)按对策表实施。

(5)定期统计巴氏杀菌乳大肠菌群情况。经过一周活动,大肠菌群超标问题解决,说明该对策表制定正确,实施有效。

表 4-12 某乳品企业解决巴氏杀菌乳大肠菌群超标问题对策

序号	主要原因	对策	执行人	检查人	期限	备注
1	工艺卫生与个人卫生差	操作人员必须体检合格才能上岗	(略)	(略)	周一至周二	由人事处配合
1	工艺卫生与个人卫生差	加强操作人员卫生培训	(略)	(略)	周一至周二	由老师讲解
1	工艺卫生与个人卫生差	操作人员工作服、鞋、帽每班次统一彻底清洗消毒	(略)	(略)	周一至周二	由后勤部配合
2	杀菌时的温度过低	严格控制杀菌温度在72~75 ℃	(略)	(略)	周一至周二	由生产部配合
2	杀菌时的温度过低	采用热交换性能更好的板式热交换器	(略)	(略)	周一至周二	由生产部配合
3	杀菌时间不够	严格控制乳液流过热交换器的时间在15~20 s	(略)	(略)	周三至周四	由原料奶采购处配合
4	生产用水大肠菌群较高	生产用水进行严格检验,合格后使用	(略)	(略)	周三至周四	由质检部配合
5	室内卫生及孳生微生物源	每班次后用消毒液进行室内消毒	(略)	(略)	周六至周日	由生产部配合
5	室内卫生及孳生微生物源	所有进出口加装防护网	(略)	(略)	周六至周日	由工程部配合

4.6.2 实训任务

收集配方奶粉在包装工序中出现的质量波动因素,并提出相应的控制措施。

实训提升技能点:收集、分析食品质量安全数据的能力。

专业技能点:"对策表法"在配方奶粉中的应用。

职业素养技能点:①调研能力;②分析、解决问题的能力。

实训组织:对学生进行分组,每个组参照"基础知识"中的内容,对配方奶粉在包装工序中出现的质量问题进行分析,并制定相应的对策。

实训提交文件:解决配方奶粉包装工序中出现质量波动问题的对策表。

实训评价:由乳制品企业或主讲教师进行评价。

学生姓名	数据收集的完整性 (20分)	数据描述的正确性 (20分)	回答质疑的准确性 (10分)	对策表 (50分)

思考题

1. 对策表法与因果图法的关系是什么?
2. 如何制作对策表?
3. 参照《数据说话:基于统计技术的质量改进》和《质量管理学》等相关资料学习对策表在食品质量中的应用。

任务7 直方图法的应用

4.7.1 基础知识

1. 直方图的概念及用途

直方图是从总体中随机抽取样本,将从样本中经过测定或收集来的数据加以整理,描绘质量分布状况,反映质量分散程度,进而判断和预测生产过程质量及不合格品率的一种常用工具。直方图是连续随机变量频率分布的一种图形表示,它以有线性刻度的轴上的连续区间来表示组,组的频率(或频数)用以相应区间为底的矩形表示,矩形的面积与各组频率(或频数)成比例,如图 4-13 所示。

Ⅰ. The Concept and Use of Histogram

A histogram is a random sampling from the whole population. The data measured or collected from the sample is collated to describe the quality distribution and the degree of dispersion of the reaction quality. Furthermore, it is a common tool to judge and predict the quality of production process and the rate of nonconforming products. A histogram is a graphical representation of frequency distribution of continuous random variables. It represents the group as a continuous interval on the axis with a linear scale, and the rate (or frequency) of the group is represented by a rectangle at the base of the corresponding interval. The area of the rectangle is proportional to the rate (or frequency) of each group, as shown in Figure 4-13.

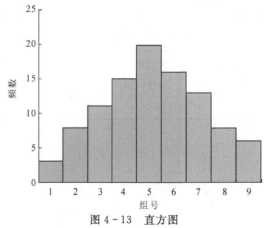

图 4-13 直方图

直方图的主要用途有以下几点。

(1) 能比较直观地看出产品质量特性值的分布状态,借此可判断生产过程是否处于稳定状态,并进行工序质量分析。

(2) 便于掌握工序能力及工序能力保证产品质量的程度,并通过工序能力来估算工序的不合格品率。

(3) 用以方便、较精确地计算质量数据的特征值。

2. 直方图的作图方法

(1) 收集数据。数据个数一般为 50 个以上,最低不少于 30 个。

(2) 求极差值。在原始数据中找出最大值和最小值,计算两者的差就是极差,即

$$R = X_{\max} - X_{\min}$$

(3) 确定分组的组数和组距。一批数据究竟分多少组,通常根据数据个数的多少来确定。具体方法参考表 4-13。

The main uses of histograms are as follows.

1. The distribution state of product quality characteristic value can be seen intuitively, which can be used to judge whether the production process is in a stable state or not and to analyze the process quality.

2. Easy to master process capability and process capability to ensure the degree of product quality, and through the process capacity to estimate the rate of nonconforming products.

3. It is used to calculate the characteristic value of quality data succinctly and accurately.

Ⅱ. Graphic Method of Histogram

1. Data collection. The number of data is generally more than 50, the minimum is not less than 30.

2. Get the extreme difference value. Find the maximum and minimum values in the raw data. The difference between the two is the extreme difference, that is

$$R = X_{\max} - X_{\min}$$

3. Determine the number of groups and the distance between groups. How many groups of data are divided into, usually according to the number of data to determine. Specific methods refer to Table 4-13.

表 4-13　数据个数与组数

数据个数	分组数	一般使用组数
50~100	6~10	10
100~250	7~12	
250 以上	10~20	

需要注意的是,如果分组数取得太多,每组里出现的数据个数就会很少,甚至为零,作出的直方图就会过于分散或呈现锯齿状;若组数取得太少,则数据会集中在少数组内,而掩盖了数据的差异。分组数 k 确定以后,组

It is important to note that if the number of groups is too many, the number of data in each group will be small, or even zero, and the histogram will be too scattered or jagged; if the number of groups is too small, the data will be concentrated in a small number of groups, and

(4)确定各组界限值。分组的组界值要比抽取的数据多一位小数,以使边界值不致落入两个组内,因此先取测量单位的1/2。例如,测量单位为0.001 mm,组界的末位数应取0.001 mm/2=0.0005 mm。然后用最小值减去测量单位的1/2,作为第一组的下界值;再将此下界值加上组距,作为第一组的上界值,依次加到最大一组的上界值(即包括最大值为止)。为了计算的需要,往往要决定各组的中心值(组中值)。每组的上下界限值相加除以2,所得数据即为组中值,组中值为各组数据的代表值。

(5)制作频数分布表。将测得的原始数据分别归入到相应的组中,统计各组的数据个数,即频数。各组频数填好以后,检查一下总数是否与数据总数相符,避免重复或遗漏。

3. 直方图的观察分析

直方图的观察、判断主要从形状进行分析。观察直方图的图形形状,看是否属于正常的分布,分析工序是否处于稳定状态,判断产生异常分布的原因。直方图有不同的形状,如图4-14所示。

masked differences in data. After the number of groups k is determined, the group distance h is also determined, $h=R/k$.

4. The limit values of each group were determined. The decimal place of the group boundary value is more than the extracted data so that the boundary value does not fall into the two groups, and the 1/2 of the unit of measurement is first taken. For example, the unit of measurement is 0.001 mm, and the last position of the group boundary should be 0.001 mm/2 = 0.0005 mm. Then 1/2 of the unit of measurement is subtracted from the minimum value as the lower bound value of the first group, and the lower bound value and the group distance are added as the upper bound value of the first group, which is added to the upper bound value (including the maximum value) of the largest group in turn. The central value of each group (group median) is often determined in order to calculate the needs. If the upper and lower bounds of each group are added and divided by 2, the resulting data is the median value of the group, and the median value of the group is the representative value of the data of each group.

5. Make the frequency distribution table. The original data are classified into the corresponding groups, and the number of data in each group is counted, that is, frequency. After filling out the frequency of each group, check if the total number is consistent with the total number of data to avoid duplication or omission.

III. Observation and Analysis of the Histogram

The observation and judgment of histogram are mainly carried out from shape analysis. Observe the shape of the histogram to see if it belongs to the normal distribution. Analyze whether the process is in a stable state and determine the cause of abnormal distribution. The histogram has different shapes, as shown in Figure 4-14.

(1)标准型,如图 4-14(a)所示。标准型又称对称型,数据的平均值与最大值和最小值的中间值相同或接近。平均值附近的数据频数最多,频数由中间值向两边缓慢下降,并且以平均值为中心左右对称。这种形状是最常见的,这时判定质量处于稳定状态。

(2)偏态型,如图 4-14(b)所示。数据的平均值位于中间值的左侧(或右侧),从左至右(或从右至左)数据分布的频数增加后突然减少,形状不对称。一些有形位公差等要求的特性值是偏态型分布,也有的是由于操作习惯而造成的。例如,由于食品添加剂称量者担心产生添加剂含量超标,称量时往往称量较少,而呈左偏型;反之,而呈右偏型。

(3)孤岛型,如图 4-14(c)所示。在直方图的左边或右边出现孤立的长方形,这是测量有误,或生产中出现异常因素而造成的。如原材料一时的变化,杀菌温度不稳定等。

(4)锯齿型,如图 4-14(d)所示。直方图如锯齿一样凹凸不平,大多是由于分组不当,或是检测数据不准而造成的。应查明原因,采取措施,重新作图分析。

(5)平顶型,如图 4-14(e)所示。直方图没有突出的顶峰,这主要是在生产过程中有缓慢变化的因素影响而造成的。

1. Standard type, see Figure 4-14(a). The standard type is also called symmetric type. The average value of the data is the same or close to the median value of the maximum and minimum values. The frequency near the average value is the most, and the frequency decreases slowly to both sides of the median value, and is symmetrical to the left and right with the average value. This shape is the most common. At this point, the quality is determined to be stable.

2. Skewness, see Figure 4-14(b). The average value of the data is on the left (or right side) of the median value, from left to right (or from right to left), the frequency of the data distribution increases then decreases suddenly and the shape is asymmetrical. Some characteristic values such as visible bit tolerances are biased distributions, and some are caused by operating habits. For example, because the food additive weighing person is worried that the additive content exceeds the standard, when weighing is often less than the weighing, and presents the left deviation type; conversely, presents the right deviation type.

3. Isolated island type, see Figure 4-14(c). An isolated rectangle appears on the left or right of the histogram. This is a measurement error, or the production of abnormal caused by cable. Such as sudden changes in raw materials, unstable sterilization temperature and so on.

4. Serrated, see Figure 4-14(d). Histograms are as rough as serrated ones, mostly due to improper grouping or inaccurate detection data. Reasons should be identified and measures taken to remap the analysis.

5. Flat-top type, see Figure 4-14(e). The histogram has no prominent pinnacle. This is mainly caused by slow changes in the production process.

(6)双峰型,如图 4-14(f)所示。靠近直方图中间值的频数较少,两侧各有一个"峰"。当有两种不同的相差较大的平均值混在一起时,常出现这种形式。这种情况往往是由于把不同材料、不同加工者、不同操作方法、不同设备生产的两批产品混在一起而造成的。

6. Doublet type, see Figure 4 – 14 (f). The frequency near the median value of the histogram is less, with a "peak" on each side. This form often occurs when there are two different distributions with large differences in mean values. This is often due to the mixing of two batches of products produced by different materials, processors, methods of operation and equipment.

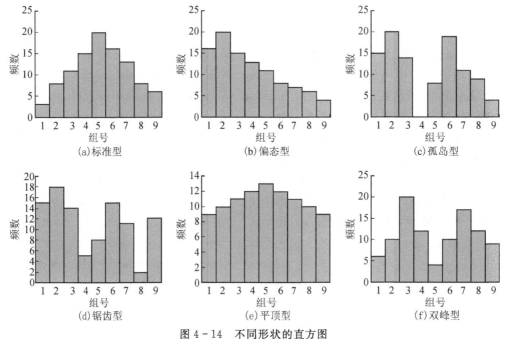

图 4-14 不同形状的直方图

4. 应用案例

从某乳品成品车间随机抽取 100 袋巴氏乳样品,分别测定其净含量,结果如表 4-14 所示。各组频数统计如表 4-15 所示。

表 4-14 某乳品厂 100 袋巴氏乳净含量数据(单位:mL)

198.2	204.0	200.3	195.1	202.5	203.4	201.1	205.2	206.8	210.0
197.3	204.0	199.9	191.2	201.7	203.2	201.1	204.4	206.3	210.0
198.4	204.1	200.5	195.4	202.5	203.5	201.2	205.3	207.0	210.2
198.7	204.2	200.6	196.0	202.6	203.7	201.3	206.0	207.2	210.3
199.2	204.2	200.7	196.2	202.7	203.7	201.3	206.0	207.2	212.8
199.8	204.3	201.0	196.7	202.9	204.0	201.4	206.2	207.2	216.1
199.9	204.3	201.1	197.2	203.0	204.0	201.4	206.2	209.0	210.0
198.0	204.0	200.3	193.4	202.5	203.3	201.1	204.9	206.6	210.2
198.6	204.1	200.5	195.7	202.6	203.5	201.2	206.0	207.1	213.3
199.7	204.2	201.0	196.4	202.8	203.9	201.4	206.1	207.3	210.2

表 4-15 各组频数统计

组号	组界	频数	组号	组界	频数
1	191.15~193.95	2	6	205.15~207.95	17
2	193.95~196.75	7	7	207.95~210.75	8
3	196.75~199.55	8	8	210.75~213.55	2
4	199.55~202.35	24	9	213.55~216.35	1
5	202.35~205.15	31			

依据直方图的作法绘出直方图,如图 4-15 所示。

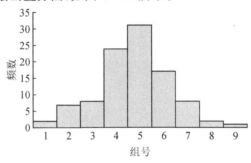

图 4-15 某乳品厂 100 袋巴氏乳净含量直方图

4.7.2 实训任务

直方图在啤酒生产中的应用。

实训提升技能点:在食品相关行业中灵活应用直方图。

专业技能点:啤酒相关知识。

职业素养技能点:①调研能力;②沟通能力。

实训组织:对学生进行分组,每个组参照"基础知识"中的内容,收集啤酒企业在生产过程中的质量安全数据,并作出直方图对其进行分析。实训结束后,每组在班级中汇报调研结论和调研实训提升了自己哪些能力。

实训提交文件:啤酒相关质量安全数据直方图。

实训评价:由啤酒企业或主讲教师进行评价。

学生姓名	数据收集的完整性 (20 分)	数据描述的正确性 (20 分)	回答质疑的准确性 (10 分)	直方图 (50 分)

思考题

1. 直方图的概念及用途是什么?
2. 如何制作直方图?作图过程中应注意什么问题?

3. 参照《数据说话:基于统计技术的质量改进》和《质量管理学》等相关资料学习直方图在食品质量安全数据中的应用。

项目 5　ISO9000 质量管理体系的建立与认证

项目概述

本项目介绍在企业内部建立 ISO9000 质量管理体系的重要性,通过实施、认证 ISO9000 质量管理体系,使企业强化管理,提高人员素质和企业文化,并使企业的形象和市场份额得到提升。

任务 1　质量管理体系

1. ISO9000 标准的产生和发展

ISO 是 International Organization for Standardization(国际标准化组织)的简称,ISO9000 质量管理体系是其制定的国际标准之一,1987 年首次提出此概念,是指"由 ISO/TC176(国际标准化组织质量管理和质量保证技术委员会)制定的所有国际标准"。该标准可帮助组织实施并有效运行质量管理体系,是质量管理体系通用的要求或指南。它不受具体的行业或经济部门的限制,可广泛适用于各种类型和规模的组织,在国内和国际贸易中促进相互理解。

最初的 ISO9000 把质量仅仅看作是技术,20 世纪 80 年代以后,人们发现这种看法是不完整的,质量不仅是技术问题,也是人类的资源问题。质量与经营是联系在一起的。值得注意的是,新版 ISO9000 族标准将全面融合国际质量宗师(如朱兰、戴明、费根堡姆等)对质量管理的经营理念和质量改进的方法以及质量管理思想,使新标准注入了更为丰富的内涵。

从 1987 年开始出版的 ISO9000 标准,至今已成为国际标准化组织的标准中应用最广泛的标准。基于多年成功的经验,它建立了一个质量管理的通用框架,并通过定义 ISO9001《质量管理体系　要求》,为组织提供了生产合格产品的基本信心,促进了全球贸易。2000 年质量管理体系发生重大修订后,在实践和技术方面都有了新的变化,反映组织在运营过程中日益加剧的复杂性、动态的环境变化和增长的需求。2012 年,国际标准化组织开始启动下一代质量管理标准新框架的研究工作,继续强化质量管理体系标准对于经济可持续增长的基础作用,以实现为未来十年或更长时间提供一个稳定的系列的核心要求;保留其通用性,适用于在任何类型、规模及行业的组织中运行;关注有效的过程管理,以便实现预期的输出。通过应用国际标准化组织的导则,增强其同其他国际标准化组织管理体系标准的兼容性和符合性,以推进其在组织内实施第一方、第二方和第三方的合格评定活动;利用简单化的语言和描述形式,便于加深理解并统一对各项要求的阐述。ISO9000:2015 版标准的主要变化在于其格式变化,以及增加了风险的重要性,包括采用与其他管理体系标准相同的新的高级结构,有利于公司执行一个以上的管理体系标准。风险识别和风险控制成为标准的要求,要求最高管理层在协调质量方针与业务需要方面采取更积极的举措。

2. ISO9000 标准的构成

1）ISO9000:2015 版标准的构成框架

ISO9000:2015 版标准的文件结构详见表 5-1 和图 5-1。

表 5-1　ISO9000:2015 版标准的文件结构

核心标准	其他标准	技术报告	小册子
ISO9000 ISO9001 ISO9004 ISO19011	ISO10012 （测量设备的质量保证要求）	ISO10006 ISO10007 ISO10013 ISO10014 ISO10015 ISO10017	·质量管理原理 选择和使用指南 ·小型企业的应用

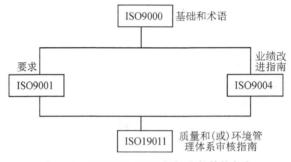

图 5-1　ISO9000:2015 版标准的结构框架

在 ISO9000:2015 版标准中，只包括 4 个核心标准：ISO9000、ISO9001、ISO9004、ISO19011。以下分别介绍这 4 项核心标准。

2）ISO9000《质量管理体系　基础和术语》

该标准表述了质量管理体系的基础知识，并确定了相关的术语。标准明确了帮助组织获得持续成功，确定质量管理七项原则是组织改进业绩的框架，也是 ISO9000 质量管理体系的基础。它表述了建立和运行质量管理体系应遵循的 12 个方面的质量管理体系基础知识，体现了七项质量管理原则的具体应用。

3）ISO9001《质量管理体系　要求》

该标准提供了质量管理体系的要求，供组织需要证实其具有稳定地提供顾客要求和适用法律法规要求产品的能力时应用。组织可通过体系的有效应用，包括持续改进体系的过程及保证符合顾客与适用的法规要求，增进顾客满意。

此标准成为用于审核和第三方认证的唯一标准。它可用于内部和外部（第二方或第三方）评价组织满足组织自身要求和顾客及法律法规要求的能力。当组织及其产品的特点不适用时，可以考虑删减。

标准应用了以过程为基础的质量管理模式结构，鼓励组织在建立、实施和改进质量管理体系及提高有效性时，采用过程方法，通过满足顾客要求，增强顾客满意。过程方法的优点是对质量管理体系中诸多单个过程之间的关系及过程的组合和相互作用进行连续的控制。

4）ISO9004《组织持续成功的管理 一种质量管理方法》

该标准以八项质量管理原则为基础，帮助组织用有效和高效的方式识别并满足顾客和其他相关方的需求和期望，实现、保持和改进组织的整体业绩和能力，从而使组织获得成功。

该标准提供了超出 ISO9001 要求的指南和建议，不用于认证或合同的目的，也不是 ISO9001 的实施指南。标准强调组织质量管理体系的设计和实施受各种需求、具体目标、所提供的产品、所采用的过程及组织的规模和结构的影响，无意统一质量管理体系的结构或文件。

5）ISO19011《质量和（或）环境管理体系审核指南》

该标准遵循"不同管理体系可以有共同管理和审核要求"的原则，为质量管理和环境管理审核的基本原则、审核方案的管理、环境和质量管理体系审核的实施，以及环境和质量管理体系审核员的资格要求提供了指南。它适用于所有运行质量和（或）环境管理体系的组织，指导其内审和外审的管理工作。

3. ISO9000 标准的作用和意义

食品企业要想长久发展，提高市场竞争力，取得 ISO9000 认证是势在必行的，其优势在于以下几方面。

1）强化品质管理，提高企业效益

通过国家认可的权威机构，对食品企业的质量管理体系进行严格的审核，使其达到国际标准化的要求。对于企业来说，可以真正达到法治化、科学化的要求，极大地提高工作效率和产品合格率，迅速提高企业的经济效益和社会效益。

2）提高了食品企业的质量信誉，增强客户信心，扩大市场份额

食品企业取得了 ISO9000 认证，对于企业外部来说，当顾客得知供方按照国际标准实行管理，拿到了 ISO9000 品质体系认证证书，并且有认证机构的严格审核和定期监督，就可以确信该企业能够稳定地提供合格产品或服务，从而放心地与企业订立供销合同，扩大了企业的市场占有率。

3）消除了国际贸易壁垒，增强了产品的竞争力

许多国家为了保护自身的利益，设置了种种贸易壁垒，包括关税壁垒和非关税壁垒。其中非关税壁垒主要是技术壁垒，技术壁垒中，又主要是产品品质认证和 ISO9000 品质体系认证的壁垒。特别是在世界贸易组织内，各成员国之间相互排除了关税壁垒，只能设置技术壁垒，所以获得认证是消除贸易壁垒的主要途径。我国"入世"以后，失去了区分国内贸易和国际贸易的严格界限，所有贸易都有可能遭遇上述技术壁垒，所以取得 ISO9000 认证对排除贸易障碍起到了十分积极的作用。

4）节省了第二方审核的精力和费用

在现代贸易实践中，经销商审核早就成为惯例，其存在很大的弊端：一个食品企业通常要为经销商供货，经销商审核无疑会给食品企业带来沉重的负担；另一方面，经销商也需支付相当的费用，同时还要考虑派出或雇佣人员的经验和水平问题，否则花了费用也达不到预期的目的。有了 ISO9000 认证就可以排除这样的弊端。因为食品企业申请了 ISO9000 认证并获得了认证证书以后，众多的经销商就不必再对食品企业进行重复审核了。这样不管是食品企业还是经销商都可以节省很多精力或费用。企业在获得了 ISO9000 认证之后，再申请 UL、CE 等产品品质认证，还可以免除认证机构对企业的质量管理体系进行重复认证的开支。

5) 有利于国际间的经济合作和技术交流

按照国际间经济合作和技术交流的惯例,合作双方必须在产品(包括服务)品质方面有共同的语言、统一的认识和共守的规范,方能进行合作与交流。ISO9000 质量管理体系认证正好提供了这样的信任,有利于双方迅速达成协议。

任务 2　质量管理认证标准

1. ISO9000 质量管理的七项原则

1) 以顾客为关注焦点

质量管理的主要关注点是满足顾客要求并且努力超越顾客的期望。组织只有赢得顾客和其他相关方的信任才能获得持续成功。与顾客相互作用的每个方面,都提供了为顾客创造更多价值的机会。理解顾客和其他相关方当前和未来的需求,有助于组织的持续成功。

2) 领导力

各层领导建立统一的宗旨和方向,并且创造全员参与的条件,以实现组织的质量目标。统一的宗旨和方向以及全员参与,能够使组织将战略、方针、过程和资源保持一致,以实现其目标。

3) 全员参与

整个组织内各级人员的胜任、授权和参与,是提高组织创造价值和提供价值能力的必要条件。为了有效和高效地管理组织,各级人员得到尊重并参与其中是极其重要的。通过表彰、授权和提高能力,促进在实现组织的质量目标过程中的全员参与。

Ⅰ. Seven Principles of ISO9000 Quality Management

A. Focus on Customers

The main focus of quality management is to meet customer requirements and strive to exceed customer expectations. An organization can only be successful if it wins the trust of customers and other interested parties. Every aspect of interaction with the customer provides the opportunity to create more value for the customer. Understanding the current and future needs of customers and other stakeholders contributes to the continued success of the organization.

B. Leadership

Leaders at all levels establish a unified purpose and direction and create conditions for full participation so as to achieve the quality goals of the organization. A unified purpose and direction, as well as full participation, enable the organization to align its strategies, policies, processes and resources so as to achieve its goals.

C. Full Participation

The competence, empowerment and participation of people at all levels throughout the organization are necessary to enhance the ability of the organization to create and deliver value. In order to manage an organization effectively and efficiently, it is essential that people at all levels be respected and involved. Promote full participation in the achievement of the organization's quality goals through recognition, delegation and enhancement of competencies.

4) 过程方法

当活动被作为相互关联、功能连贯的过程进行系统管理时，可更加有效和高效地始终得到预期的结果。质量管理体系是由相互关联的过程所组成的。理解体系如何产生结果，能够使组织尽可能地完善其体系和绩效。

5) 改进

成功的组织总是致力于持续改进。改进对于组织保持当前的业绩水平，对其内、外部条件的变化作出反应并创造新的机会都是非常必要的。

6) 决策

决策基于事实（循证决策）。基于数据和信息的分析和评价的决策更有可能产生期望的结果。决策是一个复杂的过程，并且总是包含一些不确定因素。它常涉及多种类型和来源的输入及其解释，而这些解释可能是主观的。重要的是理解因果关系和潜在的非预期后果。对事实、证据和数据的分析可使决策更加客观，因而更有信心。

7) 关系管理

为了持续成功，组织需要管理与供方等相关方的关系。相关方影响组织的绩效。组织管理与所有相关方的关系，以最大限度地发挥其在组织绩效方面的作用。对供方及合作伙伴的关系网的管理是非常重要的。

质量管理的七项原则是一个组织在质量管理方面的总体原则，这些原则需要通过具体的活动得到体现。其应用可分为质量保证和质量管理两个

D. Process Approach

When activities are systematically managed as interrelated functional coherence processes, the desired results can be achieved more effectively and efficiently at all times. The quality management system is composed of interrelated processes. Understanding how the system produces results enables the organization to improve its system and performance as much as possible.

E. Improvement

Successful organizations are always committed to continuous improvement. Improvements are necessary for the organization to maintain its current level of performance, to respond to changes in its internal and external conditions and to create new opportunities.

F. Decision-making

The decision making is based on fact (evidence-based policy making). Decision-making based on data and information analysis and evaluation is more likely to produce desired results. Decision making is a complex process and always involves some uncertainties. It often involves multiple types and sources of input and their interpretations, which may be subjective. It is important to understand causality and potential unintended consequences. Analysis of facts, evidence and data can lead to more objective and confident decision-making.

G. Relationship Management

For continued success, the organization needs to manage relationships with relevant parties, such as suppliers. Stakeholders influence the performance of the organization. Managing relationships with all relevant parties to maximize their role in organizational performance. The management of supplier and partner networks is very important.

The seven principles of quality management, which are the overall principles of an organization in terms of quality management, need to be reflected through specific activities. Its application can be

层面。

就质量保证来说,主要目的是取得足够的信任,以表明组织能够满足质量要求。因而所开展的活动主要涉及测定顾客的质量要求、设定质量方针和目标、建立并实施文件化的质量体系,最终确保质量目标的实现。

质量管理则要考虑,作为一个组织经营管理(这里说的不是营销管理)的重要组成部分,怎样保证经营目标的实现。组织要生存、要发展、要提高效率和效益,当然离不开顾客,离不开质量。因而,从质量管理的角度,要开展的活动就其深度和广度来说,远高于质量保证所需开展的活动。

2. ISO9000 质量管理体系的十二项基础

(1)质量管理体系理论说明。

(2)质量管理体系要求与产品要求。

(3)质量管理体系方法。

(4)过程方法。

(5)质量方针和质量目标。

(6)最高管理者在质量管理体系中的作用。

(7)文件。

(8)质量管理体系评价。

(9)持续改进。

(10)统计技术的作用。

(11)质量管理体系与其他管理体系的关注点。

(12)质量管理体系与组织优秀模式之间的关系。

divided into two levels: quality assurance and quality management.

In terms of quality assurance, the main objective is to gain sufficient trust to demonstrate that the organization is able to meet quality requirements. Therefore, the main activities involved measuring customer quality requirements, setting quality policies and objectives, establishing and implementing a documented quality system, and ultimately ensuring the achievement of quality objectives.

Quality management should consider, as an important part of an organization management (not marketing management), how to ensure the realization of business objectives. If an organization wants to survive, develop, improve efficiency and benefit, of course, it cannot do without customers, cannot do without quality. Thus, from a quality management perspective, the activities to be undertaken are far more profound and extensive than those required for quality assurance.

Ⅱ. Twelve Foundations of the ISO9000 Quality Management System

1. Theory of the quality management system.

2. The requirements and the product requirements of the quality management system.

3. Methods of quality management system.

4. Process approach.

5. Quality policy and quality objectives.

6. The role of the top management in the quality management system.

7. Documents.

8. Evaluation of quality management system.

9. Continuous improvement.

10. The role of statistical technology.

11. The focus of quality management system and other management system.

12. The relationship between quality management system and organizational excellence model.

3. ISO9001:2015 版标准总则

1)总则

采用质量管理体系应该是组织的一项战略性决策,可以帮助组织改进其整体绩效,并为可持续发展计划提供良好的基础。

对于根据本标准实施质量管理体系的组织来说,潜在的收益是:

(1)稳定提供满足顾客要求和法律法规要求的产品和服务的能力;

(2)获取增强顾客满意的机会;

(3)应对与组织环境和目标相关的风险;

(4)证实符合质量管理体系特定要求的能力。

本标准可用于内部和外部。

以下方面不是本标准的目的:

(1)统一不同质量管理体系的结构;

(2)统一本标准条款结构的文件;

(3)在组织中使用本标准的特定术语。

本标准所规定的质量管理体系要求是对产品要求的补充。

本标准采用过程方法,该方法结合了策划—实施—检查—改进(PDCA)循环和基于风险的思维。过程方法使组织能够策划其过程及相互作用。PDCA 循环使组织能够确保其过程得到充分的资源和管理,并确定和实施改进机会。基于风险的思维能够使组织确定可能导致其过程和质量管理体

Ⅲ. ISO9001:2015 Standard General

A. General

The adoption of a quality management system is a strategic decision for an organization that can help improve its overall performance and provide a sound basis for sustainable development initiatives.

The potential benefits to an organization of implementing a quality management system based on this International Standard are:

1. The ability to consistently provide products and services that meet customer and applicable statutory and regulatory requirements;

2. Facilitating opportunities to enhance customer satisfaction;

3. Addressing risks and opportunities associated with its context and objectives;

4. The ability to demonstrate conformity to specified quality management system requirements.

This International Standard can be used by internal and external parties.

It is not the intent of this International Standard to imply the need for:

1. Uniformity in the structure of different quality management systems;

2. Alignment of documentation to the clause structure of this International Standard;

3. The use of the specific terminology of this International Standard within the organization.

The quality management system requirements specified in this International Standard are complementary to requirements for products and services.

This International Standard employs the process approach, which incorporates the Plan-Do-Check-Act (PDCA) cycle and risk-based thinking. The process approach enables an organization to plan its processes and their interactions. The PDCA cycle enables an organization to ensure that its processes are adequately resourced and managed, and that opportunities for improvement are determined and

系偏离所策划的结果的因素,采取预防性控制,以最小化负面影响,并在机会出现时将机会最大化。在日益变化和复杂的环境中,持续满足要求和应对未来的需求和期望是组织面临的挑战。要实现这个目标,组织可能发现除了纠正和持续改进以外,变革、创新和重组也是必要的。

本标准中采用了以下动词形式:"shall"表示要求;"should"表示建议;"may"表示允许;"can"表示可能性或能力。标注为"注"的信息是理解或说明相关要求的指南。

2) 质量管理原则

本标准基于ISO9000中阐述的质量管理原则,包括每项原则的说明、对组织的重要性、与该原则相关的益处的示例,以及应用该原则时改进组织绩效所采取的典型措施的示例。

质量管理的原则是:以顾客为关注焦点;领导力;全员参与;过程方法;改进;循证决策;关系管理。

3) 过程方法

(1) 总则。本标准鼓励在建立、实施质量管理体系以及改进其有效性时采用过程方法,通过满足顾客要求,增强顾客满意。采用过程方法须考虑特定要求。

acted on. Risk-based thinking enables an organization to determine the factors that could cause its processes and its quality management system to deviate from the planned results, to put in place preventive controls to minimize negative effects and to make maximum use of opportunities as they arise. Consistently meeting requirements and addressing future needs and expectations poses a challenge for organizations in an increasingly dynamic and complex environment. To achieve this objective, the organization might find it necessary to adopt various forms of improvement in addition to correction and continual improvement, such as breakthrough change, innovation and reorganization.

In this International Standard, the following verbal forms are used: "shall" indicates a requirement; "should" indicates a recommendation; "may" indicates a permission; "can" indicates a possibility or a capability. Information marked as "NOTE" is for guidance in understanding or clarifying the associated requirement.

B. Quality Management Principles

This International Standard is based on the quality management principles described in ISO9000. The descriptions include a statement of each principle, a rationale of why the principle is important for the organization, some examples of benefits associated with the principle and examples of typical actions to improve the organization's performance when applying the principle.

The quality management principles are: customer focus; leadership; engagement of people; process approach; improvement; evidence-based decision making; relationship management.

C. Process Approach

1. General. This International Standard promotes the adoption of a process approach when developing, implementing and improving the effectiveness of a quality management system, to enhance customer satisfaction by meeting customer

将相互关联的过程作为体系进行理解和管理,会有助于组织实现其预期结果的有效性和效率。该方法能够使组织控制体系过程间的相互关系和相互依存,以使组织的整体绩效得到提高。

过程方法运用系统的定义和管理过程及其相互作用,以期实现与组织的质量方针和战略方向一致的预期结果。为了利用机会优势和预防不期望的结果,基于风险的思维,运用PDCA循环将能够实现对过程和整个体系进行管理。

在质量管理体系中应用过程方法能够:对要求的满足达成理解和一致性;从增值的角度考虑过程;实现有效的过程绩效;在评价数据和信息的基础上改进过程。

图5-2给出了过程示意,并展示了过程要素的相互作用。监督和测量检查对控制而言是必要的,应具体到每个过程,并根据相关风险而改变。

(2)策划—实施—检查—改进(PDCA)循环。PDCA循环能够应用于所有过程和整个质量管理体系。

图5-3展示了《质量管理体系 要求》中第4)至第10)点与PDCA的关联。

requirements. Specific requirements are considered essential to the adoption of a process approach.

Understanding and managing interrelated processes as a system contributes to the organization's effectiveness and efficiency in achieving its intended results. This approach enables the organization to control the interrelationships and interdependencies among the processes of the system, so that the overall performance of the organization can be enhanced.

The process approach involves the systematic definition and management of processes, and their interactions, so as to achieve the intended results in accordance with the quality policy and strategic direction of the organization. Management of the processes and the system as a whole can be achieved using the PDCA cycle with an overall focus on risk-based thinking aimed at taking advantage of opportunities and preventing undesirable results.

The application of the process approach in a quality management system enables: understanding and consistency in meeting requirements; the consideration of processes in terms of added value; the achievement of effective process performance; improvement of processes based on evaluation of data and information.

Figure 5-2 gives a schematic representation of any process and shows the interaction of its elements. The monitoring and measuring check points, which are necessary for control, are specific to each process and will vary depending on the related risks.

2. Plan-Do-Check-Act (PDCA) cycle. The PDCA cycle can be applied to all processes and to the quality management system as a whole.

Figure 5-3 illustrates how Clauses D to J can be grouped in relation to the PDCA cycle.

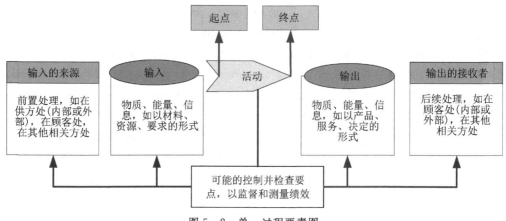

图 5-2 单一过程要素图

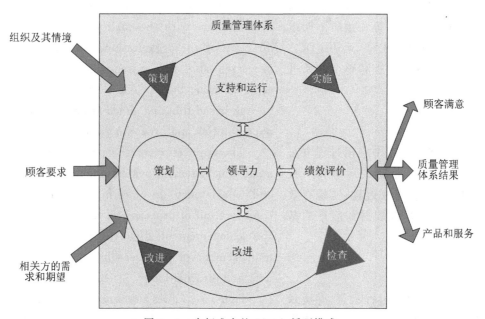

图 5-3 本标准中的 PDCA 循环模式

PDCA 模式可以简要描述如下。

①策划（Plan）：根据顾客要求和组织方针，为提供结果安排好体系目标及其过程，以及所需的资源，识别并提出风险及机会。

②实施（Do）：实施所策划的安排。

③检查（Check）：根据方针、目标和要求及策划的活动，对过程、产品和服务进行监督和测量（适用时），并报

The PDCA cycle can be briefly described as follows.

Plan：establish the objectives of the system and its processes, and the resources needed to deliver results in accordance with customers' requirements and the organization's policies; and identify and address risks and opportunities.

Do：implement what was planned.

Check：monitor and (where applicable) measure processes and the resulting products and services against policies, objectives and requirements and

告结果。

④改进(Act)：必要时采取措施，以改进过程绩效。

(3)基于风险的思维。为实现有效的质量管理体系，基于风险的思维是必要的。基于风险的思维的概念在本标准的以往版本中一直是没有言明的，如实施预防措施以消除潜在不合格，分析发生的不合格并采取与不合格的影响相适应的措施防止其再发生。

为符合本标准的要求，组织需要策划和实施应对风险和机会的措施。应对风险和机会是为提高质量管理体系有效性、实现改进的结果并防止负面影响建立基础。

机会可以形成有益于实现预期结果的状况，例如，让组织吸引顾客、开发新产品和服务、减少浪费或改进生产效率的一系列情况。应对机会的措施还可以包括对相关风险的考虑。风险是不确定的影响，这种不确定可能是正面或负面的影响。来自风险的正面偏离可以提供机会，但不是风险的所有正面影响都能产生机会。

4)与其他管理体系标准的关系

本标准采用了国际标准化组织为改进其管理体系国际标准间的一致性而开发的框架。

本标准能够使组织应用过程方

planned activities, and report the results.

Act: take actions to improve performance, as necessary.

3. Risk-based thinking. Risk-based thinking is essential for achieving an effective quality management system. The concept of risk-based thinking has been implicit in previous editions of this International Standard including, for example, carrying out preventive action to eliminate potential nonconformities, analyzing any nonconformities that do occur, and taking action to prevent recurrence that is appropriate for the effects of the nonconformity.

To conform to the requirements of this International Standard, an organization needs to plan and implement actions to address risks and opportunities. Addressing both risks and opportunities establishes a basis for increasing the effectiveness of the quality management system, achieving improved results and preventing negative effects.

Opportunities can arise as a result of a situation favourable to achieving an intended result, for example, a set of circumstances that allow the organization to attract customers, develop new products and services, reduce waste or improve productivity. Actions to address opportunities can also include consideration of associated risks. Risk is the effect of uncertainty and any such uncertainty can have positive or negative effects. A positive deviation arising from a risk can provide an opportunity, but not all positive effects of risk result in opportunities.

D. Relationship with Other Management System Standards

This International Standard applies the framework developed by ISO to improve alignment among its International Standards for management systems.

This International Standard enables an

法,加上 PDCA 循环和基于风险的思维,使其质量管理体系与其他管理体系标准的要求保持一致或整合。

本标准与 ISO9000 和 ISO9004 相关联:ISO9000《质量管理体系 基础和术语》为本标准的正确理解和实施提供了必要的背景;ISO9004《组织持续成功的管理 一种质量管理方法》为选择超越本标准要求发展的组织提供了指南。

本标准不包括针对其他管理体系的特定要求,如环境管理、职业健康安全管理或财务管理。

已开发了一部分基于本标准要求的特定行业的质量管理体系标准,这些标准中的某些标准规定了附加的质量管理体系要求,而其他标准限于为本标准在特定行业内的应用提供指南。

4.《质量管理体系 要求》
1)范围
本标准为有下列需求的组织规定了质量管理体系要求:
(1)需要证实其有能力稳定地提供满足顾客要求和适用的法律法规要求的产品或服务;
(2)通过体系的有效应用,包括体系改进过程的有效应用,以及保证符合顾客要求和适用的法律法规要求,旨在增强顾客满意。

organization to use the process approach, coupled with the PDCA cycle and risk-based thinking, to align or integrate its quality management system with the requirements of other management system standards.

This International Standard relates to ISO9000 and ISO9004 as follows. ISO9000 *Quality Management Systems — Fundamentals and Vocabulary* provides essential background for the proper understanding and implementation of this International Standard; ISO9004 *Managing for the Sustained Success of an Organization — A Quality Management Approach* provides guidance for organizations that choose to progress beyond the requirements of this International Standard.

This International Standard does not include requirements specific to other management systems, such as those for environmental management, occupational health and safety management, or financial management.

Sector-specific quality management system standards based on the requirements of this International Standard have been developed for a number of sectors. Some of these standards specify additional quality management system requirements, while others are limited to providing guidance to the application of this International Standard within the particular sector.

Ⅳ. *Quality Management Systems—Requirements*
A. Scope

This International Standard specifies requirements for a quality management system when an organization:
1. Needs to demonstrate its ability to consistently provide products and services that meet customer and applicable statutory and regulatory requirements, and
2. Aims to enhance customer satisfaction through the effective application of the system, including processes for improvement of the system and the assurance of conformity to customer and

本标准的所有要求是通用的,旨在适用于各种类型、不同规模及提供不同产品和服务的组织。

注 1 本标准中,术语"产品"或"服务"仅适用于预期提供给顾客或顾客所要求的产品和服务。

注 2 法律法规要求可称作法定要求。

2) 规范性引用文件

下列文件中的全部和部分内容在本标准中引用并在应用中不可或缺。凡是标注日期的引用文件,仅该版本适用于本标准。凡未标注日期的引用文件,引用文件的最新版(包括修订版)适用于本标准。

ISO9000:2015《质量管理体系 基础和术语》。

3) 术语和定义

本标准采用 ISO9000:2015 中的术语和定义。

4) 组织的环境

(1) 理解组织及其环境。组织应确定与其宗旨和战略方向有关且影响质量管理体系实现其预期结果的能力的内部和外部环境。

组织应监督和评审有关内部和外部环境的信息。

注 1 环境可能是正面或负面的因素,或要考虑的状况。

注 2 可以通过考虑源于国际、国家、地区或本地的法律法规、技术、竞争、市场、文化、社会和经济环境的情况,促进对外部环境的了解。

applicable statutory and regulatory requirements.

All the requirements of this International Standard are generic and are intended to be applicable to any organization, regardless of its type or size, or the products and services it provides.

NOTE 1 In this International Standard, the terms "product" or "service" only apply to products and services intended for, or required by, a customer.

NOTE 2 Statutory and regulatory requirements can be expressed as legal requirements.

B. Normative References

The following documents, in whole or in part, are normatively referenced in this document and are indispensable for its application. For dated references, only the edition cited applies. For undated references, the latest edition of the referenced document (including any amendments) applies.

ISO9000:2015, *Quality Management Systems — Fundamentals and Vocabulary*.

C. Terms and Definitions

For the purposes of this document, the terms and definitions given in ISO9000:2015 apply.

D. Context of the Organization

1. Understand the organization and its context. The organization shall determine external and internal issues that are relevant to its purpose and its strategic direction and that affect its ability to achieve the intended result(s) of its quality management system.

The organization shall monitor and review information about these external and internal issues.

NOTE 1 Issues can include positive and negative factors or conditions for consideration.

NOTE 2 Understanding the external context can be facilitated by considering issues arising from legal, technological, competitive, market, cultural, social and economic environments, whether international, national, regional or local.

注3 可以通过考虑与价值、文化、知识和组织绩效相关的情况,促进对内部环境的了解。

(2)理解相关方的需求和期望。由于对组织持续提供满足顾客和适用的法律法规要求的产品和服务的能力存在影响或潜在影响,组织应确定:

①与质量管理体系有关的相关方;
②与质量管理体系有关的相关方的要求。

组织应监督和评审相关方及其有关要求的信息。

(3)确定质量管理体系的范围。组织应确定质量管理体系的边界和适用性来建立其范围。

在确定质量管理体系范围时,组织应考虑:
①涉及的外部和内部情况;
②涉及的相关方要求;
③组织的产品与服务。

如本标准的要求在确定的范围内适用,组织应应用本标准的所有要求。

组织的质量管理体系范围应作为文件资料提供和保存。范围应说明质量管理体系所覆盖的产品和服务,并对组织确定不适用于管理体系范围的本标准的要求说明正当理由。

NOTE 3 Understanding the internal context can be facilitated by considering issues related to values, culture, knowledge and performance of the organization.

2. Understand the needs and expectations of interested parties. Due to their effect or potential effect on the organization's ability to consistently provide products and services that meet customer and applicable statutory and regulatory requirements, the organization shall determine:

a. The interested parties that are relevant to the quality management system;
b. The requirements of these interested parties that are relevant to the quality management system.

The organization shall monitor and review information about these interested parties and their relevant requirements.

3. Determine the scope of the quality management system. The organization shall determine the boundaries and applicability of the quality management system to establish its scope.

When determining this scope, the organization shall consider:

a. The external and internal issues referred to;
b. The requirements of relevant interested parties referred to;
c. The products and services of the organization.

The organization shall apply all the requirements of this International Standard if they are applicable within the determined scope of its quality management system.

The scope of the organization's quality management system shall be available and be maintained as documented information. The scope shall state the types of products and services covered, and provide justification for any requirement of this International Standard that the organization determines is not applicable to the scope of its quality management system.

当被确定为不适用的要求不影响组织确保产品和服务的符合性,以及增强顾客满意的能力或责任时,才能声称符合本标准的要求。

(4)质量管理体系及其过程。

①组织应按本标准的要求建立、实施、保持和持续改进质量管理体系,包括质量管理体系所需的过程及其相互作用。

组织应确定质量管理体系所需的过程及其在整个组织中的应用,组织应:

A. 确定这些过程所需的输入和期望的输出;

B. 确定这些过程的顺序和相互作用;

C. 确定和应用所需的准则和方法(包括监督、测量和相关的绩效指标),以确保这些过程的有效运行和控制;

D. 确定这些过程所需的资源并确保其可用性;

E. 规定这些过程的职责和权限;

F. 应对按照6)中(1)的要求确定的风险和机会;

G. 评价这些过程并实施所需的变更,以确保这些过程实现预期的结果;

H. 改进过程和质量管理体系。

②在必要的范围内,组织应:

A. 保存文件资料,为过程运行提供支持;

Conformity to this International Standard may only be claimed if the requirements determined as not being applicable do not affect the organization's ability or responsibility to ensure the conformity of its products and services and the enhancement of customer satisfaction.

4. Quality management system and its processes.

a. The organization shall establish, implement, maintain and continually improve a quality management system, including the processes needed and their interactions, in accordance with the requirements of this International Standard.

The organization shall determine the processes needed for the quality management system and their application throughout the organization, and shall:

a) Determine the inputs required and the outputs expected from these processes;

b) Determine the sequence and interaction of these processes;

c) Determine and apply the criteria and methods (including monitoring, measurements and related performance indicators) needed to ensure the effective operation and control of these processes;

d) Determine the resources needed for these processes and ensure their availability;

e) Assign the responsibilities and authorities for these processes;

f) Address the risks and opportunities as determined in accordance with the requirements of Clause F.1;

g) Evaluate these processes and implement any changes needed to ensure that these processes achieve their intended results;

h) Improve the processes and the quality management system.

b. To the extent necessary, the organization shall:

a) Maintain documented information to support the operation of its processes;

B. 保留文件,以确保过程是按计划执行的。

5)领导力

(1)领导力和承诺。

①总则。最高管理者应通过以下方面证实其领导力和对质量管理体系的承诺：

A. 对质量管理体系的有效性承担责任；

B. 确保质量方针和质量目标得到建立,并与组织的战略方向和组织环境保持一致；

C. 确保将质量管理体系要求融入组织的业务过程；

D. 促进过程方法和基于风险的思维的应用；

E. 确保获得质量管理体系所需的资源；

F. 传达有效的质量管理以及满足质量管理体系要求的重要性；

G. 确保质量管理体系实现预期的结果；

H. 鼓励、指导和支持员工为质量管理体系的有效性作出贡献；

I. 增强改进；

J. 支持其他相关管理者在其负责的区域证实其领导力。

注 本标准中的"业务"从广义上解释为对于组织的存在而言具有核心价值的活动,组织可以是公有的、私有的、营利或非营利的。

b) Retain documented information to have confidence that the processes are being carried out as planned.

E. Leadership

1. Leadership and commitment.

a. General. Top management shall demonstrate leadership and commitment with respect to the quality management system by：

a) Taking accountability for the effectiveness of the quality management system；

b) Ensuring that the quality policy and quality objectives are established for the quality management system and are compatible with the context and strategic direction of the organization；

c) Ensuring the integration of the quality management system requirements into the organization's business processes；

d) Promoting the use of the process approach and risk-based thinking；

e) Ensuring that the resources needed for the quality management system are available；

f) Communicating the importance of effective quality management and of conforming to the quality management system requirements；

g) Ensuring that the quality management system achieves its intended results；

h) Engaging, directing and supporting persons to contribute to the effectiveness of the quality management system；

i) Promoting improvement；

j) Supporting other relevant management roles to demonstrate their leadership as it applies to their areas of responsibility.

NOTE Reference to "business" in this International Standard can be interpreted broadly to mean those activities that are core to the purposes of the organization's existence, whether the organization is public, private, for profit or not for profit.

②以顾客为关注焦点。最高管理者应通过以下方面,证实其对以顾客为关注焦点的领导力和承诺：

A. 确定、理解和持续满足顾客要求和适用的法律法规要求；

B. 确定和应对影响产品和服务符合性以及增强顾客满意能力的风险与机会；

C. 维护以增强顾客满意为关注焦点。

（2）方针。

①建立质量方针。最高管理者应建立、实施和保持质量方针,方针应：

A. 与组织的宗旨和环境相适应,并支持其战略方向；

B. 提供制定质量目标的框架；

C. 包括对满足适用要求的承诺；

D. 包括对持续改进质量管理体系的承诺。

②沟通质量方针。质量方针应：

A. 为可获得的保存为文件的资料；

B. 在组织内得到沟通、理解和应用；

C. 适当时,可为相关方所获取。

（3）组织的角色、职责和权限。最高管理者应确保相关角色的职责和权限在整个组织得到分派、沟通和理解。

最高管理者应规定职责和权

b. Customer focus. Top management shall demonstrate leadership and commitment with respect to customer focus by ensuring that：

a) Customer and applicable statutory and regulatory requirements are determined, understood and consistently met；

b) The risks and opportunities that can affect conformity of products and services and the ability to enhance customer satisfaction are determined and addressed；

c) The focus on enhancing customer satisfaction is maintained.

2. Policy.

a. Establish the quality policy. Top management shall establish, implement and maintain a quality policy that：

a) Is appropriate to the purpose and context of the organization and supports its strategic direction；

b) Provides a framework for setting quality objectives；

c) Includes a commitment to satisfy applicable requirements；

d) Includes a commitment to continual improvement of the quality management system.

b. Communicate the quality policy. The quality policy shall：

a) Be available and be maintained as documented information；

b) Be communicated, understood and applied within the organization；

c) Be available to relevant interested parties, as appropriate.

3. Organizational roles, responsibilities and authorities. Top management shall ensure that the responsibilities and authorities for relevant roles are assigned, communicated and understood within the organization.

Top management shall assign the responsibility and

限以：

①确保质量管理体系符合本标准的要求；

②确保过程实现其预期的输出；

③向最高管理者报告质量管理体系绩效和改进机会；

④确保在整个组织提高以顾客为关注焦点的意识；

⑤对质量管理体系的变更进行策划和实施时，维护质量管理体系的完整性。

6）策划

（1）应对风险和机会的措施。

①策划质量管理体系时，组织应考虑4）中(1)和(2)的要求，并确定需要应对的风险和机会以：

A. 确保质量管理体系能够实现其预期的结果；

B. 增强期望的影响；

C. 预防或减少非预期的影响；

D. 实现持续改进。

②组织应策划：

A. 应对风险和机会的措施；

B. 如何在质量管理体系过程中融入和实施这些措施［见 4)中第(4)点］，及如何评价这些措施的有效性。

所采取的应对风险和机会的措施应与对产品和服务符合性的潜在影响相适应。

注 1 应对风险的选择可以包括避免风险、为获取机会而接受风险、消

authority for：

a. Ensuring that the quality management system conforms to the requirements of this International Standard；

b. Ensuring that the processes are delivering their intended outputs；

c. Reporting on the performance of the quality management system and on opportunities for improvement, in particular to top management；

d. Ensuring the promotion of customer focus throughout the organization；

e. Ensuring that the integrity of the quality management system is maintained when changes to the quality management system are planned and implemented.

F. Planning

1. Actions to address risks and opportunities.

a. When planning for the quality management system, the organization shall consider the issues referred to in Clause D.1 and the requirements referred to in Clause D.2 and determine the risks and opportunities that need to be addressed to：

a) Give assurance that the quality management system can achieve its intended result(s)；

b) Enhance desirable effects；

c) Prevent, or reduce, undesired effects；

d) Achieve improvement.

b. The organization shall plan：

a) Actions to address these risks and opportunities；

b) How to integrate and implement the actions into its quality management system processes (see Clause D.4)；evaluate the effectiveness of these actions.

Actions taken to address risks and opportunities shall be proportionate to the potential impact on the conformity of products and services.

NOTE 1 Options to address risks can include avoiding risk, taking risk in order to pursue an

除风险源、改变可能性或结果、分担风险或经过决策而保留风险。

注2 机会可以带来新实践的采用、发布新产品、打开新市场、获得新客户、建立合作关系、使用新技术,以及其他期望的或可行的可能性,以满足组织或其顾客的需求。

(2)质量目标及实现目标的策划。

①组织应在质量管理体系所需的相关职能、层次和过程上建立质量目标。

质量目标应:

A. 与质量方针保持一致;

B. 可测量;

C. 考虑适用的要求;

D. 与产品和服务的符合性以及增强顾客满意有关;

E. 受到监督;

F. 得到沟通;

G. 适当时进行更新。

组织应保存质量目标的文件资料。

②策划如何实现其质量目标时,组织应确定:

A. 要做什么;

B. 需要什么资源;

C. 由谁负责;

D. 什么时候完成;

E. 如何评价结果。

(3)变更的策划。当组织确定了质量管理体系变更的需求时[见4)中第(4)点],应按策划的方式进行变更。

组织应考虑:

①变更的目的及其潜在后果;

opportunity, eliminating the risk source, changing the likelihood or consequences, sharing the risk, or retaining risk by informed decision.

NOTE 2 Opportunities can lead to the adoption of new practices, launching new products, opening new markets, addressing new customers, building partnerships, using new technology and other desirable and viable possibilities to address the organization's or its customers' needs.

2. Quality objectives and planning to achieve them.

a. The organization shall establish quality objectives at relevant functions, levels and processes needed for the quality management system.

The quality objectives shall:

a) Be consistent with the quality policy;

b) Be measurable;

c) Take into account applicable requirements;

d) Be relevant to conformity of products and services and to enhancement of customer satisfaction;

e) Be monitored;

f) Be communicated;

g) Be updated as appropriate.

The organization shall maintain documented information on the quality objectives.

b. When planning how to achieve its quality objectives, the organization shall determine:

a) What will be done;

b) What resources will be required;

c) Who will be responsible;

d) When it will be completed;

e) How the results will be evaluated.

3. Planning of changes. When the organization determines the need for changes to the quality management system, the changes shall be carried out in a planned manner (see Clause D. 4).

The organization shall consider:

a. The purpose of the changes and their

②质量管理体系的完整性；

③资源的获得；

④职责和权限的分配与再分配。

7）支持

（1）资源。

①总则。组织应确定和提供建立、实施、维护和持续改进质量管理体系所需的资源。

组织应考虑：

A. 现有内部资源的能力和局限；

B. 需要从外部供方获取的资源。

②人。组织应确定和提供质量管理体系有效实施及过程的运行和控制所需的人员。

③基础设施。组织应确定、提供并维护为达到产品和服务符合要求的过程的运行所需的基础设施。

注 基础设施可包括：
A. 建筑物和相关的设施；
B. 设备，包括硬件和软件；
C. 运输资源；
D. 信息和通信技术。

④过程运行环境。组织应确定、提供并维护过程的运行与达到产品和服务的符合性所需的环境。

注 适宜的环境可以是人文的因素和物理的因素的组合，例如：

A. 社会的（如不歧视、平和、不对

potential consequences;

b. The integrity of the quality management system;

c. The availability of resources;

d. The allocation or reallocation of responsibilities and authorities.

G. Support

1. Resources.

a. General. The organization shall determine and provide the resources needed for the establishment, implementation, maintenance and continual improvement of the quality management system.

The organization shall consider:

a) The capabilities and constraints on existing internal resources;

b) What needs to be obtained from external providers.

b. People. The organization shall determine and provide the persons necessary for the effective implementation of its quality management system and for the operation and control of its processes.

c. Infrastructure. The organization shall determine, provide and maintain the infrastructure necessary for the operation of its processes and to achieve conformity of products and services.

NOTE Infrastructure can include:

a) Buildings and associated utilities;

b) Equipment, including hardware and software;

c) Transportation resources;

d) Information and communication technology.

d. Environment for the operation of processes. The organization shall determine, provide and maintain the environment necessary for the operation of its processes and to achieve conformity of products and services.

NOTE A suitable environment can be a combination of human and physical factors, such as:

a) Social (e. g. non-discriminatory, calm, non-

抗);

B. 心理的(如压力降低、倦怠预防、情感保护);

C. 物理的(温度、热度、湿度、照明、空气流通、卫生、噪音)。

根据所提供的产品和服务,这些因素可能有显著的差异。

⑤监督和测量资源。

A. 总则。当监督或测量用于产品和服务符合特定要求的证据时,组织应确定和提供所需的资源以确保有效和可靠的结果。

组织应确保所提供的资源:

a. 与所进行的监督和测量活动的具体类型相适应;

b. 能够维持,以确保其持续满足使用要求。

组织应保留适当的文件资料,作为监督和测量资源满足使用要求的证据。

B. 测量的追溯性。对测量有追溯性要求,或组织认为是对测量结果的有效性提供信心的必要部分时,测量设备应:

a. 根据可溯源至国际或国家标准的测量标准,按照规定的时间间隔或在使用前进行检定或校准,或两者都进行;当不存在上述标准时,应保留作为校准或检定依据的文件资料;

b. 具有标识,以确定其校准状态;

c. 防止可能使校准状态和后续测量结果失效的调整、损坏或退化。

当发现测量设备不能满足其预期

confrontational);

b) Psychological (e. g. stress-reducing, burnout prevention, emotionally protective);

c) Physical (e. g. temperature, heat, humidity, light, airflow, hygiene, noise).

These factors can differ substantially depending on the products and services provided.

e. Monitoring and measuring resources.

a) General. The organization shall determine and provide the resources needed to ensure valid and reliable results when monitoring or measuring is used to verify the conformity of products and services to requirements.

The organization shall ensure that the resources provided are:

• Suitable for the specific type of monitoring and measurement activities being undertaken;

• Maintained to ensure their continuing fitness for their purpose.

The organization shall retain appropriate documented information as evidence of fitness for purpose of the monitoring and measurement resources.

b) Measurement traceability. When measurement traceability is a requirement, or is considered by the organization to be an essential part of providing confidence in the validity of measurement results, measuring equipment shall be:

• Calibrated or verified, or both, at specified intervals, or prior to use, against measurement standards traceable to international or national measurement standards; when no such standards exist, the basis used for calibration or verification shall be retained as documented information;

• Identified in order to determine their status;

• Safeguarded from adjustments, damage or deterioration that would invalidate the calibration status and subsequent measurement results.

The organization shall determine if the validity

使用要求时,组织应确定以往测量结果的有效性是否受到不利影响,必要时采取适当的纠正措施。

⑥组织知识。组织应确定过程的运行与达到产品和服务的符合性所需的知识。这些知识应得到维护并在需要时易于获取。

在应对变化的需求和趋势时,组织应考虑现有的知识,确定如何获取必要的附加知识和所需的更新。

注1 组织知识是组织的特定知识,是获得的经验,是实现组织目标所使用和分享的信息。

注2 组织知识可以基于：

A. 内部资源(如知识产权,从经验获得的知识,从失败和成功项目中吸取的教训,获取和分享非文件化的知识和经验,过程、产品和服务的改进结果)；

B. 外部资源(如标准、学术交流、会议,以及从顾客和供方收集的知识)。

(2)能力。组织应：

①确定在组织控制下从事影响质量管理体系绩效和有效性工作的人员的必要能力；

②基于适当的教育、培训或经验,确保这些人员是胜任的；

③酌情采取措施以获取所需的能力,并评价这些措施的有效性；

of previous measurement results has been adversely affected when measuring equipment is found to be unfit for its intended purpose, and shall take appropriate action as necessary.

f. Organizational knowledge. The organization shall determine the knowledge necessary for the operation of its processes and to achieve conformity of products and services. This knowledge shall be maintained and be made available to the extent necessary.

When addressing changing needs and trends, the organization shall consider its current knowledge and determine how to acquire or access any necessary additional knowledge and required updates.

NOTE 1 Organizational knowledge is knowledge specific to the organization; it is gained by experience. It is information that is used and shared to achieve the organization's objectives.

NOTE 2 Organizational knowledge can be based on:

a) Internal sources (e. g. intellectual property, knowledge gained from experience, lessons learned from failures and successful projects, capturing and sharing undocumented knowledge and experience, the results of improvements in processes, products and services);

b) External sources (e. g. standards, academia, conferences, gathering knowledge from customers or external providers).

2. Competence. The organization shall:

a. Determine the necessary competence of person(s) doing work under its control that affects the performance and effectiveness of the quality management system;

b. Ensure that these persons are competent on the basis of appropriate education, training, or experience;

c. Where applicable, take actions to acquire the necessary competence, and evaluate the effectiveness of the actions taken;

④保留适当的文件资料作为能力的证据。

注 适用的措施可以包括为现有员工提供培训、辅导或重新分配工作，雇佣或外包胜任的人员。

(3)意识。组织应确保在组织控制下工作的有关人员意识到：

①质量方针；
②相关的质量目标；
③他们对质量管理体系有效性的贡献，包括改进绩效的益处；
④不符合质量管理体系要求可能引发的后果。

(4)沟通。组织应确定与质量管理体系有关的内部和外部沟通，包括：

①沟通的内容；
②沟通的时机；
③沟通的对象；
④如何沟通；
⑤谁负责沟通。

(5)文件资料。

①总则。组织的质量管理体系应包括：

A.本标准要求的文件资料；

B.组织确定的为确保质量管理体系有效运行所需的文件资料。

注 不同组织质量管理体系的文件资料的规模可以因以下方面而不同：组织的规模、活动、过程、产品和服务的类型；过程及其相互作用的复杂程度；人员的能力。

d. Retain appropriate documented information as evidence of competence.

NOTE Applicable actions can include, for example, the provision of training to, the mentoring of, or the reassignment of currently employed persons; or the hiring or contracting of competent persons.

3. Awareness. The organization shall ensure that persons doing work under the organization's control are aware of:

a. The quality policy;
b. Relevant quality objectives;
c. Their contribution to the effectiveness of the quality management system, including the benefits of improved performance;
d. The implications of not conforming with the quality management system requirements.

4. Communication. The organization shall determine the internal and external communications relevant to the quality management system, including:

a. On what it will communicate;
b. When to communicate;
c. With whom to communicate;
d. How to communicate;
e. Who communicates.

5. Document.

a. General. The organization's quality management system shall include:

a) Document required by this International Standard;

b) Document determined by the organization as being necessary for the effectiveness of the quality management system.

NOTE The extent of document for a quality management system can differ from one organization to another due to: the size of organization and its type of activities, processes, products and services; the complexity of processes and their interactions; the competence of persons.

②编制和更新。在编制和更新文件资料时,组织应确保适当的:

A. 标识和说明(如标题、日期、作者、文件编号等);

B. 格式(如语言、软件版本、图示)和媒介(如纸质、电子);

C. 评审和批准适宜和充分。

③文件资料的控制。

A. 应对质量管理体系和本标准所要求的文件资料进行控制,以确保:

a. 在需要使用文件之处和之时能获得适用的文件;

b. 文件得到充分的保护(如防止泄密、使用不当、缺损)。

B. 组织应酌情实施以下活动对文件资料进行控制:

a. 分发、获取、检索和使用;

b. 储存、保护,包括保存易读性;

c. 更改的控制(如版本控制);

d. 保留和处置。

组织所确定的质量管理体系的制定和运行所需的外部来源文件,应适当加以识别和控制。

作为符合性证据保留的文件资料应妥善保存,以防止其非预期性失效。

注 获取是指有关文件资料访问许可的决定,或访问和更改文件资料的许可和权利。

b. Creating and updating. When creating and updating documented information, the organization shall ensure appropriate:

a) Identification and description (e.g. a title, date, author, or reference number);

b) Format (e.g. language, software version, graphics) and media (e.g. paper, electronic);

c) Review and approval for suitability and adequacy.

c. Control of documented information.

a) Document required by the quality management system and by this International Standard shall be controlled to ensure:

• It is available and suitable for use, where and when it is needed;

• It is adequately protected (e.g. from loss of confidentiality, improper use, or loss of integrity).

b) For the control of documented information, the organization shall address the following activities, as applicable:

• Distribution, access, retrieval and use;

• Storage and preservation, including preservation of legibility;

• Control of changes (e.g. version control);

• Retention and disposition.

Documented information of external origin determined by the organization to be necessary for the planning and operation of the quality management system shall be identified as appropriate, and be controlled.

Documented information retained as evidence of conformity shall be protected from unintended alterations.

NOTE Access can imply a decision regarding the permission to view the document only, or the permission and authority to view and change the document.

8)运行

(1)运行的策划和控制。组织应策划、实施和控制满足产品和服务提供要求所需的过程[见4)中第(4)点],并实施6)中确定的措施。

①确定产品和服务的要求。

②建立以下准则:过程;产品和服务的接受。

③确定为使产品和服务符合要求所需的资源。

④按照准则实施过程控制。

⑤确定、维护和保留必要的文件资料:确信过程已按照计划予以实施;证实产品和服务与其要求相符合。

策划的输出应适合组织的运行。

组织应控制有计划的变更,评审非预期的变更的后果,必要时采取措施减轻任何不良影响。

组织应确保外包过程得到控制[见8)中第(4)点]。

(2)产品和服务要求。

①顾客沟通。与顾客沟通应包括:

A. 提供与产品和服务有关的信息;

B. 处理问询,合同或订单的处理,包括更改;

C. 获取顾客关于产品和服务的反馈,包括顾客抱怨;

D. 顾客财产的处理和控制;

H. Operation

1. Operational planning and control. The organization shall plan, implement and control the processes (see Clause D. 4) needed to meet the requirements for the provision of products and services, and to implement the actions determined in Clause F, by:

a. Determining the requirements for the products and services;

b. Establishing criteria for the processes, the acceptance of products and services;

c. Determining the resources needed to achieve conformity to the product and service requirements;

d. Implementing control of the processes in accordance with the criteria;

e. Determining, maintaining and retaining documented information to the extent necessary: to have confidence that the processes have been carried out as planned; to demonstrate the conformity of products and services to their requirements.

The output of this planning shall be suitable for the organization's operations.

The organization shall control planned changes and review the consequences of unintended changes, taking action to mitigate any adverse effects, as necessary.

The organization shall ensure that outsourced processes are controlled (see Clause H. 4).

2. Requirements for products and services.

a. Customer communication. Communication with customers shall include:

a) Providing information relating to products and services;

b) Handling enquiries, contracts or orders, including changes;

c) Obtaining customer feedback relating to products and services, including customer complaints;

d) Handling or controlling customer property;

E. 确定相关应急措施的特定要求。

②产品和服务要求的确定。在确定提供给顾客的产品和服务的要求时,组织应确保:

A. 产品和服务的要求得到确定,包括适用的法律法规要求,及组织认为必要的要求;

B. 组织能够满足其所提供产品和服务的申明要求。

③产品和服务要求的评审。

A. 组织应确保其有能力满足顾客对产品和服务的要求,组织应在向顾客承诺提供产品和服务之前实施评审,包括:

a. 顾客特定的要求,包括交付和交付后活动的要求;

b. 顾客虽然没有明示,但规定用途或已知的预期用途所必需的要求;

c. 组织规定的要求;

d. 适用于产品和服务的法律法规要求;

e. 与以前表述不一致的合同或订单的要求。

组织应确保与以前规定不一致的合同或订单要求已得到解决。

顾客没有提供其要求的文件化说明时,组织应在事前对顾客要求进行确认。

e) Establishing specific requirements for contingency actions, when relevant.

b. Determination the requirements for products and services. When determining the requirements for the products and services to be offered to customers, the organization shall ensure that:

a) The requirements for the products and services are defined, including any applicable statutory and regulatory requirements; those considered necessary by the organization;

b) The organization can meet the claims for the products and services it offers.

c. Review of the requirements for products and services.

a) The organization shall ensure that it has the ability to meet the requirements for products and services to be offered to customers. The organization shall conduct a review before committing to supply products and services to a customer, to include:

• Requirements specified by the customer, including the requirements for delivery and postdelivery activities;

• Requirements not stated by the customer, but necessary for the specified or intended use, when known;

• Requirements specified by the organization;

• Statutory and regulatory requirements applicable to the products and services;

• Contract or order requirements differing from those previously expressed.

The organization shall ensure that contract or order requirements differing from those previously defined are resolved.

The customer's requirements shall be confirmed by the organization before acceptance, when the customer does not provide a documented statement of their requirements.

注 在某些情况下,如网络销售,对每一个订单进行正式的评审可能是不实际的。作为替代方法,可对提供给顾客的有关的产品信息进行评审,如产品目录。

B. 组织应酌情保留以下方面的文件资料:

a. 评审结果;

b. 产品和服务的任何新要求。

④产品和服务要求的变更。若产品和服务要求发生变更,组织应确保相关的文件资料得到修改,并确保相关人员知道已变更的要求。

(3)产品和服务的设计和开发。

①总则。组织应建立、实施和维护设计和开发过程,以确保后续产品和服务的提供。

②设计和开发策划。在确定设计和开发的阶段和控制时,组织应考虑:

A. 设计和开发活动的性质、持续时间和复杂性;

B. 所需的过程阶段,包括适用的设计和开发的评审方法;

C. 所需的设计和开发的验证和确认方法;

D. 设计和开发过程涉及的职责和权限;

E. 产品和服务的设计和开发所需的内部和外部资源;

F. 控制设计和开发过程涉及的个人之间的接口需求;

NOTE In some situations, such as internet sales, a formal review is impractical for each order. Instead, the review can cover relevant product information, such as catalogues.

b) The organization shall retain documented information, as applicable:

• On the results of the review;

• On any new requirements for the products and services.

d. Change to requirements for products and service. The organization shall ensure that relevant documented information is amended, and that relevant persons are made aware of the changed requirements, when the requirements for products and services are changed.

3. Design and development of products and services.

a. General. The organization shall establish, implement and maintain a design and development process that is appropriate to ensure the subsequent provision of products and services.

b. Design and development planning. In determining the stages and controls for design and development, the organization shall consider:

a) The nature, duration and complexity of the design and development activities;

b) The required process stages, including applicable design and development reviews;

c) The required design and development verification and validation activities;

d) The responsibilities and authorities involved in the design and development process;

e) The internal and external resource needs for the design and development of products and services;

f) The need to control interfaces between persons involved in the design and development process;

G. 顾客和用户参与设计和开发过程的需求;

H. 提供后续产品和服务的要求;

I. 顾客和其他相关方对设计和开发过程所期望的控制程度;

J. 证实设计和开发要求已得到满足所需的文件资料。

③设计和开发输入。组织应确定有关产品和服务设计和开发的具体类型的基本要求,组织应考虑:

A. 功能和性能要求;
B. 来自以前类似设计和开发的信息;
C. 法律法规要求;
D. 组织承诺执行的标准或行业规则;
E. 产品和服务的性质引起的潜在失效后果。

输入应满足设计和开发目的,完整并且清楚。设计和开发输入的矛盾应予以解决。组织应保留设计和开发输入的文件资料。

④设计和开发控制。组织对设计和开发过程进行控制,以确保:

A. 要实现的结果得到确定;
B. 实施评审,以评价设计和开发结果满足要求的能力;

C. 实施验证活动,以确保设计和开发的输出满足设计和开发输入的要求;

D. 实施确认活动,以确保形成的产品和服务能够满足规定的应用或预

g) The need for involvement of customers and users in the design and development process;

h) The requirements for subsequent provision of products and services;

i) The level of control expected for the design and development process by customers and other relevant interested parties;

j) The documented information needed to demonstrate that design and development requirements have been met.

c. Design and development Inputs. The organization shall determine the requirements essential for the specific types of products and services to be designed and developed. The organization shall consider:

a) Functional and performance requirements;

b) Information derived from previous similar design and development activities;

c) Statutory and regulatory requirements;

d) Standards or codes of practice that the organization has committed to implement;

e) Potential consequences of failure due to the nature of the products and services.

Inputs shall be adequate for design and development purposes, complete and unambiguous. Conflicting design and development inputs shall be resolved. The organization shall retain documented information on design and development inputs.

d. Design and development controls. The organization shall apply controls to the design and development process to ensure that:

a) The results to be achieved are defined;

b) Reviews are conducted to evaluate the ability of the results of design and development to meet requirements;

c) Verification activities are conducted to ensure that the design and devclopment outputs meet the input requirements;

d) Validation activities are conducted to ensure that the resulting products and services meet the

期用途；

E. 对评审或验证和确认活动中确定的问题采取必要的措施；

F. 保留这些活动的文件资料。

注 设计和开发的评审、验证和确认具有不同的目的，它们可以按适合组织的方式单独或任意组合。

⑤设计和开发输出。组织应确保设计和开发的输出：

A. 满足输入的要求；

B. 对于所提供的产品和服务的后续过程是充分的；

C. 适当时，参照监督和测量要求以及接受标准；

D. 具体说明产品和服务的特性，这些特性对于它们的预期用途并以安全和适当方式提供是必不可少的。

组织应保留设计和开发输出的文件资料。

⑥设计和开发的更改。在产品和服务的设计和开发期间或后续过程中，组织应识别、评审对其设计和开发的更改并进行必要的控制，以确保其符合要求，没有负面影响。

组织应保留以下文件资料：

A. 设计和开发的变更；

B. 评审结果；

C. 变更的授权；

D. 所采取的预防负面影响的措施。

requirements for the specified application or intended use;

e) Any necessary actions are taken on problems determined during the reviews, or verification and validation activities;

f) Documented information of these activities is retained.

NOTE Design and development reviews, verification and validation have distinct purposes. They can be conducted separately or in any combination, as is suitable for the products and services of the organization.

e. Design and development outputs. The organization shall ensure that design and development outputs:

a) Meet the input requirements;

b) Are adequate for the subsequent processes for the provision of products and services;

c) Include or reference monitoring and measuring requirements, as appropriate, and acceptance criteria;

d) Specify the characteristics of the products and services that are essential for their intended purpose and their safe and proper provision.

The organization shall retain documented information on design and development outputs.

f. Design and development changes. The organization shall identify, review and control changes made during, or subsequent to, the design and development of products and services, to the extent necessary to ensure that there is no adverse impact on conformity to requirements.

The organization shall retain documented information on:

a) Design and development changes;

b) The results of reviews;

c) The authorization of the changes;

d) The actions taken to prevent adverse impacts.

(4)外部提供的过程、产品和服务的控制。

①总则。组织应确定对外部提供的过程、产品和服务符合要求。

在以下情况下,组织应确定对外部提供的过程、产品和服务的控制:

A. 外部供方提供的产品和服务并入组织自己的产品和服务;

B. 由外部供方代表组织直接向顾客提供产品和服务;

C. 由外部供方提供的作为组织决定结果的过程或过程的一部分。

组织应根据外部供方按照规定的要求提供过程或产品和服务的能力,确定和应用对外部供方的评价、选择、绩效监督和重新评价的准则。组织应保留这些活动和由评价引起的必要措施的文件资料。

②控制类型和程度。组织应确保外部提供的过程、产品和服务对组织稳定地向顾客交付符合要求的产品和服务的能力没有负面影响。

组织应:
A. 确保将外部提供的过程保持在其质量管理体系控制范围内;
B. 确定拟对外部供方及其形成的输出实施的控制;

C. 考虑外部提供的过程、产品和服务对组织持续满足顾客和适用的法律法规要求的能力的潜在影响,以及

4. Control of externally provided processes, products and services.

a. General. The organization shall ensure that externally provided processes, products and services conform to requirements.

The organization shall determine the controls to be applied to externally provided processes, products and services when:

a) Products and services from external providers are intended for incorporation into the organization's own products and services;

b) Products and services are provided directly to the customer(s) by external providers on behalf of the organization;

c) A process, or part of a process, is provided by an external provider as a result of a decision by the organization.

The organization shall determine and apply criteria for the evaluation, selection, monitoring of performance, and re-evaluation of external providers, based on their ability to provide processes or products and services in accordance with requirements. The organization shall retain documented information of these activities and any necessary actions arising from the evaluations.

b. Type and extent of control. The organization shall ensure that externally provided processes, products and services do not adversely affect the organization's ability to consistently deliver conforming products and services to its customers.

The organization shall:

a) Ensure that externally provided processes remain within the control of its quality management system;

b) Define both the controls that it intends to apply to an external provider and those it intends to apply to the resulting output;

c) Take into consideration: the potential impact of the externally provided processes, products and services on the organization's ability to

外部供方实施的控制的有效性;

D. 确定验证或其他必要活动,以确定外部提供的过程、产品和服务满足要求。

③外部供方信息。在与外部供方沟通前,组织应确保要求的充分性。

组织应与外部供方沟通以下要求:

A. 将要提供的过程、产品和服务;

B. 批准的产品和服务,方法、过程和设备,以及产品和服务的发放;

C. 能力,包含所要求的人员资格;

D. 外部供方和组织之间的相互作用;

E. 组织实施的对外部供方绩效的控制和监督;

F. 组织或其顾客拟在外部供方现场实施的验证或确认活动。

(5)生产和服务的提供。

①提供生产和服务的控制。组织应在受控条件下实施产品和服务的提供。

适用时,受控条件应包括:

A. 获得规定的文件资料,如生产的产品、提供的服务或执行的活动的特性,以及要达到的结果;

consistently meet customer and applicable statutory and regulatory requirements; the effectiveness of the controls applied by the external provider;

d) Determine the verification, or other activities, necessary to ensure that the externally provided processes, products and services meet requirements.

c. Information for external providers. The organization shall ensure the adequacy of requirements prior to their communication to the external provider.

The organization shall communicate to external providers its requirements for:

a) The processes, products and services to be provided;

b) The approval of products and services; methods, processes and equipment; the release of products and services;

c) Competence, including any required qualification of persons;

d) The external providers' interactions with the organization;

e) Control and monitoring of the external providers' performance to be applied by the organization;

f) Verification or validation activities that the organization, or its customer, intends to perform at the external providers' premises.

5. Production and service provision.

a. Control of production and service provision. The organization shall implement production and service provision under controlled conditions.

Controlled conditions shall include, as applicable:

a) The availability of documented information that defines: the characteristics of the products to be produced, the services to be provided, or the activities to be performed; the results to be achieved;

B. 获得和使用适宜的监督和测量资源;

C. 在适当阶段进行监督和测量,以验证过程或输出的控制准则及产品和服务的接受准则已得到满足;

D. 利用适当的设施和环境来运作过程;

E. 指派胜任的人员,包括所要求的资格;

F. 当生产和服务提供过程形成的输出不能由后续的监督或测量加以验证,组织应对这些过程实现计划结果的能力进行确认和定期再确认;

G. 实施防止人为错误的措施;

H. 实施产品和服务的发放、交付和交付后活动。

②标识和可追溯性。必要时,组织应使用适宜的方法识别输出,以确保产品和服务的符合性。

组织应在提供产品和服务的全过程中,针对监督和测量要求识别输出的状态。

在有可追溯性要求的场合,组织应控制输出的唯一性标识,并应保留可追溯性所需的文件资料。

③顾客或者外部供方财产。组织应爱护在组织控制下或组织使用的顾客或外部供方的财产。

组织应识别、验证、保护和维护供其使用或构成产品和服务的顾客或外

b) The availability and use of suitable monitoring and measuring resources;

c) The implementation of monitoring and measurement activities at appropriate stages to verify that criteria for control of processes or outputs, and acceptance criteria for products and services, have been met;

d) The use of suitable infrastructure and environment for the operation of processes;

e) The appointment of competent persons, including any required qualification;

f) The validation, and periodic revalidation, of the ability to achieve planned results of the processes for production and service provision, where the resulting output cannot be verified by subsequent monitoring or measurement;

g) The implementation of actions to prevent human error;

h) The implementation of release, delivery and post-delivery activities.

b. Identification and traceability. The organization shall use suitable means to identify outputs when it is necessary to ensure the conformity of products and services.

The organization shall identify the status of outputs with respect to monitoring and measurement requirements throughout production and service provision.

The organization shall control the unique identification of the outputs when traceability is a requirement, and shall retain the documented information necessary to enable traceability.

c. Property belonging to customers or external providers. The organization shall exercise care with property belonging to customers or external providers while it is under the organization's control or being used by the organization.

The organization shall identify, verify, protect and safeguard customers' or external

部供方的财产。

若顾客或外部供方财产丢失、损坏或发现其他不适合使用的情况,组织应报告顾客或外部供方,并保留有关已发生状况的文件资料。

注 顾客或外部供方财产可以包括材料、组件、工具和设备、顾客现场、知识产权和个人数据。

④防护。组织应在产品和服务提供期间对输出进行必要的防护,以确保其符合要求。

注 防护可以包括标识、搬运、污染控制、包装、贮存、传送或运输以及保护。

⑤交付后的活动。组织应满足与产品和服务有关的交付后活动的要求。

在确定所需的交付后活动的程度时,组织应考虑:

A. 法律法规要求;
B. 与产品和服务有关的潜在不期望的后果;
C. 产品和服务的性质、用途和预期寿命;
D. 顾客要求;
E. 顾客反馈。

注 交付后活动可包括诸如担保条件下的措施、合同规定的维护服务、附加服务如回收或最终处置。

⑥变更的控制。组织应对生产或服务提供的变更进行评审和控制,以确保持续符合要求的必要范围。

providers' property provided for use or incorporation into the products and services.

When the property of a customer or external provider is lost, damaged or otherwise found to be unsuitable for use, the organization shall report this to the customer or external provider and retain documented information on what has occurred.

NOTE A customer's or external provider's property can include material, components, tools and equipment, premises, intellectual property and personal data.

d. Preservation. The organization shall preserve the outputs during production and service provision, to the extent necessary to ensure conformity to requirements.

NOTE Preservation can include identification, handling, contamination control, packaging, storage, transmission or transportation, and protection.

e. Post-delivery activities. The organization shall meet requirements for post-delivery activities associated with the products and services.

In determining the extent of post-delivery activities that are required, the organization shall consider:

a) Statutory and regulatory requirements;
b) The potential undesired consequences associated with its products and services;
c) The nature, use and intended lifetime of its products and services;
d) Customer requirements;
e) Customer feedback.

NOTE Post-delivery activities can include actions under warranty provisions, contractual obligations such as maintenance services, and supplementary services such as recycling or final disposal.

f. Control of changes. The organization shall review and control changes for production or service provision, to the extent necessary to ensure

组织应保留描述变更的评审结果、有权变更的人员以及评审引起的任何必要措施的文件资料。

（6）产品和服务的发放。组织应在适当阶段实施策划的安排，以验证产品和服务要求已得到满足。

在策划的安排圆满完成之前，不应向顾客发放产品和服务。除非得到有关授权人员的批准，并视顾客的申请而定。

组织应保留产品和服务发放的文件资料，这些资料应包括：

①符合接受准则的证据；

②有权发放人员的可追溯性。

（7）对不合格过程输出的控制。

①组织应确保不符合要求的输出得到识别和控制，以防止非预期的使用或交付。

组织应根据不合格的性质及其对产品和服务符合性的影响采取适当的措施。这也适用于产品交付后、服务提供期间或之后发现的不合格产品和服务。

组织应通过以下一种或几种途径处置不合格的输出：

A. 纠正；

B. 隔离、遏制、退回或暂停提供产品和服务；

continuing conformity with requirements.

The organization shall retain documented information describing the results of the review of changes, the person(s) authorizing the change, and any necessary actions arising from the review.

6. Release of products and services. The organization shall implement planned arrangements, at appropriate stages, to verify that the product and service requirements have been met.

The release of products and services to the customer shall not proceed until the planned arrangements have been satisfactorily completed, unless otherwise approved by a relevant authority and, as applicable by the customer.

The organization shall retain documented information on the release of products and services. The documented information shall include：

a. Evidence of conformity with the acceptance criteria；

b. Traceability to the person(s) authorizing the release.

7. Control of nonconforming process outputs.

a. The organization shall ensure that outputs that do not conform to their requirements are identified and controlled to prevent their unintended use or delivery.

The organization shall take appropriate action based on the nature of the nonconformity and its effect on the conformity of products and services. This shall also apply to nonconforming products and services detected after delivery of products, during or after the provision of services.

The organization shall deal with nonconforming outputs in one or more of the following ways：

a)Correction；

b) Segregation, containment, return or suspension of provision of products and services；

C. 告知顾客；

D. 获得让步接受的授权。

当不合格的输出得到纠正时，应验证与要求的符合性。

②组织应保留以下文件资料：

A. 描述不合格；

B. 描述所采取的措施；

C. 描述所获得的让步；

D. 确定就不合格的措施作出决定的权限。

9）绩效评价

（1）监督、测量、分析和评价。

①总则。组织应确定：

A. 所需要的监督和测量；

B. 所需的监督、测量、分析和评价方法，以确保有效的结果；

C. 实施监督和测量的时机；

D. 分析和评价监督和测量结果的时机。

组织应评价质量管理体系的绩效和有效性。

组织应保留适当的文件资料作为结果的证据。

②顾客满意。组织应监测顾客对其要求和期望得到满足的程度的感受。组织应确定获取、监测和评审这种信息的方法。

注 监测顾客感受的示例可以包括顾客调查、顾客对已交付产品或服务的反馈、顾客交流会议、市场占有率分析、赞扬、索赔担保和经销商报告。

c）Informing the customer；

d）Obtaining authorization for acceptance under concession.

Conformity to the requirements shall be verified when nonconforming outputs are corrected.

b. The organization shall retain documented information that：

a）Describes the nonconformity；

b）Describes the actions taken；

c）Describes any concessions obtained；

d）Identifies the authority deciding the action in respect of the nonconformity.

I. Performance Evaluation

1. Monitoring, measurement, analysis and evaluation.

a. General. The organization shall determine：

a）What needs to be monitored and measured；

b）The methods for monitoring, measurement, analysis and evaluation needed to ensure valid results；

c）When the monitoring and measuring shall be performed；

d）When the results from monitoring and measurement shall be analysed and evaluated.

The organization shall evaluate the performance and the effectiveness of the quality management system.

The organization shall retain appropriate documented information as evidence of the results.

b. Customer satisfaction. The organization shall monitor customers' perceptions of the degree to which their needs and expectations have been fulfilled. The organization shall determine the methods for obtaining, monitoring and reviewing this information.

NOTE Examples of monitoring customer perceptions can include customer surveys, customer feedback on delivered products and services, meetings with customers, market-share analysis, compliments, warranty claims and dealer reports.

③分析和评价。组织应分析、评价来自监测的适当数据和信息。

应将分析结果用于评价：

A. 产品和服务的符合性；
B. 顾客满意程度；
C. 质量管理体系的绩效和有效性；
D. 策划是否得到有效实施；
E. 所采取的应对风险和机会的措施是否有效；
F. 外部供方的绩效；
G. 质量管理体系改进的需求。

注 分析数据的方法可以包括统计技术。

（2）内部审核。

①组织应按策划的时间间隔进行内部审核，以确定质量管理体系是否符合组织和本标准的要求，并得到有效实施和维护。

②组织应当：

A. 策划、建立、实施和维护一个或多个审核方案，包括审核的频次、方法、职责、策划要求和报告编制，审核方案应考虑有关过程的重要性、影响组织的变化和以往审核的结果；

B. 确定每次审核的准则和范围；

C. 选择审核员和实施审核，确保审核过程的客观性和公正性；

D. 确保将审核结果报告给有关管理者；

c. Analysis and evaluation. The organization shall analyse and evaluate appropriate data and information arising from monitoring and measurement.

The results of analysis shall be used to evaluate：

a) Conformity of products and services；
b) The degree of customer satisfaction；
c) The performance and effectiveness of the quality management system；
d) If planning has been implemented effectively；
e) The effectiveness of actions taken to address risks and opportunities；
f) The performance of external providers；
g) The need for improvements to the quality management system.

NOTE Methods to analyse data can include statistical techniques.

2. Internal audit.

a. The organization shall conduct internal audits at planned intervals to provide information on whether the quality management system conforms to the requirements of the organization and this International Standard, and is effectively implemented and maintained.

b. The organization shall：

a) Plan, establish, implement and maintain an audit programme(s) including the frequency, methods, responsibilities, planning requirements and reporting, which shall take into consideration the importance of the processes concerned, changes affecting the organization, and the results of previous audits；

b) Define the audit criteria and scope for each audit；

c) Select auditors and conduct audits to ensure objectivity and the impartiality of the audit process；

d) Ensure that the results of the audits are reported to relevant management；

E. 及时采取适当的纠正和纠正措施；

F. 保留文件资料，作为审核方案实施和审核结果的证据。

注 参见 ISO19011《质量和（或）环境管理体系审核指南》。

（3）管理评审。

①总则。最高管理者应按策划的时间间隔评审组织的质量管理体系，以确保其持续的适宜性、充分性和有效性，并与组织的战略方向保持一致。

②管理评审输入。策划和实施管理评审时应考虑：

A. 以往管理评审的措施的状况。

B. 与质量管理体系有关的外部和内部的变更。

C. 质量管理体系绩效和有效性的信息，包括以下方面的趋势：

a. 顾客满意和来自相关方的反馈；

b. 质量目标的实现程度；

c. 过程绩效及产品和服务的符合性；

d. 不合格和纠正措施；

e. 监督和测量结果；

f. 审核结果；

g. 外部供方的绩效。

D. 资源的充分性。

E. 所采取的应对风险和机会的措施的有效性[见 6)中第(1)点]。

F. 改进机会。

③管理评审输出。管理评审的输出应包括与以下方面有关的决定和

e) Take appropriate correction and corrective actions without undue delay;

f) Retain documented information as evidence of the implementation of the audit programme and the audit results.

NOTE See ISO19011 for guidance.

3. Management review.

a. General. Top management shall review the organization's quality management system, at planned intervals, to ensure its continuing suitability, adequacy, effectiveness and alignment with the strategic direction of the organization.

b. Management review input. The management review shall be planned and carried out taking into consideration:

a) The status of actions from previous management reviews.

b) Changes in external and internal issues that are relevant to the quality management system.

c) Information on the performance and effectiveness of the quality management system, including trends in:

• Customer satisfaction and feedback from relevant interested parties;

• The extent to which quality objectives have been met;

• Process performance and conformity of products and services;

• Nonconformities and corrective actions;

• Monitoring and measurement results;

• Audit results;

• The performance of external providers.

d) The adequacy of resources.

e) The effectiveness of actions taken to address risks and opportunities (see Clause F.1).

f) Opportunities for improvement.

c. Management review output. The outputs of the management review shall include decisions and

措施：

　　A. 改进的机会；

　　B. 质量管理体系变更的需求；

　　C. 资源需求。

　组织应保留文件资料，作为管理评审的结果的证据。

10）改进

（1）总则。组织应确定和选择改进机会并实施必要的措施，以满足顾客要求和增强顾客满意。这些应包括：

①改进产品和服务，以满足要求及应对未来的需求和期望；

②纠正、预防或减少非预期的影响；

③改进质量管理体系的绩效和有效性。

注 改进的示例可以包括纠正、纠正措施、持续改进、变革、创新或重组。

（2）不合格和纠正措施。

①发生不合格（包括来自投诉的不合格）时，组织应：

　　A. 酌情对不合格作出反应：

　　a. 采取措施控制和纠正不合格；
　　b. 处理后果。

　　B. 评价是否需要采取措施以消除不合格的原因，通过以下方面使其不再发生或不在其他地方发生：

　　a. 审查和分析不合格项；
　　b. 确定不合格的原因；
　　c. 确定是否存在或可能潜在发生类似的不合格。

actions related to：

a) Opportunities for improvement；

b) Any need for changes to the quality management system；

c) Resource needs.

The organization shall retain documented information as evidence of the results of management reviews.

J. Improvement

1. General. The organization shall determine and select opportunities for improvement and implement any necessary actions to meet customer requirements and enhance customer satisfaction. These shall include：

a. Improving products and services to meet requirements as well as to address future needs and expectations；

b. Correcting, preventing or reducing undesired effects；

c. Improving the performance and effectiveness of the quality management system.

NOTE Examples of improvement can include correction, corrective action, continual improvement, breakthrough change, innovation and reorganization.

2. Nonconformity and corrective action.

a. When a nonconformity occurs, including any arising from complaints, the organization shall：

a) React to the nonconformity and, as applicable：

• Take action to control and correct it；

• Deal with the consequences.

b) Evaluate the need for action to eliminate the cause(s) of the nonconformity, in order that it does not recur or occur elsewhere, by：

• Reviewing and analysing the nonconformity；

• Determining the causes of the nonconformity；

• Determining if similar nonconformities exist, or could potentially occur.

C. 实施所需的措施。

D. 审查所采取的纠正措施的有效性。

E. 必要时，更新在策划期间确定的风险和机会。

F. 必要时，对质量管理体系进行变更。

纠正措施应与所遇到的不合格项的影响程度相适应。

②组织应保留文件资料作为以下方面的证据：

A. 不合格项的性质以及随后所采取的任何措施；

B. 纠正措施的结果。

（3）持续改进。组织应持续改进质量管理体系的适宜性、充分性和有效性。

组织应考虑分析和评价的结果，以及管理评审的结果，以确定是否有成为持续改进部分的需要或机会。

c) Implement any action needed．

d) Review the effectiveness of any corrective action taken．

e) Update risks and opportunities determined during planning, if necessary．

f) Make changes to the quality management system, if necessary．

Corrective actions shall be appropriate to the effects of the nonconformities encountered．

b. The organization shall retain documented information as evidence of：

a) The nature of the nonconformities and any subsequent actions taken；

b) The results of any corrective action．

3. Continual improvement. The organization shall continually improve the suitability, adequacy and effectiveness of the quality management system．

The organization shall consider the results of analysis and evaluation, and the outputs from management review, to determine if there are needs or opportunities that shall be addressed as part of continual improvement.

任务3 质量管理体系的建立

5.3.1 基础知识

建立、完善质量管理体系一般要经历质量管理体系的策划与设计、质量管理体系文件的编制、质量管理体系的试运行、质量管理体系的审核与评审四个阶段，每个阶段又可分为若干具体步骤。

There are four stages to establish and perfect the quality management system, that is, the planning and design of the quality management system, the compilation of the quality management system documents, the trial operation of the quality management system, the audit and evaluation of the quality management system. Each stage can be divided into a number of specific steps.

1. 质量管理体系的策划与设计

该阶段主要是做好各种准备工

Ⅰ. Planning and Design of Quality Management System

This stage is mainly concerned with various

作,包括教育培训、统一认识,组织落实、拟定计划,确定质量方针、制定质量目标,调查和分析现状,调整组织结构、配备资源等。

1)教育培训,统一认识

质量管理体系建立和完善的过程,是始于教育,终于教育的过程,也是提高认识和统一认识的过程。教育培训要分层次,循序渐进地进行。

(1)第一层次为决策层,包括党、政、技(术)领导,主要培训以下内容。

①通过介绍质量管理和质量保证的发展和本单位的经验教训,说明建立、完善质量管理体系的迫切性和重要性。

②通过 ISO9000 标准的总体介绍,提高按国家(国际)标准建立质量管理体系的认识。

③通过质量管理体系要素讲解(重点应讲解"管理职责"等总体要素),明确决策层领导在质量管理体系建设中的关键地位和主导作用。

(2)第二层次为管理层,重点是管理、技术和生产部门的负责人,以及与建立质量管理体系有关的工作人员。这一层次的人员是建设、完善质量管理体系的骨干力量,起着承上启下的作用,要使他们全面接受 ISO9000 标准有关内容的培训,在方法上可采取讲解与研讨相结合。

preparations, including education and training, unifying understanding, organizing implementation, drawing up plans; defining quality policies and setting quality objectives; investigating and analyzing the current situation; adjusting the organizational structure and allocating resources, etc.

A. Education and Training, Unifying Understanding

The process of establishing and improving the quality management system is beginning with education and finally education, and it is also a process of raising awareness and unifying understanding. Education and training should be carried out step by step.

1. The first level is the decision-making level, including the party, government, technology (academic) leadership. Main training:

a. By introducing the development of quality management and quality assurance and the experience and lessons of our unit, the urgency and importance of establishing and perfecting quality management system are explained;

b. Through the general introduction of ISO9000 standard, improve the understanding of establishing quality management system according to national (international) standard.

c. Through the explanation of quality management system elements (emphasis should be placed on general elements such as "management responsibilities"), the key position and leading role of decision-making level leaders in the construction of quality system are clarified.

2. The second level is management, with a focus on heads of management, technology and production, as well as staff involved in the establishment of quality management systems. These two levels of personnel are the backbone of the construction and improvement of the quality management system, and play a role in connecting the preceding and the following. They should be

(3)第三层次为执行层,即与产品质量形成全过程有关的作业人员。对这一层次人员主要培训与本岗位质量活动有关的内容,包括在质量活动中应承担的任务,完成任务应赋予的权限,以及造成质量过失应承担的责任等。

2)组织落实,拟定计划

尽管质量管理体系建设涉及一个组织的所有部门和全体职工,但对多数单位来说,成立一个精干的工作班子可能是需要的。根据一些单位的做法,这个班子也可分三个层次。

(1)第一层次:成立以最高管理者(厂长、总经理等)为组长,质量主管领导为副组长的质量管理体系建设领导小组(或委员会)。其主要任务包括:

①体系建设的总体规划;
②制定质量方针和目标;

③按职能部门进行质量职能的分解。

(2)第二层次:成立由各职能部门领导(或代表)参加的工作班子。这个工作班子一般由质量部门和计划部门的领导共同牵头,其主要任务是按照体系建设的总体规划具体组织实施。

(3)第三层次:成立要素工作小组。根据各职能部门的分工明确质量管理体系要素的责任单位。例如,"设计控制"一般应由设计部门负责,"采

fully trained in the relevant contents of the ISO9000 standard, which can be explained and discussed in terms of methods.

3. The third level is the executive layer, that is, the operator that is related to the whole process of product quality formation. The main training for this level of personnel is related to the quality activities of this post, including the tasks to be undertaken in the quality activities, the authority to complete the tasks, and the responsibility for causing the quality negligence.

B. Organize Implementation and Draw up Plans

Although the construction of a quality management system involves all departments and staff of an organization, for most units it may be necessary to set up a competent working group. According to the practice of some units, this team can also be divided into three levels.

1. The first level: set up the leading group (or committee) on quality management system construction with the top management (director, general manager, etc.) as the leader and the quality supervisor as the deputy leader. Its main tasks include:

a. The overall plan of system construction;

b. Development of quality guidelines and objectives;

c. Decompose the quality function according to the functional department.

2. The second level: set up a working group attended by the leaders (or representative) of various functional departments. This team is generally led by the leadership of the quality department and the planning department, and its main task is to organize and implement according to the overall plan of the system construction.

3. The third level: the establishment of a working group on elements. According to the division of labor among the functional departments, the responsible units for the

购"要素由物资采购部门负责。

组织和责任落实后,按不同层次分别制订工作计划。在制订工作计划时应注意:

①目标要明确。明确要完成什么任务,解决哪些主要问题,要达到什么目的。

②要控制进程。建立质量管理体系的主要阶段要规定完成任务的时间表、主要负责人和参与人员,以及他们的职责分工及相互协作关系。

③要突出重点。重点主要是体系中的薄弱环节及关键的少数。少数可能是某个或某几个要素,也可能是要素中的一些活动。

3)确定质量方针,制定质量目标

质量方针体现了一个组织对质量的追求、对顾客的承诺,是职工质量行为的准则和质量工作的方向。

制定质量方针的要求是:

(1)与总方针相协调;
(2)应包含质量目标;
(3)结合组织的特点;
(4)确保各级人员都能理解和坚持执行。

elements of the quality management system should be defined. For example, "design control" should generally be the responsibility of the design department, and the "procurement" element should be the responsibility of the material procurement department.

When the organization and responsibility are implemented, work plans are drawn up at different levels. When developing a work plan, attention should be paid to:

a. Be clear about your goals. What are the tasks to be accomplished, what are the major problems to be solved, and what are the objectives to be achieved?

b. Control the process. The main phase of establishing a quality management system is to specify the time frame for completion of the task, the principal person in charge and the participants, as well as their division of duties and relationships of collaboration.

c. Highlight the key points. The focus is mainly on the weak links and the key minority in the system. This minority may be one or more elements, or it may be some of the activities in the element.

C. Determine the Quality Policy and Establish the Quality Target

The quality policy embodies the pursuit of the quality of an organization and the commitment to the customer, which is the criterion for the quality and behavior of the workers and the direction of the quality work.

The requirements for the development of the quality policy are:

1. Coordinating with the general policy;
2. Including quality objectives;
3. Combining with the characteristics of the organization;
4. Ensuring that implementation is understood and sustained at all levels.

4)调查和分析现状

调查和分析现状的目的是为了合理地选择体系要素,内容包括:

(1)体系情况分析,即分析本组织的质量管理体系情况,以便根据所处的质量管理体系情况选择质量管理体系要素的要求。

(2)产品特点分析,即分析产品的技术密集程度、使用对象、产品安全特性等,以确定要素的采用程度。

(3)组织结构分析,即组织的管理机构设置是否适应质量管理体系的需要。应建立与质量管理体系相适应的组织结构,并确立各机构间的隶属关系、联系方法。

(4)生产设备和检测设备能否适应质量管理体系的有关要求。

(5)技术、管理和操作人员的组成、结构及水平状况的分析。

(6)管理基础工作情况分析,即标准化、计量、质量责任制、质量教育和质量信息等工作的分析。

对以上内容可采取与标准中规定的质量管理体系要素要求进行对比性分析。

5)调整组织结构,配备资源

因为在一个组织中除质量管理外,还有其他各种管理。组织机构设置由于历史沿革,多数并不是按质量形成的客观规律来设置相应的职能部

D. Survey and Analysis of the Current Situation

The purpose of the survey and analysis is to reasonably select the elements of the system, including:

1. System situation analysis. The quality management system of the organization is analyzed in order to select the requirements of the quality management system elements according to the quality management system.

2. Product characteristics analysis. The technology intensive degree, the using object, the product safety characteristic and so on of the product are analyzed to determine the use degree of the elements.

3. Organizational structure analysis. Whether or not the organization's management structure meets the needs of the quality management system. Organizational structures adapted to quality management systems should be established and reporting relationships and methods of communication among agencies established.

4. Whether production equipment and testing equipment meet the requirements of the quality management system.

5. The analysis of the composition, structure and level of the technology, management and operating personnel.

6. Analysis of management basic work situation. Analysis of standardization, measurement, quality responsibilitysystems, quality education and quality information.

The above contents can be compared with the requirements of the quality management system elements specified in the standard.

E. Adjustment of the Organizational Structure and Allocate Resources

In an organization, there are all kinds of management apart from quality management. Since the historical evolution of organizational setup is not based on the objective law of quality, so after the

门的,所以在完成落实质量管理体系要素并展开对应的质量活动以后,必须将活动中相应的工作职责和权限分配到各职能部门。一方面是客观展开的质量活动,一方面是人为的现有的职能部门。处理两者之间的关系,一般地讲,一个质量职能部门可以负责或参与多个质量活动,但不要让一项质量活动由多个职能部门来负责。

目前我国企业现有职能部门对质量管理活动所承担的职责、所起的作用普遍不够理想,总的来说应该加强。

在活动展开的过程中,必须涉及相应的硬件、软件和人员配备,根据需要应进行适当的调配和充实。

2. 质量管理体系文件的编制

质量管理体系文件是描述质量管理的一整套文件,是质量管理体系运行的依据。

1)质量管理体系包括的文件

(1)形成文件的质量方针和质量目标。

(2)质量手册。

(3)本标准所要求的形成文件的程序和记录。

(4)组织确定的为确保其过程有效策划、运行和控制所需的文件,包括记录。

注1 本标准出现"形成文件的程序"之处,即要求建立该程序,形成文件,并加以实施和维护。一个文件可包括对一个或多个程序的要求。一个

completion of the implementation of the elements of the quality management system and the corresponding quality activities, the corresponding job responsibilities and competencies in the activity must be assigned to the functional departments. On the one hand, it is an objective quality activity, on the other hand, it is a man-made existing functional department. When the relationship between the two is handled, generally speaking, a quality functional department can be responsible for or participate in multiple quality activities, but do not allow a quality activity to be held by multiple functional departments.

At present, the responsibilities and functions of the of Chinese enterprises to quality management activities are generally not ideal and should be strengthened.

In carrying out the activities, the corresponding hardware, software and staffing must be involved, and the appropriate deployment and enrichment should be carried out as needed.

Ⅱ. Preparation of Quality Management System Documents

The documents of the quality management system are a whole set of documents describing quality management, and they are the basis of the operation of quality management system.

A. The Quality Management System Documents

1. The quality policy and quality goal of the documented.

2. Quality manual.

3. The procedures and records of the documented required by this standard.

4. Documents, including records, identified by the organization to ensure that its processes are effectively planned, operated, and controlled.

NOTE 1 Where a "procedure of documented" occurs in this standard, it is required that the procedure be established, form the file, implemented and maintained. A document may

程序的要求可以被包含在多个文件中。

注 2 不同组织的质量管理体系文件的多少与详略程度可以不同,取决于:

①组织的规模和活动的类型;

②过程及其相互作用的复杂程度;

③人员的能力。

注 3 文件可采用任何形式或类型的媒介。

2)质量管理体系文件编写的原则

(1)系统性。质量管理体系文件反映一个组织质量管理体系运行的全过程。文件的各个层次间、文件与文件之间应做到层次清楚、接口明确、结构合理。

(2)法规性。质量管理体系文件是必须执行的法规性文件,应保持其相对的稳定性和连续性。

(3)协调性。应保证质量管理体系文件与其他管理性文件的协调统一,保证质量管理体系文件之间的协调一致。

(4)高增值性。质量管理体系文件不是质量管理体系现状的简单写实,质量管理体系文件随着质量管理体系的不断改进而完善。

(5)继承性。在编制质量管理体系文件时,要继承以往有效的经验和做法。

(6)可操作性。应发动各部门有

include requirements for one or more procedures. The requirements of a program that forms a document can be included in multiple files.

NOTE 2 The documents of the quality management system of different organizations can be different from the degree of detail, which depending on:

a. The scale of the organization and the type of activity;

b. The complexity of the process and its interaction;

c. The ability of person.

NOTE 3 The documents can be used in any form or type of media.

B. Principles for Document Preparation of the Quality Management System

1. Systematic. The quality management system documents are the whole process that reflects the operation of an organization quality management system. There should be clear levels, clear interfaces and reasonable structure among all levels of documents and between documents and documents.

2. Statutory. Quality management system documents is a statutory document that must be enforced, and they should maintain their relative stability and continuity.

3. Coordination. The quality management system documents should be harmonized with other management documents, and the quality management system documents should be harmonized.

4. High value increment. The quality management system document is not a simple reality of the current situation of the quality management system, and the quality management system document is perfected along with the continuous improvement of the quality management system

5. Succession. In the preparation of quality management system document, the practice of effective past experience should be inherited.

6. Maneuverability. Personnel with practical

实践经验的人员,集思广益、共同参与,确保文件的可操作性,切忌照搬照抄,闭门造车。

(7)唯一性。对一个组织来说,质量管理体系只有一个,因此质量管理体系文件也应该是唯一的,要杜绝不同版本并存的现象。

(8)见证性。质量管理体系文件可作为本组织质量管理体系有效运行并得到维护的客观证据,向顾客、第三方证实本组织质量管理体系的运行情况。

(9)适宜性。质量管理体系文件的编制和形式应考虑企业的产品特点、规模、管理经验等。文件的详略程度应与人员的素质、技能和培训等因素相适宜。

为了确保质量手册的有效实施,使企业取得良好的效益,在编制质量手册的过程中,应注意以下问题:

①要注重从企业自身需要出发编制手册,防止走形式;

②注意总结本企业质量管理的经验;

③充分利用现有管理标准、工作标准;

④要注意让职工积极参与;

⑤注意使用符合本国文化传统的语言。

experience in various departments should be mobilized to brainstorm and participate together to ensure the maneuverability of the documents and avoid copying them as well as to shut oneself up in a room making a cart.

7. Uniqueness. For an organization, there is only one quality management system, so the quality management system document should also be unique, which should eliminate the coexistence of different versions.

8. Witness. The quality management system documents can be used as an objective evidence for the effective operation and maintenance of the quality management system of the organization, and confirm to customers and third parties the performance of the quality management system of the organization.

9. Suitability. The preparation and form of quality management system documents should take into account the product characteristics, scale, management experience and so on. The level of detail of the documents should be commensurate with factors such as the quality, skills and training of the personnel.

In order to ensure the effective implementation of the quality manual and make enterprises obtain good benefits, the following problems should be noted in the process of compiling the quality manual:

a. Pay attention to the need of the enterprise to compile the book of words to prevent the form of absences;

b. Pay attention to summing up the experience of quality management in our enterprise;

c. Make full use of existing management standards and working standards;

d. Pay attention to the active participation of employees;

e. Pay attention to the use of language consistent with the cultural traditions of the country.

3)质量管理体系文件编写的准备

文件编制前应完成质量管理体系结构的设计(包括质量方针、质量目标的制定,ISO9001条款的确定,企业现状的诊断,质量责任分配及资源配备等),同时应进行下列准备:

(1)确定文件编写的主管机构(一般为ISO9001推进小组),指导和协调文件编写工作;

(2)收集整理企业现有文件;

(3)对编写人员进行培训,明确编写的要求、方法、原则和注意事项;

(4)编写指导性文件。

为了使质量管理体系文件统一协调,达到规范化和标准化要求,应编制指导性文件,就文件的要求、内容、格式等作出规定。

4)质量管理体系文件的编制内容和要求

从质量管理体系的建设角度讲,应强调以下几个问题。

(1)质量管理体系文件一般应在第一阶段工作完成后才正式编制,必要时也可交叉进行。如果前期工作不做,直接编制体系文件,就容易产生系统性、整体性不强,以及脱离实际等弊病。

(2)除质量手册需统一组织编制外,其他体系文件应按分工由归口职能部门分别编制。先提出草案,再组织审核,这样做有利于今后文件的执行。

C. Preparation of Quality Management System Document

Prior to the preparation of the document, the design of the quality management system structure (including the establishment of the quality policy and quality objectives, the determination of ISO9001 terms, the diagnosis of the current situation of the enterprise, the allocation of quality responsibility and the allocation of resources, etc.) shall be completed, and the following preparations shall be made:

1. Develop unwritten authorities (In general, the ISO9001 propulsion group) to guide and coordinate the preparation of documents;

2. Collect and organize the existing documents of the enterprise;

3. Train the writers and make clear the requirements, methods, principles and precautions;

4. Preparation of guidance documents.

In order to harmonize the quality management system documents and achieve the requirements of normalization and standardization, we should draw up guidance documents, and stipulate the requirements, contents and formats of documents.

D. The Content and Requirements of the Quality Management System Documents

Several problems should be emphasized from the perspective of the construction of the quality management system.

1. Quality management system documents should generally be formally formulated after the first stage of work and can be crossed when necessary. If the earlier work is not done, it is easy to produce the lack of systematicness, integrity, and disengagement from reality.

2. In addition to the need for unified organization and formulation of the quality manual, other system documents shall be formulated separately by the functional departments concerned according to the

（3）质量管理体系文件的编制应结合本单位的质量职能分配进行。按所选择的质量管理体系要求，逐个展开为各项质量活动（包括直接质量活动和间接质量活动），将质量职能分配落实到各职能部门。质量活动项目和分配可采用矩阵图的形式表述，质量职能矩阵图也可作为附件附于质量手册之后。

（4）为了使所编制的质量管理体系文件作到协调、统一，在编制前应制订"质量管理体系文件明细表"，将现行的质量手册（如果已编制）、企业标准、规章制度、管理办法以及记录表收集在一起，与质量管理体系要素进行比较，从而确定新编、增编或修订质量管理体系文件的项目。

（5）为了提高质量管理体系文件的编制效率，减少返工，在文件编制过程中要加强文件层次间、文件与文件间的协调。尽管如此，一套质量好的质量管理体系文件也要经过自上而下和自下而上的多次反复。

（6）编制质量管理体系文件的关键是讲求实效，不走形式。既要从总体上和原则上满足 ISO9000 标准，又要在方法上和具体做法上符合本单位的实际。

division of labour, first proposing drafts and then organizing for examination and verification, which will be conducive to the implementation of future documents.

3. The compilation of quality management system documents shall be carried out in conjunction with the distribution of quality functions of the unit. According to the requirements of the selected quality management system, the quality activities (including direct quality activities and indirect quality activities) are carried out one by one, and the distribution of quality functions is carried out to the various functional departments. Quality activity items and assignments may be expressed in the form of a matrix chart, and the quality function matrix chart may be attached to the quality manual as an annex.

4. In order to harmonize and unify the quality management system documents, a "detailed list of quality management system documents" should be prepared prior to preparation, collecting existing quality manuals (if prepared), enterprise standards, regulations, management practices and record sheets to compare with the elements of a quality management system, so as to identify new, addenda or revisions to the quality management system document project.

5. In order to improve the efficiency of document preparation and reduce the rework, we should strengthen the coordination among the levels of documents, between documents and documents in the process of document preparation. Nevertheless, a set of high-grade quality management system documents have to go through the multiple iterations from top-down and bottom-up.

6. The key to compiling quality management system documents is to pay more attention to actual effect and not to take the form. It is necessary to meet the standards of ISO9000 on the whole and in principle, and also to accord with the reality of our unit in terms of methods and specific practices.

3. 质量管理体系的试运行

质量管理体系文件编制完成后,质量管理体系将进入试运行阶段。其目的是通过试运行,考验质量管理体系文件的有效性和协调性,并对暴露出的问题采取改进措施和纠正措施,以达到进一步完善质量管理体系文件的目的。

在质量管理体系试运行过程中,要重点抓好以下工作。

(1)有针对性地宣传质量管理体系文件,使全体职工认识到新建立或完善的质量管理体系是对过去质量管理体系的变革,是为了向国际标准接轨,要适应这种变革就必须认真学习、贯彻质量管理体系文件。

(2)实践是检验真理的唯一标准。质量管理体系文件通过试运行必然会出现一些问题,全体职工应将实践中出现的问题和改进意见如实反映给有关部门,以便采取纠正措施。

(3)将质量管理体系试运行中暴露出的问题,如体系设计不周、项目不全等进行协调、改进。

(4)加强信息管理,不仅是质量管理体系试运行本身的需要,也是保证试运行成功的关键。所有与质量活动有关的人员都应按质量管理体系文件的要求,做好质量信息的收集、分析、

Ⅲ. Trial Operation of the Quality Management System

After the completion of the quality management system documents, the quality management system will enter the trial operation stage. The purpose is to test the effectiveness and coordination of quality management system documents through trial operation, and take corrective actions and corrective actions to solve the problems, so as to further improve the purpose of quality management system documents.

During the trial operation of the quality management system, we should focus on the following tasks.

1. Publicize the quality management system documents in a targeted manner. Make all staff realize that the newly established or perfect quality management system is the change of the past quality management system, in order to meet the international standard, we must study and implement the quality management system documents seriously to adapt to this kind of change.

2. Practice is the only criterion for testing truth. Through the trial operation of the quality management system documents, there will inevitably be some problems, and all the staff and workers will truthfully reflect the problems and suggestions for improvement in practice to the relevant departments in order to take corrective measures.

3. Coordinate and improve the problems exposed in the quality management system trial operation, such as inadequate system design, incomplete project, etc.

4. Strengthen the information management is not only the need of the quality management system trial operation itself, but also the key to ensure the success of the trial operation. All personnel concerned with quality activities shall do

传递、反馈、处理和归档等工作。

4. 质量管理体系的审核与评审

质量管理体系审核在体系建立的初始阶段往往更加重要。在这一阶段,质量管理体系审核的重点,主要是验证和确认体系文件的适用性和有效性。

1)审核与评审的主要内容

(1)规定的质量方针和质量目标是否可行。

(2)体系文件是否覆盖了所有主要质量活动,各文件之间的接口是否清楚。

(3)组织管理结构能否满足质量管理体系运行的需要,各部门、各岗位的质量职责是否明确。

(4)质量管理体系要素的选择是否合理。

(5)规定的质量记录是否能起到见证作用。

(6)所有职工是否养成了按体系文件操作或工作的习惯,执行情况如何。

2)该阶段体系审核的特点

(1)体系正常运行时的体系审核,重点在符合性。在试运行阶段,通常是将符合性与适用性结合起来进行的。

(2)为使问题尽可能地在试运行阶段暴露无遗,除组织审核组进行正式审核外,还应有广大职工的参与,鼓励他们通过试运行的实践,发现和提出问题。

well in the collection, analysis, transmission, feedback, processing and archiving of quality information in accordance with the requirements of the quality management system documents.

IV. Audit and Review of Quality Management System

Quality management system audit is often more important in the initial stage of system establishment. In this stage, the focus of the quality management system audit is to verify and confirm the applicability and validity of the system documents.

A. The Main Contents of the Audit and Review

1. The feasibility of the specified quality policy and quality objectives.

2. Whether the system documents cover all major quality activities and whether the interface between the documents is clear.

3. Whether the organizational structure can meet the needs of the operation of the quality management system, and whether the quality responsibilities of the departments and posts are clear.

4. Whether the selection of quality management system elements is reasonable.

5. Does the specified quality record serve as a witness.

6. Whether all the workers have formed the habit of operating or working according to the system documents, and how the implementation is conducted.

B. The Characteristics of the System Review at This Stage

1. In the normal operation of the system, the system audit focuses on the conformity, and in the trial operation stage, it usually combines the conformity with the applicability.

2. In order to make the problem appear as far as possible during the trial run phase, besides the formal audit of the audit team, there should be a wide range of workers' participation, and encourage them to find and ask questions through

(3)在试运行的每一阶段结束后，一般应正式安排一次审核，以便及时对发现的问题进行纠正。对一些重大问题也可根据需要，适时地组织审核。

(4)在试运行中要对所有要素审核一遍。

(5)充分考虑对产品的保证作用。

(6)在内部审核的基础上，由最高管理者组织一次体系评审。

应当强调，质量管理体系是在不断改进中得以完善的。质量管理体系进入正常运行后，仍然要采取内部审核、管理评审等各种手段以使质量管理体系能够保持和不断完善。

trial run practice.

3. After the end of each stage of trial operation, a general audit should be formally arranged so as to correct the discovered problems in time, and to organize some important problems according to their needs.

4. All the factors should be reviewed in the trial operation.

5. Give full consideration to the guarantee of the product.

6. On the basis of internal audit, the highest manager organizes a system review.

It should be emphasized that the quality management system is improving continuously, and the quality management system is going to run normally. All kinds of means, such as internal audit, management review and so on, should be adopted in order to keep and improve the quality management system.

5.3.2 实训任务

某食品企业为通过ISO9001:2015质量管理体系的认证，要在企业内部建立质量管理体系。请你按照质量管理体系建立的四个步骤，给企业安排具体实施步骤。

请按照ISO9001:2015质量管理体系的要求给上述食品企业编写质量管理体系文件。

任务4 内审和管理评审

5.4.1 基础知识

1. 质量管理体系内部审核

1)内部审核控制程序

(1)目的。确保本公司质量管理体系的有效性和符合性，并及时采取纠正预防措施，以实现体系的持续改进。

(2)适用范围。适用于本公司质量管理体系覆盖的所有过程、部门和场所的审核。

(3)定义。第一方审核又称内部

Ⅰ. Internal Audit of the Quality Management System

A. Internal Audit Control Procedure

1. Purposes. Ensure the effectiveness and compliance of our quality management system, and take corrective and preventive actions in time to achieve continuous improvement of the system.

2. Scope of application. Applicable to all processes, departments and sites covered by the company's quality management system.

3. Definitions. The first party audit, also

审核,用于内部的目的,由组织自己或以组织的名义进行,可作为组织自我合格声明的基础。内部审核是质量管理体系的一种质量体系认证审核。

(4)职责:

①管理者代表及品质部负责内部质量审核(以下简称内审)的组织、管理、领导工作;

②内审小组负责内审的准备、实施、报告、验证等工作;

③公司各部门积极配合,接受内审,并对内审中出现的不合格项采取纠正与预防措施。

(5)作业内容。

①内审的策划。

A. 每年年初,管理者代表应根据标准和实际需要编制一份该年度审核计划,呈报总经理批准后组织实施。内审应覆盖质量管理体系的所有过程、部门和场所,每年至少2次。

B. 在以下特殊情况可提出内审要求:

a. 当合同要求或客户需要评价质量管理体系时;

b. 当机构和职能有重大变更时;

c. 发现严重不合格而需要审查时;

d. 第三方审核认证或监督审核前;

e. 总经理提出要求时。

C. 管理者代表依据以上情况适时提出内审建议,总经理批准后实施。

called internal audit, is used for internal purposes and is carried out by the organization itself or in the name of the organization. It can be used as the basis of the organization's self-qualification statement. Internal audit is a quality system certification audit of quality management system.

4. Duties:

a. Management Representative and Quality Department shall be responsible for the organization, management and leadership of internal quality audit (hereinafter referred to as internal audit);

b. The Internal Audit Panel shall be responsible for the preparation, implementation, report and verification of the internal audit;

c. The company departments actively cooperate and accept internal audit, and take corrective and preventive measures to the unqualified items in internal audit.

5. Job content.

a. Planning of internal audit.

a) At the beginning of each year, the management representative shall prepare a plan of review for the year, based on standards and actual needs, internal audit shall cover all processes, departments and sites of the quality management system, at least twice a year.

b) Internal audit requests may be made in the following exceptional circumstances:

• When contractual requirements or customer needs evaluate the quality management system;

• When there is a significant change in structure and function;

• Where serious disqualification is found and needs to be reviewed;

• Before the third party audit certification or supervision audit;

• When requested by the general manager.

c) According to the above circumstances, the management representative shall make timely proposals for internal audit, which shall be implemented after the approval of the general manager.

② 成立内审小组。

A. 根据内审活动的目的、范围、部门、过程及日程安排,管理者代表提出内审小组名单,总经理批准后组成内审小组。

B. 内审人员资格条件:

a. 内审人员应是所在部门负责人或主要骨干;

b. 内审人员须通过质量管理体系内审课程培训并考试合格;

c. 内审人员须经总经理确认授权。

C. 内审人员职责。

a. 内审组长职责:

• 协商并制订审核活动计划,准备工作文件,布置审核组成员工作;

• 主持审核会议,控制现场审核的实施,使审核按计划和要求进行;

• 确认内审员审核发现的不合格项报告;

• 向管理者代表和总经理提交审核报告,报告内容包括对受审核方提出的改进建议和要求;

• 做好审核实施过程中文件的分发工作。

b. 内审员职责:

• 根据审核要求编制检查表;

• 按审核计划完成审核任务;

• 将审核发现形成书面资料,编制不合格项报告;

b. Establishment of an internal audit panel.

a) According to the purpose, scope, department, process and schedule of the internal audit activities, the management representative proposes the list of the internal audit team, which is formed after the approval of the general manager.

b) Qualifications of internal auditors:

• The internal audit personnel shall be the head or the main backbone of the department in which they belong;

• Internal audit personnel shall be trained and passed the internal audit course in the quality management system;

• The internal audit personnel shall be authorized by the general manager.

c) Responsibilities of internal auditors.

1) Duties of internal audit team leader:

• Negotiate and draw up the audit activity plan, prepare the working documents, and arrange the work of the audit team members;

• Lead the audit meeting, control the implementation of the on-site audit, make the audit according to the plan and requirements;

• Confirm the reports of nonconformities found by the internal auditors;

• Submit audit reports to management representatives and general managers, including recommendations and requirements for improvement to the auditee;

• The distribution of documents in the course of audit and implementation.

2) Internal auditor's duty:

• Preparation of checklists in accordance with audit requirements;

• Complete the audit task according to the audit plan;

• The findings of the audit shall be formed into written materials and reports of nonconformities shall be prepared;

・支持配合内审组长工作，协助受审核方制定纠正措施，并实施跟踪审核。

③编制审核工作计划。

A. 由内审组长编制审核工作计划，经管理者代表批准后组织召开审核小组会议，明确各成员分工和要求，确保每位内审员清楚了解审核任务。

B. 审核工作计划包括以下内容：

a. 被审部门的目的、范围、日期；

b. 依据的文件。

2)内审检查表的编制要求

(1)按部门写。按照标准和体系文件的要求，然后根据受审核部门的职能来编写。如检查文件管理部门，先根据手册职能分配表中与文件管理部门相关的要素，将标准条款文件控制写上，然后再逐步列出要查的具体内容。这样要素和部门职能就不容易遗漏。

(2)按流程写。根据企业的主要流程，一项项查流程的接口是否明确，流程职能中的规定是否都做到了等。

• Support the internal audit team leader, assist the auditee in formulating corrective actions, and follow up the audit.

c. Preparation of audit work plan.

a) The internal audit team leader shall prepare the audit work plan, and after the approval of the management representative, organize the meeting of the audit group to clarify the division of labor and requirements of each member, and ensure that each internal auditor clearly understands the audit task.

b) The audit work plan includes the following:

• The purpose, scope and date of the department to be tried;

• Document basis.

B. Requirements for the Preparation of Internal Audit Checklists

1. By department: according to the requirements of the standards and system documents, and then according to the functions of the audited department. If a document management department is checked, it will be based on the elements related to the document management department in the functional distribution table of the manual, the standard article document control is written first, and then listed step by step the specific contents to be looked up. It is not easy to omit elements and departmental functions.

2. Write according to the process: according to the main process of the enterprise, a check interface of the process is clear, the process functions are specified whether to do.

2. 如何组织好 ISO9000 的管理评审工作

管理评审是 ISO9001:2015 标准对组织最高管理者提出的重要要求之一，是组织的最高管理者为了解、促进、改进本组织质量管理体系运行的主动行为，是实施质量管理和质量控制的重要手段，是对质量管理体系运行的整体效果以及现状的适宜性进行综合评价的方法之一，是发现质量管理体系存在的问题并进行质量改进的主要依据。最高管理者可以通过管理评审全面检查和评价组织的质量方针、质量目标，以及质量体系的适宜性、有效性和充分性，找出质量体系运行中需要提高和改进的方面和环节，制定切实可行的纠正措施，不断提高组织在市场的竞争能力。

最高管理者的重视是搞好管理评审的关键。管理评审的重点是评审输入文件，如果出现管理评审输入文件内容单一,管理评审信息输入质量不高,避重就轻,没有关键问题和环节,那么管理评审的效果往往不理想,评审只能是流于形式。

1) 管理评审的目的

组织的最高管理者应明确管理评审的上述目的,切忌评审仅仅停留在汇报、分析标准要素和程序文件的执行上,而对组织的质量方针、质量目标及质量体系是否适应组织内部、外部的环境变化,在市场中所处的水平,是否适合本组织的实际情况不作分析和评价。从而无法对组织的质量方针、质量目标和质量体系是否持续适宜、充分和有效作出结论,也很难针对评审的目的提出需要改进的问题,更谈不上达到管理评审的目的。所以,最高管理者和参加管理评审的所有人员,都必须明确管理评审的目的,围绕目的进行评审。

2) 管理评审的内容

组织所处的客观环境(内部的、外部的)不断变化,如新的法律、法规要求,新的市场要求,组织的人事变动、产品变换,以及工艺路线调整、设备更新等,都可能影响到质量管理体系。因此,要针对内、外部环境变化及时评价质量方针、质量目标及质量管理体系的持续适宜性。

组织针对变化了的环境对质量管理体系作出变更,采取持续改进活动,并对质量管理体系采取相应的调整。在这些过程中,难免出现这样或那样考虑不周全的问题,有的可能是过程未得到充分展开,有的可能是职责或接口关系规定得不明确,有的是资源配置不合理,有的是控制要求不落实,使相关过程未能协调有效地受控和运行。因此,应针对上述问题的出现及时评价质量方针、质量目标及质量管理体系的充分性。

组织应通过质量管理体系建立后的产品质量、过程质量的符合性,顾客满意的程度,质量管理体系是否已得到有效运行,内审、纠正和预防措施是否都正常实施,来及时评价质量方针、质量目标及质量管理体系的有效性。

3) 管理评审输入文件

为使管理评审有计划、有步骤地进行,并达到预期的效果和目的,在管理评审前,由主管部门在征求最高管理者意见的前提下,列出本次评审的内容、时间、地点及各部门需输入的信息资料。各部门接到任务后,准备管理评审输入文件。一般情况下应有以下输入信息:

(1) 内、外部审核结果;

(2) 质量管理体系运行情况及改进意见(质量方针、质量目标的实现情况,质量目标修订建议和依据,质量管理体系运行中长期存在的问题或系统性问题,质量管理体系文件的符合性和可操作性,文件修改的依据和建议等);

(3) 产品实物质量信息;

(4) 不合格品分布情况及处理情况信息;

(5) 纠正和预防措施的分析、实施及验证;

(6) 服务信息(合同履约率、顾客满意度及不满意度、顾客投诉的处理等);

(7) 组织机构设置、资源配置状况信息(人员、设备、办公环境等);

(8) 上次管理评审提出的改进措施的实施情况及验证信息。

上述信息应由组织的各职能部门在汇总、分析正常管理资料的基础上编写,提供的信息资料不能就事论事,要抓住问题的关键和实质。

例如,在提交产品实物质量信息资料时,不能简单地列出数据,应作加工和分析。应提供合格产品、优良产品的比例,说明产品质量是否能实现质量方针和质量目标。同时还应与前期产品质量对比,发现需要调整和改进建议的依据。只有用这些数据作依据,最高管理者才能准确判断组织产品实物质量的好坏,以及质量目标的适宜性和有效性。

又如,提交不合格品分布情况及处理情况信息时,应将不合格品进行分类、排序,找出产生不合格品较多的单位或生产过程,必要时计算出不合格品的频次,应用统计技术找出产生不合格品频次高的主要原因,并制定纠正措施。通过以上工作,总结出评审期内不合格品的数量和分布情况,产生不合格品的主要原因和采取的纠正措施,纠正措施是否实施,实施后效果如何,哪些单位或生产过程对不合格品的控制还存在问题,提出改进建议等上交管理评审,使最高管理者就此作出评价,并作出质量改进决策。

4) 管理评审应评审的重要问题

最高管理者应把管理评审的重点放在影响产品质量的关键问题、长期存在的质量问题、质量管理体系运行中的系统问题上。这些问题,有些是受内、外部因素的影响而难以解决的,有些是需要较大投入才能解决的问题,最高管理者应根据组织自身的能力和需要,协调外部因素,采取纠正措施,解决内部存在的问题。

5) 管理评审提出纠正措施及验证

针对管理评审输入文件的信息,及对各种信息的分析,找出问题并制定纠正措施。管理评审中提出的纠正措施应切合实际,具有可操作性,应确保纠正措施得到有效的实施。对于那些长期存在的问题或系统性问题,有时不能急于求成,应分期采取纠正措施。对于纠正措施的效果也应分期验证,以便及时发现问题,予以调整。管理评审提出的纠正措施实施后应予以验证,并报下次管理评审,使最高管理者及时掌握纠正措施实施效果的信息。只有解决了体系中存在的系统问题和长期存在的问题,组织的质量管理体系才能得以完善和提高,才能更好地反映出企业质量管理体系的自我完善和自我提高能力,才能反映最高管理者质量管理和质量提高的决心。

总之,管理评审是由组织的最高管理者定期(一般一年一次)进行的较高层次的对质量管理体系的正式评价,采用会议形式进行,时间有限。组织的各职能部门必须认真对待管理评审,积极参与管理评审,使管理评审更加有效。

5.4.2 实训任务

按企业内审要求,为某食品企业编写内审文件,包括内审检查表、内审报告、内审不合格报告单。

按照质量评审的要求和步骤为某食品企业安排和编写某一年度的质量评审文件。

思考题

1. 简述 ISO9000 认证的优势。
2. ISO9000 质量管理的七项原则是什么?
3. ISO9000 质量管理体系的十二项基础都包括什么?
4. 质量管理体系建立的四个阶段是什么?
5. 质量手册编写的原则及所包括的内容是什么?
6. 内审不符合(不合格)项报告的内容是什么?
7. 企业申请 ISO9000 认证前必须具备哪些基本条件?

8. 现场审核的主要程序包括哪些？
9. 质量管理体系认证的实施步骤是什么？

拓展学习网站

1. 国家市场监督管理总局(http://samr.saic.gov.cn)
2. 各省市场监督管理局

项目6　食品安全管理基础知识

项目概述

本项目通过学习食品安全危害及控制手段、良好的操作规范（GMP）、卫生标准操作程序（SSOP）等内容，掌握食品安全管理的基础知识。

任务1　食品安全危害及控制手段

1. 食品安全的基本概念

《中华人民共和国食品安全法》中对食品安全的定义："食品安全，指食品无毒、无害，符合应当有的营养要求，对人体健康不造成任何急性、亚急性或者慢性危害。"世界卫生组织（WHO）在《加强国家级食品安全计划指南》中认为食品安全是"对食品按其原定用途进行制作和食用时不会使消费者受害的一种担保"。这种担保具有两个方面的含义：一是指食品在生产和加工以及分发销售直至最终消费的全过程中，保证都没有限定剂量的有毒有害物质的介入；第二是指在整个过程中，如果存在对营养成分的损害、破坏或引起各成分间比例变化的话，这些变化也保证在可接受的幅度范围内。

2. 食品加工中的安全危害

1）食品加工中的生物性危害

食品加工中的生物性危害主要是食品中微生物的污染。食品的微生物污染不仅降低食品质量，而且对人体健康产生危害。食品的微生物污染占

Ⅰ. Basic Concepts of Food Safety

The definition of food safety in *Food Safety Law of the People's Republic of China* is: "food safety refers to non-toxic, harmless food according with the nutritional requirements that should be required, and does not cause any acute, sub-acute or chronic harm to human health." In the *Guideline for Strengthening the National Food Safety Plan*, WHO thought so: "there is a guarantee that food will not be victimized when it is made and consumed for its intended purpose". This guarantee has the following two meanings: first, the whole process of food production and processing, distribution and sale until final consumption, the guarantee is not subject to a limited dose of toxic and harmful substances involved; the second is that, throughout the process, if there is damage to the nutrient component or a change in the proportion of components is caused, these changes are also guaranteed to be within an acceptable range.

Ⅱ. Safety Hazards in Food Processing

A. Biological Hazards in Food Processing

Biological hazards in food processing are mainly caused by microbial contamination in food. Microbial contamination of food not only reduces the quality of food, but also causes harm to human

整个食品污染的比重很大,危害也很大。

食品微生物污染的来源有食品原料本身的污染、食品加工过程中的污染,以及食品贮存、运输及销售中的污染。

(1)细菌性危害。

①致病菌。致病菌一般是指肠道致病菌和致病性球菌,主要包括沙门氏菌、志贺氏菌、金黄色葡萄球菌、致病性链球菌等四种。致病菌不允许在食品中被检出。

②常见的食品细菌。

A.肠杆菌科(enterobacteriaceae),为革兰氏阴性,需氧及兼性厌氧,包括志贺氏菌属及沙门氏菌属、耶尔森氏菌属等致病菌。

B.乳杆菌属(lactobacillus),为革兰氏阳性杆菌,厌氧或微需氧,在乳品中多见。

C.微球菌属(micrococcus)和葡萄球菌属(staphylococcus),为革兰氏阳性细菌,嗜中温,营养要求较低,在肉、水产食品和蛋品上常见,有的能使食品变色。

D.芽孢杆菌属(bacillus)与芽孢梭菌属(clostridium),分布较广泛,尤其多见于肉和鱼。前者需氧或兼性厌氧,后者厌氧,属中温菌者多,间或嗜热菌,是罐头食品中常见的腐败菌。

E.假单胞菌属(pseudomonas),为革兰氏阴性无芽孢杆菌,需氧,嗜冷,在pH值为5.0～5.2下发育,是典型的腐败细菌,在肉和鱼上易繁殖,多见冷冻食品。

health. Microbial contamination of food occupies a large proportion of the whole food pollution, and the harm is also great.

The sources of food microbial contamination are the contamination of food raw materials, the contamination of food in the process of processing and the pollution of food in storage, transportation and sale.

1. Bacterial hazard.

a. Pathogenic bacteria. Pathogenic bacteria generally refer to pathogenic entero bacteria and pathogenic cocci. Which mainly include the following four kinds: Salmonella, Shigella, Staphylococcus aureus, Pathogenic Streptococcus. Pathogenic bacteria are not allowed to be detected in food.

b. Common food bacteria.

a) Enterobacteriaceae. Gram-negative, aerobic and facultative anaerobic, including Shigella and Salmonella, Yersinia and other pathogenic bacteria.

b) Lactobacillus. Gram-positive bacilli, anaerobic or micro aerobic, are common in dairy products.

c) Micrococcus and staphylococcus. The genus is Gram-positive bacteria, with moderate temperature and low nutritional requirements. Common in meat, aquatic food and eggs, and some can make food discoloration.

d) Bacillus and clostridium. More widely distributed, especially in meat and fish. The former aerobic or facultative anaerobic, the latter anaerobic, a medium temperature bacteria, occasionally or thermophilic bacteria, which is a common spoilage bacteria in canned food.

e) Pseudomonas. It is a Gram-negative bacilli, aerobic, cold, and developed under pH value 5.0～5.2. It is a typical spoilage bacteria. It is easy to reproduce in meat and fish. Frozen food is more common.

(2)病毒性危害。

①肝炎病毒。我国食品的病毒污染以肝炎病毒最为严重,主要为甲型肝炎病毒和戊型肝炎病毒。甲型肝炎病毒可以通过食品传播。1987年12月至1988年1月上海因食用含甲肝病毒的毛蚶(贝壳类水产),引起甲型肝炎的暴发流行。究其原因是沿海或靠近湖泊居住的人们喜食毛蚶、蛏子、蛤蜊等贝壳,尤其上海人讲究取其味。因此,食用毛蚶时仅用开水烫一下,然后取贝肉蘸调味料食用,这种吃法固然味道鲜美,但其中的甲肝病毒并没有杀死,结果引起食源性病毒病。戊型肝炎病毒不稳定,容易被破坏。

②朊病毒。朊病毒是一种不含核酸的蛋白感染因子,能引起哺乳动物中枢神经组织病变。朊病毒能引起人和动物的可转移性神经退化疾病,如牛海绵脑病(BSE,俗称疯牛病)、克雅氏病(CJD)等疾病。目前英国已知至少有70人死于新型克雅氏病,而医学界怀疑克雅氏病可能和食用牛海绵脑病病牛制成的肉制品有关。

(3)寄生虫危害。

①猪囊虫。猪囊虫俗称"米猪肉",是指带囊尾蚴的猪肉。人如果食用了没有死亡的猪肉囊虫,由于肠液和胆汁的刺激,其头节即可伸出包囊,以带钩的吸盘牢固地吸附在人的肠壁上,从中吸取营养并发育为成虫,即绦虫,使人患绦虫病。

2. Viral hazard.

a. Hepatitis virus. Hepatitis virus is the most serious virus contamination in food in China, mainly hepatitis A virus and hepatitis E virus. Hepatitis A virus can be transmitted through food. From December 1987 to January 1988, the outbreak of hepatitis A in Shanghai was caused by the consumption of scapharca (shellfish) containing hepatitis A virus. The reason is that the people living in coastal or near lakes like to eat the scapharca, clams, through shell and other shellfish, especially Shanghai people pay attention to the taste, therefore, when eating the scapharca, only use boiling water to burn, and then take out the shellfish meat, dip in seasoning to eat. This kind of food is delicious, but the hepatitis A virus is not killed, resulting in foodborne virus disease. Hepatitis E virus is unstable and vulnerable to destruction.

b. Prion virus. Prion is a non-nucleic acid protein infection factor, which can cause lesions in mammalian central nervous tissue. Prions can cause transferable neurodegenerative diseases in humans and animals, such as BSE, CJD and so on. At least 70 people are known to have died of new Creutzfeldt-Jakob disease in the UK, and the medical profession suspects that it may be related to eating meat made from cattle with BSE disease.

3. Parasitic harm.

a. Cysticercus cellulosae. Cysticercus cellulosae, commonly known as "rice pork", refers to the pork with measle. If a person eats a pork cysticercosis that is not dead, due to intestinal fluid and bile, the head node can extend to the hook of the cyst, then the hook of the sucker firmly adsorbed on the intestinal wall, so as to absorb nutrients and develop into imago, that is cestode, and cause the people suffering from the taeniasis.

②旋毛虫。旋毛虫是一种很小的线虫,肉眼不易看见。当人误食含旋毛虫幼虫的食品后,幼虫则从囊内伸出进入十二指肠和空肠,并迅速发育为成虫,每条成虫可产1500个以上幼虫。幼虫穿过肠壁,随血液循环到全身,主要寄生在横纹肌肉内,使被寄生的肌肉发生变性。患者初期呈恶心、呕吐、腹痛和下痢等症状,随后体温升高。由于在肌肉内寄生,肌肉发炎,疼痛难忍。根据寄生的部位,出现声音嘶哑、呼吸和吞咽困难等症状。

2)食品加工中的化学性危害
(1)食品中天然存在的化学危害。
①真菌毒素。霉菌能引起农作物的病害和食品霉变,并产生有毒的代谢产物——霉菌毒素。目前已知的霉菌毒素有200多种,主要有黄曲霉毒素、镰刀菌毒素(T-2毒素、脱氧雪腐镰刀菌烯醇、玉米赤霉烯酮、伏马菌素等)、赭曲霉毒素、杂色曲霉素、展青霉素、3-硝基丙酸等。

A. 黄曲霉毒素(aflatoxins)是由黄曲霉和寄生曲霉产生的次生代谢产物。已发现的黄曲霉毒素有20多种,其中以黄曲霉毒素 B_1 的毒性和致癌性最强,在食品中的污染也最普遍。

B. 赭曲霉毒素(ochratoxin)是由曲霉毒属和青霉属的一些菌种产生的二次代谢产物。该毒素是异香豆素的系列衍生物,包括赭曲霉毒素 A、B 和 C,其中赭曲霉毒素 A 是植物性食品中的主要污染物,是谷物、大豆、咖啡豆和可可豆的污染物。

b. Trichinella spiralis. Trichinella spiralis is a small nematode that is not visible to the naked eye. When the food containing the larvae of Trichinella spiralis is eaten by mistake, the larvae escape into the duodenum and jejunum from the sac and develop rapidly into imago, each of which can produce more than 1500 larvae. The larva passes through the wall of the intestine and circulates to the whole body with the blood, mainly parasitic in the transverse muscles, so that the parasitic muscles degenerate. The patient developed nausea, vomiting, abdominal pain and dysentery at the initial stage, and then elevated body temperature. Because of parasitism in the muscle, muscle inflammation, pain is unbearable. According to the parasitic site, symptoms such as hoarseness, dyspnea and dysphagia occur.

B. Chemical Hazards in Food Processing
1. Natural chemical hazards in food.

a. Mycotoxins. Mould can cause crop diseases and food mildew, and produce toxic metabolites—mycotoxins. At present, there are 200 kinds of known mycotoxins, mainly aflatoxin, fusarium toxin (T-2 toxin, DON, zearalenone, fumonisin), aspergillus germanium toxin, versicolorin, patulin, 3-nitropropionic acid and so on.

a) Aflatoxins is composed of aspergillus flavus and aspergillus parasites. Aflatoxins have been found in more than 20, of which aflatoxin B_1 is the most toxic and carcinogenic, and the most common in food.

b) Ochratoxin is the two metabolite produced by some strains of Aspergillus and Penicillium. This toxin is isocoumarin derivatives including pigment, ochratoxin A, B and C, which are the main pollutants of ochratoxin A in vegetable foods, cereals, beans, pollutants, coffee beans and cocoa beans.

C. 单端孢霉烯族化合物（trichothecenes）是一组生物活性和化学结构相似的有毒代谢产物，大多数单端孢霉烯族化合物是由镰刀菌属的菌种产生的，其中最重要的菌种是产生 DON 和 NIV 的禾谷镰刀菌。单端孢霉烯族化合物的主要毒性作用为细胞毒性、免疫抑制和致畸作用，可能有弱致癌性，是污染谷物和饲料的污染物。

②植物食品中的天然毒素。

A. 红细胞凝集素和皂素。红细胞凝集素又称外源凝集素，是一种糖蛋白，存在于大豆、四季豆、豌豆、小扁豆、蚕豆和花生等食物原料中。四季豆又称菜豆、扁豆、刀豆、芸豆和豆角等。由四季豆等引起的食物中毒事件时有发生。

B. 生物碱。生物碱是一类含氮的有机化合物，有类似碱的性质，遇酸可生成盐。存在于食用植物中的生物碱主要有龙葵碱、秋水仙碱和咖啡碱等。龙葵碱又称茄碱、龙葵毒素和马铃薯毒素，是由葡萄糖残基和茄啶组成的一种弱碱性糖苷。它存在于马铃薯、番茄及茄子等茄科植物中。马铃薯中龙葵碱的含量随品种、部位和季节的不同而不同。发芽马铃薯的幼芽和芽眼部分含量最高，绿色马铃薯和出现黑斑的马铃薯块茎中含量也较高。当食入 0.2～0.4 g 茄碱时即可发生中毒。

③动物食品中的天然毒素。

A. 河鲀毒素。河鲀是一种味道极鲜美但含剧毒的鱼类。河鲀中的有毒成分是河鲀毒素（TTX），其毒性比氰化钾高 1000 倍。因此河鲀中毒是世界上最严重的动物性食品中毒，其死亡率居食物中毒死亡率的首位。河鲀毒

c) Trichothecenes is a group of toxic metabolites with similar biological activity and chemical structure. Most monosporin compounds are produced by fusarium species, the most important of which are fusarium graminearum, which produces DON and NIV. The main toxic effects of monosporin compounds are cytotoxicity, immunosuppression and teratogenicity, which may have weak carcinogenicity and are pollutants that pollute cereals and feed.

b. Natural toxins in plant foods.

a) Hemaglutinin and saponin. Hemaglutinin also called lectin, which is a glycoprotein found in soybeans, green bean, peas, lentils, vicia beans and peanuts and other food raw materials. Green beans also known as beans, lentils, sword beans, kidney beans and long beans. Food poisoning caused by green bean and so on occurs from time to time.

b) Alkaloid. Alkaloids are a class of organic compounds containing nitrogen, which are similar to alkali, and can produce salt with acid. Alkaloid from edible plants are mainly solanine, colchicine and caffeine. Solanine also called solasonine, solanum nigrum toxin and potato toxin, which is a kind of alkaline glycoside composed of glucose residues and solanidine. It is found in solanaceae plants such as potatoes, tomatoes and eggplant. The content of solanine in potato varied with variety, position and season. The content of sprout and bud eye part of germinated potato was the highest, and the content of green potato and potato tuber with black spot was also higher. Poisoning can occur when the cannibalism is 0.2 g or 0.4 g solanine.

c. Natural toxins in animal food.

a) Tetrodotoxin. Puffer fish is a delicious but highly toxic fish. The toxic component of puffer fish is tetrodotoxin (TTX), which is 1000 times more toxic than potassium cyanide. Therefore, globefish poisoning is the most serious animal food poisoning in the world, and the mortality rate is the first in food poisoning.

素是一种神经毒素,能阻断神经传导,使神经麻痹,病死率高达40%。河鲀毒素性质比较稳定,盐腌、日晒均不被破坏,在100 ℃下加热24 h或120 ℃下加热60 min才能被完全破坏。因此,一般家庭烹调难以去除毒性,所以严禁擅自经营、加工和销售河鲀。

B. 动物腺体和内脏中的毒素。动物腺体和内脏中的毒素包括甲状腺素、肾上腺分泌的激素、变性淋巴腺、动物肝脏中的毒素以及胆囊毒素等。为安全起见,防止甲状腺素中毒,建议烹调前应注意摘除甲状腺;无论淋巴结有无病变,消费者应将其除去为宜;要食用健康的新鲜动物肝脏,食用前充分清洗、煮熟煮透;一次摄入不能太多;如果在摘除胆囊时不小心弄破胆囊,应用清水充分洗涤、浸泡以便去除残留的胆囊毒素。

④毒蘑菇中的天然毒素。我国已知食用蘑菇约有700多种,毒蘑菇约为190多种。食用蘑菇和有毒蘑菇在外观上很难分辨,因此,因误食毒蘑菇而引起的中毒事件频频发生。蘑菇毒素从化学结构上可分为生物碱类、肽类(毒环肽)及其他化合物(如有机酸等),根据中毒时出现的临床症状可分为胃肠毒素、神经精神毒素、血液毒素、原浆毒素和其他毒素五类。

鉴于蘑菇种类繁多,难以识别,所以在采集野蘑菇时,要在专业人员或有识别能力的人员指导下进行,以便

Tetrodotoxin is a neurotoxin that blocks nerve conduction and paralyzes the nerves with a mortality of up to 40%. The nature of tetrodotoxin is relatively stable, salted and sunburned are not damaged. It can be completely destroyed by heating at 100 ℃ for 24 h or 120 ℃ for 60 min. Therefore, the general domestic cooking is difficult to remove toxicity, so it is strictly prohibited to operate, process and sell puffer fish.

b) Toxins in the glands and viscera of animals. Toxins in animal glands and viscera include thyroxine, hormones secreted by the adrenal gland, degenerated lymph glands, toxins in the animal liver, and cholecyst toxins. In order to prevent thyroxine poisoning, it is recommended that thyroid gland should be removed before cooking; consumers should remove thyroid gland whether or not lymph nodes are diseased; to eat healthy fresh animal liver, wash, cook and cook thoroughly before eating; do not ingest too much at a time; if you accidentally break the gallbladder when you remove the gallbladder, wash and soak in clear water to remove the remaining gallbladder toxins.

d. Natural toxins in poisonous mushrooms. There are about 700 kinds of edible mushrooms and 190 kinds of poisonous mushrooms in China. It is difficult to distinguish edible mushrooms from poisonous mushrooms in appearance, therefore, poisoning events caused by accidental eating of poisonous mushrooms occur frequently. Mushroom toxins can be classified into alkaloids, peptides (toxic cyclic peptides) and other compounds (such as organic acids, etc.). According to the clinical symptoms of poisoning, they can be classified into gastrointestinal toxins, neuropsychotoxins, blood toxins, and so on. Five classes of protoplasmic and other toxins.

In view of the variety of mushrooms, it is difficult to identify, so in the collection of wild mushrooms, under the guidance of professionals or

剔除毒蕈。对一般人来说,最有效的措施是绝对不采摘不认识的野蘑菇,也不食用没有吃过的蘑菇。

(2)环境污染导致的化学危害。

①重金属污染。重金属是指相对密度大于4或5的金属,约45种,如铜、铅、锌、铁、钴、镍、钒、铌、钽、钛、锰、镉、汞、钨、钼、金、银等。大部分重金属如汞、铅、镉等并非生命活动所必需,而且所有重金属超过一定浓度都对人体有毒。

A. 汞对食品的污染。汞分为无机汞和有机汞,有机汞曾用作杀菌剂,用以拌种或田间喷粉,目前已禁止使用。通过食物进入人体的甲基汞可以直接进入血液,与红细胞血红蛋白的硫基结合,随血液分布于各组织器官,并可以透过血脑屏障侵入脑组织,严重损害小脑和大脑两半球,致使中毒患者视觉、听觉产生严重障碍。严重者出现精神错乱、痉挛死亡。

B. 砷对食品的污染。砷分为无机砷和有机砷。无机砷多数为3价砷和5价砷化合物,有机砷主要为5价砷。长期摄入少量的砷化物可导致慢性砷中毒,症状为进行性衰弱、食欲不振、恶心、呕吐等,同时出现皮肤色素沉着、角质增生、末梢神经炎等特有体征。患者出现末梢多发性神经炎,四肢感觉异常、麻木、疼痛,行走困难,直至肌肉萎缩。

have the ability to recognize, in order to eliminate toadstools. For the average person, the most effective measure is to never pick wild mushrooms that you don't know, or eat mushrooms that you haven't eaten.

2. Chemical hazards from environmental pollution.

a. Heavy metal pollution. Heavy metals refer to metals with relative density greater than 4 or 5. There are about 45 kinds of metals, such as copper, lead, zinc, iron, cobalt, nickel, vanadium, niobium, tantalum, titanium, manganese, cadmium, mercury, tungsten, molybdenum, gold, silver and so on. Most heavy metals such as mercury, lead and cadmium are not necessary for life activities, and all heavy metals over a certain concentration are toxic to the human body.

a) Contamination of food by mercury. Mercury is divided into inorganic mercury and organic mercury, organic mercury used as fungicides for seed or field spraying powder, has been banned. Methylmercury, which enters the human body through food, can enter the blood directly, bind to the thiol group of erythrocyte hemoglobin, distribute with the blood in various tissues and organs, and invade the brain through the blood-brain barrier, causing serious damage to the cerebellar and cerebral hemispheres, causes the poisoning patient vision, the hearing produces the serious obstacle. In severe cases, insanity and spasmodic death has occurred.

b) Arsenic contamination of food. Arsenic is divided into inorganic arsenic and organic arsenic. Inorganic arsenic is mainly composed of 3-valent arsenic and 5-valent arsenic compounds, while organic arsenic is mainly 5-valent arsenic. Long-term intake of a small amount of arsenic can lead to chronic arsenism, symptoms for progressive weakness, loss of appetite, nausea, vomiting, and other unique signs such as skin pigmentation, keratinosis, peripheral neuritis. Patients with peripheral polyneuritis, limbs feel

C. 镉对食品的污染。镉广泛存在于自然界,但含量很低。一般食品中均可以检出镉。金属镉一般无毒,而化合物有毒。急性镉中毒出现流涎、恶心、呕吐等消化道症状。慢性镉中毒可使钙代谢失调,引起肾结石所致的肾绞痛,骨软化症或骨质疏松所致的骨骼症状。镉有致突变和致畸作用,对 DNA 的合成有强抑制作用,并可诱发肿瘤。

②二噁英对食品的污染。二噁英是一类三环芳香族化合物。它属于脂溶性化合物,难以生物降解。二噁英具有强烈的致肝癌毒性,主要来源是含氯化合物的生产和使用。垃圾的焚烧、煤、石油、汽油、沥青等的燃烧也会产生二噁英。一般人群接触的二噁英 90% 以上来源于膳食,尤其是鱼、肉、蛋、奶等高脂肪食物。

③N-亚硝基化合物对食品的污染。N-亚硝基化合物是一类具有—N—N=O 结构的有机化合物,对动物有较强的致癌作用,能诱发多种器官和组织的肿瘤。我国某些地区食管癌高发,被认为与当地食品中亚硝胺检出率较高有关。

(3)农药残留。农药按其用途可分为杀虫剂、杀菌剂、除草剂、杀螨剂、植物生长调节剂、粮食防虫剂、灭鼠药和昆虫不育剂等。按其化学组成又可分为有机氯、有机磷、氨基甲酸酯和拟

abnormal, numbness, pain, walking difficulties, until muscle atrophy.

c) Cadmium contamination of food. Cadmium exists widely in nature, but its content is very low. Cadmium can be detected in general food. Metal cadmium is generally non-toxic, while compounds are toxic. Acute cadmium poisoning appeared salivation, nausea, vomiting and other gastrointestinal symptoms. Chronic cadmium poisoning can cause calcium metabolism disorders, causing kidney stones caused by renal colic, osteomalacia or osteoporosis caused by bone symptoms. Cadmium has mutagenicity and teratogenicity, which can inhibit the synthesis of DNA and induce tumors.

b. Dioxin's pollution to food. Dioxin is a class of tricyclic aromatic compounds. Dioxin is a fat soluble compound, which is difficult to be biodegraded. Dioxin has a strong toxicity to liver cancer. The main source of dioxin is the production and use of chlorinated compounds. The burning of garbage, coal, oil, gasoline, asphalt and other burning will also produce dioxin. More than 90% dioxins from the general population are derived from dietary, especially high fat foods such as fish, meat and egg milk.

c. Pollution of N-nitroso compounds in food. N-nitroso compounds are a class of organic compounds with the —N—N=O structure, strong carcinogenic effects on animal, can induce a variety of organs and tissues of the tumor. The high incidence of esophageal cancer in some parts of China is considered to be related to the high detection rate of nitrosamines in local food.

3. Pesticide residues. Pesticide can be divided into insecticide, fungicide, herbicide, acaricide, plant growth regulator, food insecticide, rodenticide and insect sterility. According to its chemical composition, it can be divided into organochlorine, organophosphorus,

除虫菊酯等类型。

①有机氯农药。有机氯农药是指在组成上含氯的有机杀虫、杀菌剂。有机氯农药包括滴滴涕（DDT，二氯二苯三氯乙烷）和六六六（BHC，六氯环己烷）、氯丹、林丹、艾氏剂和狄氏剂等。虽然此类农药于1983年就已停止生产和使用，但毕竟此类农药有30多年的使用历史，而且有机氯农药化学性质稳定、不易降解，因此，其对食品的污染和残留仍普遍存在。

②有机磷农药。有机磷农药是指在组成上含磷的有机杀虫、杀菌剂等。多数有机磷农药化学性质不稳定，遇光和热易分解，在碱性环境中易水解。在作物中经过一段时间的自然分解转化为毒性较小的无机磷。有机磷农药对食品的污染普遍存在，主要污染植物性食品，尤其是含有芳香物质的植物，如水果、蔬菜等。主要的污染方式是直接施用农药或来自土壤的农药污染。

③氨基甲酸酯类农药。氨基甲酸酯类为氨基甲酸的N-甲基取代酯类，是含氮类农药。用于农业生产的主要有杀虫剂、杀菌剂和除草剂。氨基甲酸酯类杀虫剂具有致畸、致突变、致癌的可能。

④拟除虫菊酯类农药。拟除虫菊酯类农药是近年发展较快的一类农药，是模拟天然菊酯的化学结构而合成的有机化合物。中毒者可出现头痛、乏力、流涎、惊厥、抽搐、痉挛、呼吸困难、血压下降、恶心、呕吐等症状。该类农药还具有致突变作用。

carbamate and pyrethroid.

a. Organochlorine pesticides. Organochlorine pesticides are organic insecticides and fungicides containing chlorine in their composition. Organochlorine pesticides including DDT and BHC, chlordane, lindane, AI reagent and dieldrin etc. Although the production and use of such pesticides had ceased in 1983, they had been used for more than 30 years, and organochlorine pesticides were stable in chemical properties and difficult to degrade. Therefore, their contamination and residues of food were still widespread.

b. Organophosphorus pesticides. Organophosphorus pesticides are organic insecticides and fungicides with phosphorus in composition. Most organophosphorus pesticides are unstable in chemical properties, easily decomposed by light and heat, and easily hydrolyzed in alkaline environment. After a period of natural decomposition in crops into less toxic inorganic phosphorus. Organophosphorus pesticides pollute food widely. They mainly pollute plant foods, especially plants containing aromatic substances, such as fruits, vegetables and so on. The main form of pollution is the direct application of pesticides or pesticide pollution from the soil.

c. Carbamate pesticides. Carbamate is N-methyl-substituted carbamate, which is a nitrogen-containing pesticide. Insecticides, fungicides and herbicides are mainly used in agricultural production. Carbamate insecticides have the possibility of teratogenicity, mutagenicity and carcinogenesis.

d. Pyrethroid pesticides. Pyrethroid pesticide is a kind of pesticide which has been developed rapidly in recent years. It is an organic compound synthesized by mimicking the chemical structure of natural permethrin. Poisoning can appear headache, fatiguc, salivation, convulsion, hyperspasmia, spasm, dyspnea, blood pressure drop, nausea, vomiting and other symptoms. These pesticides also have mutagenicity.

(4)兽药残留。兽药是指用于预防和治疗畜禽疾病的药物。一些促进畜禽生长、提高生产性能、改善动物性食品品质的药用成分被开发为饲料添加剂,也属于兽药的范畴。常见兽药残留的种类有抗生素类、合成抗菌素类、抗寄生虫类、杀虫剂和激素类药物。兽药残留的危害主要表现在急性中毒、过敏反应、致癌、致畸、致突变、激素(样)作用等方面。

(5)加工过程中加入的化学品。全世界批准使用的食品添加剂有25000种,中国允许使用的品种有近千种。食品添加剂的使用对食品产业的发展起着重要作用,但如果不按要求科学地使用食品添加剂,也会带来很大的负面影响。

3)食品加工中的物理性危害

物理危害主要是由于食品中存在玻璃、金属、木头、首饰、塑料等硬物,食用时易引起口腔、牙齿,甚至消化道的损伤。物理危害是顾客投诉最多的问题。需要说明的是,这里所讲的危害不包括发现头发、昆虫等异物。控制物理危害的措施有金属检测器检测,可查看并剔除掺有金属片的小包装食品,X光机可查出非铁硬物等。

3. 食品企业常用的质量安全控制体系

食品企业产品的质量安全状况不仅关系企业自身的生存与发展,更重要的是关系到国计民生与社会稳定。

4. Veterinary drug residues. Veterinary drugs are used to prevent and treat animal diseases. Some medicinal ingredients that promote the growth of livestock and poultry, improve the performance of production and improve the quality of animal food are developed as feed additives, they also belong to the category of veterinary drugs. Common veterinary drug residues include antibiotics, synthetic antibiotics, anti-parasites, insecticides and hormone drugs. The harm of veterinary drug residue is mainly manifested in acute poisoning, allergic reaction, carcinogenesis, teratogenicity, mutagenesis, hormones and so on.

5. Chemicals added in the process. There are 25000 kinds of food additives approved for use in the world and nearly 1000 kinds of food additives are permitted in China. The use of food additives plays an important role in the development of the food industry, but if the scientific use of food additives does not meet the requirements, it will also bring a great negative impact.

C. Physical Hazards in Food Processing

Physical hazards are mainly due to the existence of glass, metal, wood, jewelry, plastic and other hard objects in food. Physical hazards are the most frequently complained about by customers. It should be noted that the hazards described here do not include the discovery of foreign bodies such as hair and insects. The measures to control the physical hazards are metal detector detection, the inspection and elimination of small packaging food with metal chips, X-ray machine can detect non-iron hard objects, and so on.

Ⅲ. Quality and Safety Control System Commonly Used in Food Enterprises

The quality and safety of food products is not only related to the survival and development of the enterprise itself, but also to the national economy

企业要取得公众信任就必须有公众认可的管理模式及值得相信的证明材料,这就产生了各种认证;在一个庞大的生产加工体系中,实现科学管理,保证产品的质量安全,如果没有科学的管理方法是不可能实现的。因此,从20世纪90年代以来,在国际组织的努力下,形成了一些国际公认的质量控制体系。食品企业常用的质量安全控制体系如下。

and people's livelihood and social stability. In order to gain public trust, enterprises must have the approved management model and trustworthy proof materials, which has produced various kinds of certification. In a huge production and processing system, scientific management is realized to ensure the quality and safety of products. It is impossible to achieve without scientific management methods. Therefore, since the 1990s, with the efforts of international organizations, a number of internationally recognized quality control systems have been formed. The commonly used quality and safety control systems for food enterprises are as follows.

1)良好的操作规范(GMP)

它是保证食品具有高度安全性的良好生产管理系统。它要求食品企业应具有合理的生产过程、良好的生产设备、正确的生产知识、完善的质量控制和严格的管理体系。因此,这是食品工业实现生产工艺合理化、科学化和现代化的必备条件。

A. Good Manufacture Practice, GMP

It is a good production management system to ensure high safety of food. It requires food enterprises to have reasonable production process, good production equipment, correct production knowledge, perfect quality control and strict management system. Therefore, GMP is a necessary condition for food industry to realize rationalization, scientization and modernization of production technology.

2)卫生标准操作程序(SSOP)

企业为了使其所加工的食品符合此要求,制定了在食品加工过程中如何具体实施清洗、消毒和保持卫生的作业指导文件,把每一种卫生操作具体化、程序化,对某人执行的任务提供足够详细的规范,并在实施过程中进行严格的检查和记录,实施不力要及时纠正。

B. Sanitation Standard Operation Procedures, SSOP

In order to make the food processed by the enterprise comply with the requirements of GMP, how to carry out the operation guidance document of cleaning, disinfection and hygiene maintenance in the process of food processing is formulated, and each kind of sanitary operation is concretized and programmed. Provide sufficient detailed specifications for the tasks performed by someone, and conduct strict checks and records during the implementation process, and the lack of implementation should be corrected in a timely manner.

3)危害分析与关键控制点(HACCP)

该体系强调在食品加工的全过程中,对各种危害因素进行系统和全面

C. Hazard Analysis Critical Control Point, HACCP

The system emphasizes the systematic and comprehensive analysis of various hazard factors

的分析,然后确定关键控制点(CCP),进而确定控制、检测、纠正的方案。这是目前食品行业有效预防食品安全事故最先进的管理方案。

4)国际标准化组织(ISO)质量管理体系

ISO/TC176是国际标准化组织中的质量管理和质量保证技术委员会,负责制定世界通用的质量管理和质量保证标准。

ISO9000系列标准是ISO/TC176成立以来第一次向全世界发布的第一项管理标准,适用于所有组织。ISO22000标准《食品安全管理体系 食品链中各类组织的要求》是ISO/TC176针对食品企业制定的食品安全管理体系。

during the whole process of food processing, and then determines the critical control point (CCP), and then determines the control, detection, and correction programs. It is the most advanced management scheme to effectively prevent food safety accidents in the food industry at present.

D. ISO Quality Management System

ISO/TC176 is a universal quality management and quality assurance standard established by the Quality Management and Quality Assurance Technical Committee of the International Organization for Standardization.

The ISO9000 series standard is the first management standard issued to the world for the first time since the establishment of ISO/TC176, which is applicable to all organizations. ISO22000 standard *Food Safety Management Systems—Requirements for any Organization in the Food Chain* is a food safety management system formulated by ISO/TC176 for food enterprises.

任务2 良好的操作规范(GMP)

"GMP"是英文Good Manufacturing Practice的缩写,中文的意思是"良好的操作规范",或是"优良制造标准",是一种特别注重在生产过程中实施的对产品质量与卫生安全的自主性管理制度。它是一套适用于制药、食品等行业的强制性标准,要求企业从原料、人员、设施设备、生产过程、包装运输、质量控制等方面按国家有关法规达到卫生质量标准,形成一套可操作的作业规范,帮助企业改善企业卫生环境,及时发现生产过程中存在的问题并加以改善。简要地说,良好的操作规范要求食品生产企业应具备良好的生产设备、合理的生产过程、完善的质量管理和严格的检测系统,确保最终产品的质量(包括食品安全卫生)符合法规要求。良好的操作规范所规定的内容,是食品加工企业必须达到的最基本的条件。

1. 良好的操作规范的由来和发展

良好的操作规范原较多应用于制药工业,现许多国家将其用于食品工业,制定出相应的法规。美国最早将良好的操作规范用于工业生产,1969年其食品药物管理局(FDA)发布了食品制造、加工、包装和保存的良好的操作规范,简称GMP或FGMP基本法,并陆续发布各类食品的良好生产规范。目前,美国已立法强制实施食品良好的操作规范。良好的操作规范自20世纪70年代初在美国提出以来,已在全球不少发达国家和发展中国家得到认可并采纳。1969年,世界卫生组织向全世界推荐良好的操作规范。1972年,欧洲共同体14个成员国公布了良好的操作规范的总则。日本、英国、新加坡和很多工业先进国家引进食品良好的操作规范。日本厚生省于1975年开始制定各类食品卫生规范。我国已颁布药品生产良好操作规范的标准,并实行企业良好操作规范认证,使药品的生产及管理水平有了较大程度的提高。我国食品企业质量管理规范的制定开始于20世纪80年代中期。从1988年开始,我国先后颁布了包括 GB 14881—2013《食品生产通用卫生规范》在内的21个食品企业卫生规范。重点对厂房、设备、设施和企业自身卫生管理等方面提出卫生要求,以促进我国食品卫生状况的改善,预防和控制各种有害因素对食品的污染。

2. 良好操作规范的分类

1)从适用范围分

(1)具有国际性质的良好操作规范。如世界卫生组织的良好操作规范、北欧七国自由贸易联盟制定的良好操作规范、东南亚国家联盟的良好操作规范等。

(2)国家权力机构颁发的良好操作规范。如中华人民共和国卫生部及后来的国家药品监督管理局、美国食品药物管理局、英国卫生和社会保险部、日本厚生省等政府机关制定的良好操作规范。

(3)工业组织制定的良好操作规范。如美国制药工业联合会制定的良好操作规范(其标准不低于美国政府制定的良好操作规范)、中国医药工业公司制定的良好操作规范及其实施指南,甚至包括药厂或公司自己制定的良好操作规范。

2)从良好操作规范的性质分

(1)良好操作规范作为法典规定,如中国、美国和日本的良好操作规范。

(2)将良好操作规范作为建议性的规定,对药品生产和质量管理起到指导性作用,如世界卫生组织的良好操作规范。

3)按良好操作规范的权威性和法律效力分

按良好操作规范的权威性和法律效力可分为强制性和指导性(推荐性)的良好操作规范。

3. 良好操作规范的三大目标要素

实施良好操作规范的目标在于将人为的差错控制在最低的限度,防止对药品的污染,保证高质量产品的质量管理体系。

1)将人为的差错控制在最低的限度

(1)管理方面。例如,质量管理部门从生产管理部门独立出来;建立相互监督检查制度;指定各部门责任者;制订规范的实施细则和作业程序,各生产工序严格复核,如称量、材料贮存领用等;在各生产工序,对用于生产的运送容器、主要机械,要标明正在生产的药品名称、规格、批号等状态标志;整理和保管好记录(一般按产品有效期终止后1年,未规定有效期的药品应保存3年);人员的配备、教育和管理。

(2)装备方面。例如,各工作间要保持宽敞,消除妨碍生产的障碍;不同品种的操作必须有一定的间距,严格分开。

2)防止对药品的污染和降低质量

(1)管理方面。例如,操作室清扫和设备洗净的标准及实施;对生产人员进行严格的卫生教育;操作人员定期进行身体检查,以防止生产人员带有病菌、病毒而污染药品;限制非生产人员进入工作间等。

(2)装备方面。例如,防止粉尘对药品的污染,要有相应的机械设备(空调净化系统等);

操作室专用化;对直接接触药品的机械设备、工具、容器,选用对药物不发生变化的材质制造,如使用 L316 型不锈钢材等,注意防止机械润滑油对药品的污染;操作室的结构及天花板、地面、墙壁等清扫容易;对无菌操作区要进行微粒检查和浮游菌、沉降菌的检查,定期灭菌等。

3) 保证高质量产品的质量管理体系

(1) 管理方面。例如,质量管理部门独立行使质量管理职责;机械设备、工具、量具定期维修校正;检查生产工序各阶段的质量,包括工程检查;有计划的合理的质量控制,包括质量管理实施计划、试验方案、技术改造、质量攻关要适应生产计划要求;追踪药品批号,并作好记录;在适当条件下保存出厂后的产品质量检查留下的样品;收集消费者对药品投诉的情报信息,随时完善生产管理和质量管理等。

(2) 装备方面。例如,操作室和机械设备的合理配备,采用先进的设备及合理的工艺布局;为保证质量管理的实施,配备必要的实验、检验设备和工具等。

4. 良好操作规范的主要内容

企业要建立良好的操作规范,就需要了解良好的操作规范的内容。食品企业实施良好的操作规范有利于食品质量控制和企业的长远发展,提高食品企业和产品的声誉,促进竞争力;有利于食品进入国际市场,促进食品企业质量管理的科学化和规范化;有利于提高卫生行政部门对食品企业进行监督检查的水平,为企业提供生产和质量遵循的基本原则和必需的标准组合。参照 GB 14881—2013《食品生产通用卫生规范》,良好操作规范的内容如下。

1) 范围

本标准规定了食品生产过程中原料采购、加工、包装、贮存和运输等环节的场所、设施、人员的基本要求和管理准则。

本标准适用于各类食品的生产,如确有必要制定某类食品生产的专项卫生规范,应当以本标准作为基础。

2) 术语和定义

(1) 污染:在食品生产过程中发生的生物、化学、物理污染因素传入的过程。

(2) 虫害:由昆虫、鸟类、啮齿类动物等生物(包括苍蝇、蟑螂、麻雀、老鼠等)造成的不良影响。

(3) 食品加工人员:直接接触包装或未包装的食品、食品设备和器具、食品接触面的操作人员。

A. Range

This standard sets out the basic requirements and management guidelines for the places, facilities, personnel and so on in the process of raw material purchasing, processing, packaging, storage and transportation.

This standard is applicable to the production of all kinds of food. If it is necessary to formulate a special hygienic standard for the production of a certain kind of food, it shall be based on this standard.

B. Terms and Definitions

1. Pollution: the introduction of biological, chemical and physical pollution factors in the process of food production.

2. Insect attack: adverse effects caused by insects, birds, rodents and other organisms (including flies, cockroaches, sparrows, mice, etc.).

3. Food processing personnel: the operators of direct contact with packaged or unpackaged food, food equipment, utensils, and food contact surfaces.

(4)接触表面:设备、工器具、人体等可被接触到的表面。

(5)分离:通过在物品、设施、区域之间留有一定空间,而非通过设置物理阻断的方式进行隔离。

(6)分隔:通过设置物理阻断如墙壁、卫生屏障、遮罩或独立房间等进行隔离。

(7)食品加工场所:用于食品加工处理的建筑物和场地,以及按照相同方式管理的其他建筑物、场地和周围环境等。

(8)监控:按照预设的方式和参数进行观察或测定,以评估控制环节是否处于受控状态。

(9)工作服:根据不同生产区域的要求,为降低食品加工人员对食品的污染风险而配备的专用服装。

3) 选址及厂区环境

(1)选址。

①厂区不应选择对食品有显著污染的区域。如某地对食品安全和食品宜食用性存在明显的不利影响,且无法通过采取措施加以改善,应避免在该地建厂。

②厂区不应选择有害废弃物以及粉尘、有害气体、放射性物质和其他扩散性污染源不能有效清除的地址。

③厂区不宜选择易发生洪涝灾害的地区,难以避开时应设计必要的防范措施。

④厂区周围不宜有虫害大量孳生的潜在场所,难以避开时应设计必要的防范措施。

4. Contact surface: a surface that is accessible to equipment, tools, human bodies, etc.

5. Separation: isolation is done by leaving space between objects, facilities, and areas, rather than by setting up physical blockages.

6. Division: isolation is performed by setting up physical blocks such as walls, sanitary barriers, shades, or independent rooms.

7. Food processing place: buildings and sites used for food processing, as well as other buildings, sites and surroundings managed in the same manner.

8. Monitor: observed or measured in a preset manner and parameter to assess whether the control link is in a controlled state.

9. Work clothes: special clothing for reducing the risk of food contamination by food workers in accordance with the requirements of different production areas.

C. Site Selection and Site Environment

1. Site selection.

a. The site areas chosen should not be areas that are significantly contaminated with food. If there are obvious adverse effects on food safety and food suitability in a certain place and cannot be improved by taking measures, the establishment of a factory at that address should be avoided.

b. The site areas selected should not be addresses where the following diffuse sources of pollution cannot be effectively removed, such as hazardous wastes and dust, harmful gases, radioactive substances and others.

c. The site area should not be set up in the area prone to flood and waterlogging, and the necessary preventive measures should be designed when it is difficult to avoid.

d. It is not appropriate to establish the site area in the potential place where a large number of pests are breeding, and necessary precautions

(2)厂区环境。

①应考虑环境给食品生产带来的潜在污染风险,并采取适当的措施将其降至最低水平。

②厂区应合理布局,各功能区域划分明显,并有适当的分离或分隔措施,防止交叉污染。

③厂区内的道路应铺设混凝土、沥青或者其他硬质材料;空地应采取必要措施,如铺设水泥、地砖或铺设草坪等方式,保持环境清洁,防止正常天气下扬尘和积水等现象的发生。

④厂区绿化应与生产车间保持适当距离,植被应定期维护,以防止虫害的孳生。

⑤厂区应有适当的排水系统。

⑥宿舍、食堂、职工娱乐设施等生活区应与生产区保持适当距离或分隔。

4)厂房和车间
(1)设计和布局。

①厂房和车间的内部设计和布局应满足食品卫生操作要求,避免食品生产中发生交叉污染。

②厂房和车间的设计应根据生产工艺合理布局,预防和降低产品受污染的风险。

③厂房和车间应根据产品特点、生产工艺、生产特性以及生产过程对清洁

should be designed when it is difficult to avoid.

2. Site environment.

a. The potential pollution risks posed by the environment to food production should be taken into account and appropriate measures should be taken to minimize them.

b. The site area should be rationally distributed, each functional area should be clearly divided, and appropriate separation or division measures should be put in place to prevent cross-contamination.

c. The roads in the site area shall be paved with concrete, asphalt, or other hard materials; the opening spaces shall take necessary measures, such as laying cement, floor tiles or laying lawns to keep the environment clean, and to prevent the occurrence of dust and water in normal weather.

d. The site greening should keep an appropriate distance from the production workshop and vegetation should be maintained regularly to prevent the breeding of insect pests.

e. The site areas should have an appropriate drainage system.

f. Dormitories, canteens, staff and workers entertainment facilities and other living areas should maintain appropriate distance or separation from the production area.

D. Factory Building and Workshop

1. Design and layout.

a. The interior design and layout of factory building and workshop shall meet the requirements of food hygiene operation and avoid cross-contamination in food production.

b. The design of factory building and workshop should be rationally distributed in accordance with the production process to prevent and reduce the risk of product contamination.

c. According to the product characteristics, production process, production characteristics and

程度的要求合理划分作业区,并采取有效分离或分隔。如通常可划分为清洁作业区、准清洁作业区和一般作业区,或清洁作业区和一般作业区等。一般作业区应与其他作业区域分隔。

④厂房内设置的检验室应与生产区域分隔。

⑤厂房的面积和空间应与生产能力相适应,便于设备安置、清洁消毒、物料存储及人员操作。

(2)建筑内部结构与材料。

①内部结构。建筑内部结构应易于维护、清洁或消毒。应采用适当的耐用材料建造。

②顶棚。

A. 顶棚应使用无毒、无味,与生产需求相适应,易于观察清洁状况的材料建造;若直接在屋顶内层喷涂涂料作为顶棚,应使用无毒、无味、防霉、不易脱落、易于清洁的涂料。

B. 顶棚应易于清洁、消毒,在结构上不利于冷凝水垂直滴下,防止虫害和霉菌孳生。

C. 蒸汽、水、电等配件管路应避免设置于暴露食品的上方;如确需设置,应有能防止灰尘散落及水滴掉落的装置或措施。

③墙壁。

A. 墙面、隔断应使用无毒、无味的防渗透材料建造,在操作高度范围

production process, the factory building and workshop should be reasonably divided into operation areas on the requirements of cleanliness, and take effective separation or division. For example: generally, areas can be divided into clean operation area, quasi-clean operation area and general operation area; or clean operation area and general operation area and so on. The general operation area should be separated from other operation areas.

d. The inspection room set up in the factory building shall be separated from the production area.

e. The area and space of the factory building should be adapted to the production capacity, which is convenient for equipment placement, cleaning and disinfection, material storage and personnel operation.

2. Internal structure and materials of buildings.

a. Internal structure. The internal structure of the building should be easy to maintain, clean or disinfect. It should be constructed with appropriate durable material.

b. Ceiling.

a) The ceiling should be constructed of non-toxic, tasteless, suitable for production requirements and easy to observe the cleaning condition; if the coating is directly sprayed on the roof as the ceiling, it should use a non-toxic, tasteless, mildew resistant, not easy to shed and easy cleaning coating.

b) The ceiling should be easy to clean and disinfect, in the structure, it should not be conducive to the vertical drop of condensate to prevent the breeding of pests and mold.

c) Steam, water, electricity and other fittings should not be placed above exposed food; if necessary, the devices or measures should be provided to prevent dust and water droplets from falling.

c. Wall.

a) The construction of wall surface and partitions should use non-toxic and tasteless

内的墙面应光滑、不易积累污垢且易于清洁;若使用涂料,应无毒、无味、防霉、不易脱落、易于清洁。

B. 墙壁、隔断和地面交界处应结构合理、易于清洁,能有效避免污垢积存。例如设置漫弯形交界面等。

④门窗。

A. 门窗应闭合严密。门的表面应平滑、防吸附、不渗透,并易于清洁、消毒。应使用不透水、坚固、不变形的材料制成。

B. 清洁作业区和准清洁作业区与其他区域之间的门应能及时关闭。

C. 窗户玻璃应使用不易碎材料。若使用普通玻璃,应采取必要的措施防止玻璃破碎后对原料、包装材料及食品造成污染。

D. 窗户如设置窗台,其结构应能避免灰尘积存且易于清洁。可开启的窗户应装有易于清洁的防虫害窗纱。

⑤地面。

A. 地面应使用无毒、无味、不渗透、耐腐蚀的材料建造。地面的结构应有利于排污和清洗的需要。

B. 地面应平坦防滑、无裂缝,并易于清洁、消毒,并有适当的措施防止积水。

5)设施与设备
(1)设施。

impermeable material, the wall surface within the operating height should be smooth, not easy to accumulate dirt and easy to clean; if the paint is used, it should be non-toxic, tasteless, mildew resistant, not easy to fall off, and easy to clean.

b) The junction of walls, partitions, and ground should be properly constructed and easily cleaned to avoid dirt accumulation. For example, set the diffuse curved interface and so on.

d. Doors and windows.

a) Doors and windows should be tightly closed. The surface of the door should be smooth, anti-adsorption, non-permeable, and easy to clean, and disinfect. It should be made of impervious, sturdy, non-deformable material.

b) The doors between the clean operation area and the quasi-clean operation area as well as other areas shall be closed in a timely manner.

c) Window glass should be made of breakage-proof material. If ordinary glass is used, necessary measures should be taken to prevent the contamination of raw materials, packaging materials and food after the breakage of glass.

d) If windows are provided with windowsill, their structure should be able to avoid dust accumulation and easy to clean. Windows that can be opened should be fitted with an easily cleaned anti-pest screen.

e. Ground.

a) The ground shall be constructed of non-toxic, odorless, impermeable and corrosion-resistant materials. The structure of the ground should be conducive to the need for sewerage and cleaning.

b) The ground should be smooth and anti-skidding, free from cracks, easy to clean and disinfect, and appropriate measures should be taken to prevent stagnant water.

E. Facilities and Equipment
1. Facilities.

①供水设施。

A. 应能保证水质、水压、水量及其他要求符合生产需要。

B. 食品加工用水的水质应符合 GB 5749 的规定，对加工用水水质有特殊要求的食品应符合相应规定。间接冷却水、锅炉用水等食品生产用水的水质应符合生产需要。

C. 食品加工用水与其他不与食品接触的用水（如间接冷却水、污水或废水等）应以完全分离的管路输送，避免交叉污染。各管路系统应明确标识以便区分。

D. 自备水源及供水设施应符合有关规定。供水设施中使用的涉及饮用水卫生安全产品还应符合国家相关规定。

②排水设施。

A. 排水系统的设计和建造应保证排水畅通，便于清洁维护；应适应食品生产的需要，保证食品及生产、清洁用水不受污染。

B. 排水系统入口应安装带水封的地漏等装置，以防止固体废弃物进入及浊气逸出。

C. 排水系统出口应有适当措施以降低虫害风险。

D. 室内排水的流向应由清洁程度要求高的区域流向清洁程度要求低的

a. Water facility.

a）Water quality, water pressure, water volume and other requirements shall be guaranteed to meet the production needs.

b）The water quality used for food processing shall conform to the provisions of GB 5749, and the food with special requirements for the quality of water used for processing shall comply with the corresponding provisions. The water quality of indirect cooling water, boiler water and other food production water should meet the production needs.

c）Water for food processing and other water not in contact with food (such as indirect cooling water, sewage or waste water) should be transported in a completely separate pipeline to avoid cross-contamination. The piping systems should be clearly marked in order to distinguish between them.

d）Self-provided water sources and water supply facilities should comply with the relevant requirements. The products used in water supply facilities involving the sanitation and safety of drinking water shall also comply with the relevant provisions of the State.

b. Drainage facilities.

a）The design and construction of the drainage system shall ensure that the drainage is smooth and easy for cleaning and maintenance, and shall meet the needs of food production and ensure that the food, production and clean water are not polluted.

b）Drainage system entrances should be fitted with the floor drain with a water seal and other devices to prevent solid waste from entering and escaping from turbid gas.

c）Drainage system exits should have appropriate measures to reduce pest risk.

d）The flow direction of indoor drainage should flow from the area with high degree of cleanliness

区域,且应有防止逆流的设计。

E. 污水在排放前应经适当方式处理,以符合国家污水排放的相关规定。

③清洁消毒设施。应配备足够的食品、工器具和设备的专用清洁设施,必要时应配备适宜的消毒设施。应采取措施避免清洁、消毒工器具带来的交叉污染。

④废弃物存放设施。应配备设计合理、防止渗漏、易于清洁的存放废弃物的专用设施;车间内存放废弃物的设施和容器应标识清晰。必要时应在适当地点设置废弃物临时存放设施,并依废弃物特性分类存放。

⑤个人卫生设施。

A. 生产场所或生产车间入口处应设置更衣室;必要时特定的作业区入口处可按需要设置更衣室。更衣室应保证工作服与个人服装及其他物品分开放置。

B. 生产车间入口及车间内必要处,应按需设置换鞋(穿戴鞋套)设施或工作鞋靴消毒设施。如设置工作鞋靴消毒设施,其规格尺寸应能满足消毒需要。

C. 应根据需要设置卫生间,卫生间的结构、设施与内部材质应易于保持清洁;卫生间内的适当位置应设置洗手设施。卫生间不得与食品生产、

to the area with low degree of cleanliness and should be designed to prevent countercurrent.

e) The sewage shall be treated in an appropriate manner before discharge in order to comply with the relevant regulations of the State on sewage discharge.

c. Cleaning and disinfection facilities. Special cleaning facilities shall be provided with adequate food, utensils and equipment, and if necessary, suitable disinfection facilities also should be provided. Measures should be taken to avoid cross-contamination caused by cleaning and disinfecting utensils.

d. Waste storage facilities. Special facilities shall be provided for the storage of wastes which are reasonably designed, prevented leakage and easy to clean; facilities and containers for the storage of waste in the workshop shall be clearly marked. If necessary, temporary storage facilities shall be set up in appropriate locations and stored in accordance with the characteristics of the waste.

e. Personal hygiene facilities.

a) A dressing room shall be set up at the entrance to the production place or workshop; if necessary, the dressing room may be set up at the entrance of the specific work area. The dressing room should be guaranteed to be placed separately the work clothes from personal clothes and other items.

b) The entrance of the workshop and the necessary places in the workshop shall be provided with a shoe changing (wearing shoe covers) facility or a disinfection facility for the work shoes as required. Such as the installation of working shoes and boots disinfection facilities, its size should be able to meet the disinfection needs.

c) The toilet should be set according to the need, toilet structure, facilities and internal materials should be easy to keep clean; the appropriate location in the toilet should be

包装或贮存等区域直接连通。

D. 应在清洁作业区入口设置洗手、干手和消毒设施；如有需要，应在作业区内适当位置加设洗手和（或）消毒设施；与消毒设施配套的水龙头的开关应为非手动式。

E. 洗手设施的水龙头数量应与同班次食品加工人员数量相匹配，必要时应设置冷热水混合器。洗手池应采用光滑、不透水、易清洁的材质制成，其设计及构造应易于清洁消毒。应在临近洗手设施的显著位置标识简明易懂的洗手方法。

F. 根据对食品加工人员清洁程度的要求，必要时应可设置风淋室、淋浴室等设施。

⑥通风设施。
A. 应具有适宜的自然通风或人工通风措施；必要时应通过自然通风或机械设施有效控制生产环境的温度和湿度。通风设施应避免空气从清洁度要求低的作业区域流向清洁度要求高的作业区域。

B. 应合理设置进气口位置，进气口与排气口和户外垃圾存放装置等污染源保持适宜的距离和角度。进、排气口应装有防止虫害侵入的网罩等设施。通风排气设施应易于清洁、维修或更换。

equipped with hand-washing facilities. The toilet shall not be directly connected to the food production, packaging or storage areas.

d) Hand washing, hand drying and disinfection facilities shall be provided at the entrance to the cleaning operation area; if necessary, hand washing and (or) disinfection facilities shall be provided at appropriate locations in the working area; and the switch of the faucet matching the disinfection facility shall be of a non-manual type.

e) The number of faucets in hand washing facilities shall be matched with the number of food processors on the same shift, and if necessary, the cold and hot water mixers shall be provided. Wash basins should be smooth, impermeable and easy to clean material, their design and construction should be easy to clean and disinfect. Simple and understandable hand-washing methods should be marked in prominent locations near hand-washing facilities.

f) According to the requirements for the cleanliness of food processing personnel, it should be possible to set up air shower room, shower and other facilities when necessary.

f. Ventilation facilities.

a) Appropriate natural or manual ventilation measures should be in place, and if necessary, the temperature and humidity of the production environment should be effectively controlled through natural ventilation or mechanical facilities. Ventilation facilities should avoid the flow of air from low cleanliness areas to areas with high cleanliness requirements.

b) The position of air inlets should be set reasonably, and each air inlet should maintain the appropriate distance and angle with the exhaust port and outdoor garbage storage device. Intake and exhaust ports should be equipped with a net cover to prevent insect pests from invading.

C. 若生产过程需要对空气进行过滤净化处理,应加装空气过滤装置并定期清洁。

D. 根据生产需要,必要时应安装除尘设施。

⑦照明设施。

A. 厂房内应有充足的自然采光或人工照明,光泽和亮度应能满足生产和操作需要,光源应使食品呈现真实的颜色。

B. 如需在暴露食品和原料的正上方安装照明设施,应使用安全型照明设施或采取防护措施。

⑧仓储设施。

A. 应具有与所生产产品的数量、贮存要求相适应的仓储设施。

B. 仓库应以无毒、坚固的材料建成;仓库地面应平整,便于通风换气。仓库的设计应易于维护和清洁,防止虫害藏匿,并应有防止虫害侵入的装置。

C. 原料、半成品、成品、包装材料等应依据性质的不同分设贮存场所或分区域码放,并有明确标识,防止交叉污染。必要时仓库应设有温、湿度控制设施。

D. 贮存物品应与墙壁、地面保持适当距离,以利于空气流通及物品搬运。

E. 清洁剂、消毒剂、杀虫剂、润滑剂、燃料等物质应分别安全包装,明确标识,并应与原料、半成品、成品、包装

Ventilation and exhaust facilities should be easy to clean, maintain or replace.

c) If the production process needs to filter and purify the air, the air filter should be installed and cleaned regularly.

d) Dust removal facilities should be installed when necessary according to production needs.

g. Lighting facilities.

a) There should be sufficient natural lighting or artificial lighting in the factory building, the luster and brightness should meet the production and operation needs, and the light source should make the food appear real color.

b) If it is necessary to install lighting over exposed food and raw materials, use safe lighting or take protective measures.

h. Storage facilities.

a) There shall be storage facilities commensurate with the quantity and storage requirements of the products produced.

b) The warehouse shall be constructed of non-toxic and solid materials; the ground of the warehouse shall be flat so as to facilitate ventilation. Warehouses should be designed to be easy to maintain and clean, to prevent pest hiding, and to be equipped to prevent pest incursions.

c) According to the different properties of raw materials, semi-finished products and finished products, packaging materials should be divided into storage places, or stacked in different areas, and clearly marked to prevent cross-contamination. The warehouse shall be equipped with temperature and humidity control facilities when necessary.

d) Storage items should be kept at an appropriate distance from the walls and floors to facilitate air circulation and the handling of goods.

e) Cleaners, disinfectants, insecticides, lubricants, fuels and other substances should be separately packaged, clearly marked, and should

材料等分隔放置。

⑨温控设施。

A. 应根据食品生产的特点,配备适宜的加热、冷却、冷冻等设施,以及用于监测温度的设施。

B. 根据生产需要,可设置控制室温的设施。

(2)设备。

①生产设备。

A. 一般要求。应配备与生产能力相适应的生产设备,并按工艺流程有序排列,避免引起交叉污染。

B. 材质。

a. 与原料、半成品、成品接触的设备与用具,应使用无毒、无味、抗腐蚀、不易脱落的材料制作,并应易于清洁和保养。

b. 设备、工器具等与食品接触的表面应使用光滑、无吸收性、易于清洁保养和消毒的材料制成,在正常生产条件下不会与食品、清洁剂和消毒剂发生反应,并应保持完好无损。

C. 设计。

a. 所有生产设备应从设计和结构上避免零件、金属碎屑、润滑油或其他污染因素混入食品,并应易于清洁消毒,易于检查和维护。

b. 设备应不留空隙地固定在墙壁或地板上,或在安装时与地面和墙壁间保留足够空间,以便清洁和维护。

be separated from raw materials, semi-finished products, finished products, packaging materials and so on.

i. Temperature control facility.

a) Appropriate heating, cooling, freezing and other facilities, as well as facilities for temperature monitoring should be provided in accordance with the characteristics of food production.

b) According to production needs, the facilities can be installed to control the room temperature.

2. Equipment.

a. Production equipment.

a) General requirements. It should be equipped with production equipment suitable for production capacity and arranged in an orderly manner according to the process flow to avoid cross-contamination.

b) Material.

• Equipment and appliances in contact with raw materials, semi-finished products and finished products shall be made of non-toxic, tasteless, corrosion-resistant and non-exfoliating materials, and shall be easy to clean and maintain.

• The surfaces of equipment and utensils in contact with food shall be made of smooth, non-absorbable, easy cleaning, maintainable and disinfectant friendly materials, and shall not react with food, detergents and disinfectants under normal production conditions and shall be kept intact.

c) Design.

• From design and structure, all production equipment should avoid parts, metal detritus, lube oil or other pollution factors to mix into food, and it should be easy to clean, disinfect, easy to check and maintain.

• The equipment shall be fixed to the wall or floor without space or shall be installed with sufficient space between the floor and the wall for cleaning and maintenance.

②监控设备。用于监测、控制、记录的设备,如压力表、温度计、记录仪等,应定期校准、维护。

③设备的保养和维修。应建立设备保养和维修制度,加强设备的日常维护和保养,定期检修,及时记录。

6)卫生管理
(1)卫生管理制度。
①应制定食品加工人员和食品生产卫生管理制度以及相应的考核标准,明确岗位职责,实行岗位责任制。

②应根据食品的特点以及生产、贮存过程的卫生要求,建立对保证食品安全具有显著意义的关键控制环节的监控制度,良好实施并定期检查,发现问题及时纠正。

③应制定针对生产环境、食品加工人员、设备及设施等的卫生监控制度,确立内部监控的范围、对象和频率。记录并存档监控结果,定期对执行情况和效果进行检查,发现问题及时整改。

④应建立清洁消毒制度和清洁消毒用具管理制度。清洁消毒前后的设备和工器具应分开放置妥善保管,避免交叉污染。

(2)厂房及设施卫生管理。

①厂房内各项设施应保持清洁,

b. Monitoring equipment. Equipment for monitoring, control and recording, such as pressure gauges, thermometers, recorders, etc., they all should be calibrated and maintained regularly.

c. Maintenance and repair of equipment. Establish equipment maintenance and repair system, strengthen the daily maintenance and service of equipment, regular maintenance, timely recording.

F. Sanitary Control
1. Sanitary control system.
a. Food processing personnel and food production sanitary control system as well as the corresponding assessment standards should be formulated, the job responsibilities should be clearly defined and the post responsibility system should be carried out.

b. According to the characteristics of food and hygiene requirements of production and storage process, we should establish a monitoring system for critical control links that have significance for food safety, perform well and check regularly, then find out the problem and correct it in time.

c. It is necessary to establish a sanitary control system for the production environment, food processing personnel, equipment and facilities, and to establish the scope, object and frequency of internal monitoring. Record and file the monitoring results, check the implementation and effect regularly, then find out the problem and correct it in time.

d. The system of cleaning and disinfection and the management system of cleaning and disinfecting utensils should be established. Equipment and appliances before and after cleaning and disinfection should be kept separately and properly to avoid cross-contamination.

2. Sanitary control of factory building and facilities.
a. All facilities in the factory building should

出现问题及时维修或更新;厂房地面、屋顶、天花板及墙壁有破损时,应及时修补。

②生产、包装、贮存等设备及工器具、生产用管道、裸露食品接触表面等应定期清洁消毒。

(3)食品加工人员健康管理与卫生要求。

①食品加工人员健康管理。

A.应建立并执行食品加工人员健康管理制度。

B.食品加工人员每年应进行健康检查,取得健康证明;上岗前应接受卫生培训。

C.食品加工人员如患有痢疾、伤寒、甲型病毒性肝炎、戊型病毒性肝炎等消化道传染病,以及患有活动性肺结核、化脓性或者渗出性皮肤病等有碍食品安全的疾病,或有明显皮肤损伤未愈合的,应当调整到其他不影响食品安全的工作岗位。

②食品加工人员卫生要求。

A.进入食品生产场所前应整理个人卫生,防止污染食品。

B.进入作业区域应规范穿着洁净的工作服,并按要求洗手、消毒;头发应藏于工作帽内或使用发网约束。

C.进入作业区域不应配戴饰物、手表,不应化妆、染指甲、喷洒香水;不得携带或存放与食品生产无关的个人用品。

D.使用卫生间、接触可能污染食

be kept clean and repaired or updated in time if any problems arise. If there is any damage to the floor, roof, ceiling and wall of the factory building, the repair should be made in time.

b. Equipment and utensils, production pipes, exposed food surfaces for production, packaging and storage shall be cleaned and disinfected regularly.

3. Health management and hygiene requirements of food processors.

a. Health management of food processors.

a) Health management systems for food processors should be established and implemented.

b) Food processors shall conduct annual health checks and obtain health certificates; they shall receive hygiene training prior to taking up their posts.

c) Food processors who suffer from infectious diseases such as dysentery, typhoid fever, viral hepatitis A and viral hepatitis E, as well as diseases such as active tuberculosis, suppurative or exudative skin diseases that hinder food safety, or if there are obvious skin injuries that have not healed, they shall be adjusted to other jobs that do not affect food safety.

b. Hygienic requirements for food processors.

a) Personal hygiene should be cleaned before entering food production sites to prevent contamination of food.

b) When entering the work area, wear clean work clothes and wash hands and sterilize as required; hair should be hidden in the work cap or restricted by the use of hairnets.

c) No ornaments or watches shall be worn in the working area, nor should make up, dying of nails or perfume shall be sprayed, and personal belongings not related to food production shall not be carried or stored.

d) After the use of bathrooms, contact with

品的物品或从事与食品生产无关的其他活动后,再次从事接触食品、食品工器具、食品设备等与食品生产相关的活动前应洗手消毒。

③来访者。非食品加工人员不得进入食品生产场所,特殊情况下进入时应遵守和食品加工人员同样的卫生要求。

(4)虫害控制。

①应保持建筑物完好、环境整洁,防止虫害侵入及孳生。

②应制定和执行虫害控制措施,并定期检查。生产车间及仓库应采取有效措施(如纱帘、纱网、防鼠板、防蝇灯、风幕等),防止鼠类昆虫等侵入。若发现有虫鼠害痕迹时,应追查来源,消除隐患。

③应准确绘制虫害控制平面图,标明捕鼠器、粘鼠板、灭蝇灯、室外诱饵投放点、生化信息素捕杀装置等放置的位置。

④厂区应定期进行除虫灭害工作。

⑤采用物理、化学或生物制剂进行处理时,不应影响食品安全和食品应有的品质,不应污染食品接触表面、设备、工器具及包装材料。除虫灭害工作应有相应的记录。

⑥使用各类杀虫剂或其他药剂前,应做好预防措施,避免对人身、食品、设备工具造成污染;不慎污染时,应及时将被污染的设备、工具彻底清洁,消除污染。

articles likely to contaminate food, or engage in other activities unrelated to food production, hand washing and disinfection shall be carried out before engaging in activities related to food production, such as food, food appliances, food equipment, etc.

c. Visitors. Non-food processors shall not enter food production sites and shall comply with the same sanitary requirements as food processors when entering under special circumstances.

4. Pest control.

a. Keep buildings in good condition and clean environment to prevent insect pests from invading and breeding.

b. Pest control measures should be developed and implemented and regularly inspected. The production workshop and warehouse should take effective measures (such as gauze curtains, gauze net, anti-rat board, anti-fly lamp, wind curtain and so on) to prevent rodent insects and so on. If there are signs of insect and rodent damage, the source should be traced to eliminate hidden dangers.

c. The plan of pest control should be accurately drawn, indicating the position of rat traps, stick board, fly killing lamp, outdoor bait drop point, biochemical pheromone catching and killing device and so on.

d. The factory building should regularly carry out pest control work.

e. The use of physical, chemical or biological agents shall not affect food safety and the proper quality of food, and shall not contaminate food contact surfaces, equipment, workmanship and packaging materials. The work of insect control and pest control should be recorded accordingly.

f. Before using all kinds of insecticides or other pesticides, preventive measures shall be taken to avoid contamination of personal, food and equipment tools. When contaminated accidentally, the contaminated equipment and tools shall be

(5)废弃物处理。

①应制定废弃物存放和清除制度,有特殊要求的废弃物的处理方式应符合有关规定。废弃物应定期清除;易腐败的废弃物应尽快清除;必要时应及时清除废弃物。

②车间外废弃物放置场所应与食品加工场所隔离,防止污染;应防止不良气味或有害有毒气体逸出;应防止虫害孳生。

(6)工作服管理。

①进入作业区域应穿着工作服。

②应根据食品的特点及生产工艺的要求配备专用工作服,如衣、裤、鞋靴、帽和发网等,必要时还可配备口罩、围裙、套袖、手套等。

③应制定工作服的清洗保洁制度,必要时应及时更换;生产中应注意保持工作服干净完好。

④工作服的设计、选材和制作应适应不同作业区的要求,降低交叉污染食品的风险;应合理选择工作服口袋的位置、使用的连接扣件等,降低内容物或扣件掉落污染食品的风险。

7)食品原料、食品添加剂和食品相关产品

(1)一般要求。应建立食品原料、食品添加剂和食品相关产品的采购、

thoroughly cleaned in time to eliminate pollution.

5. Waste disposal.

a. Waste storage and removal systems shall be formulated, and the disposal of wastes with special requirements shall comply with relevant provisions. Wastes should be removed regularly; wastes liable to corruption should be removed as soon as possible; and waste should be removed in time when necessary.

b. Waste sites outside the workshop should be isolated from food processing sites to prevent pollution; undesirable odors or harmful toxic gases should be prevented; and pest breeding should be prevented.

6. Work clothes management.

a. Work clothes should be worn when entering the work area.

b. Special work clothes should be provided according to the characteristics of food and production technology, such as clothing, trousers, shoes, hats and hair nets, as well as masks, aprons, sleeves, gloves and so on.

c. The washing and cleaning system of working clothes should be made and should be replaced in time when necessary; it should be paid attention to keeping the work clothes clean and perfect in production.

d. The design, selection and manufacture of work clothes should be adapted to the requirements of different operation areas to reduce the risk of cross-contamination of food, and the location of the pocket of the work clothes and the used connection fasteners should be reasonably selected to reduce the risk of contaminated food by the falling contents or fasteners.

G. Food Raw Materials, Food Additives and Food-related Products

1. General requirements. Systems for the procurement, acceptance, transportation and

验收、运输和贮存管理制度,确保所使用的食品原料、食品添加剂和食品相关产品符合国家有关要求。不得将任何危害人体健康和生命安全的物质添加到食品中。

(2)食品原料。

①采购的食品原料应当查验供货者的许可证和产品合格证明文件;对无法提供合格证明文件的食品原料,应当依照食品安全标准进行检验。

②食品原料必须经过验收合格后方可使用。经验收不合格的食品原料应在指定区域与合格品分开放置并明显标记,并应及时进行退、换货等处理。

③加工前宜进行感官检验,必要时应进行实验室检验;检验发现涉及食品安全项目指标异常的,不得使用;只应使用确定适用的食品原料。

④食品原料运输及贮存中应避免日光直射、备有防雨防尘设施;根据食品原料的特点和卫生需要,必要时还应具备保温、冷藏、保鲜等设施。

⑤食品原料运输工具和容器应保持清洁、维护良好,必要时应进行消毒。食品原料不得与有毒、有害物品同时装运,避免污染食品原料。

⑥食品原料仓库应设专人管理,建立管理制度,定期检查质量和卫生

storage of food raw materials, food additives and food-related products shall be established to ensure that the food raw materials, food additives and food-related products used meet the relevant requirements of the State. No substance harmful to human health and safety shall be added to food.

2. Food raw materials.

a. The purchased food raw materials shall be inspected for the license of the supplier and the certification documents for the conformity of the products; the raw materials for food that cannot provide the documents of conformity shall be inspected in accordance with the food safety standards.

b. Food raw materials must pass the acceptance test before they can be used. The unqualified food raw materials shall be placed and marked separately from the qualified products in the designated area, and shall be returned and exchanged in time.

c. Sensory tests should be carried out before processing, and laboratory tests should be carried out when necessary; if abnormal indexes of food safety items are found to be involved in the inspection, they may not be used; and only food raw materials that are suitable for use shall be used.

d. In the transportation and storage of food raw materials, direct sunlight should be avoided, the rain and dust prevention facilities should be provided. According to the characteristics and hygiene needs of food raw materials, there should also be such facilities as heat preservation, cold storage, fresh keeping and so on.

e. The transportation tools and containers of food raw materials should be kept clean, maintained and disinfected if necessary. Food raw materials shall not be shipped at the same time as toxic or harmful substances to avoid contamination of food raw materials.

f. The warehouse of food raw materials shall be managed by special personnel, and a

情况,及时清理变质或超过保质期的食品原料。仓库出货顺序应遵循先进先出的原则,必要时应根据不同食品原料的特性确定出货顺序。

(3)食品添加剂。

①采购食品添加剂应当查验供货者的许可证和产品合格证明文件。食品添加剂必须经过验收合格后方可使用。

②运输食品添加剂的工具和容器应保持清洁、维护良好,并能提供必要的保护,避免污染食品添加剂。

③食品添加剂的贮藏应有专人管理,定期检查质量和卫生情况,及时清理变质或超过保质期的食品添加剂。仓库出货顺序应遵循先进先出的原则,必要时应根据食品添加剂的特性确定出货顺序。

(4)食品相关产品。

①采购食品包装材料、容器、洗涤剂、消毒剂等食品相关产品应当查验产品的合格证明文件,实行许可管理的食品相关产品还应查验供货者的许可证。食品包装材料等食品相关产品必须经过验收合格后方可使用。

②运输食品相关产品的工具和容器应保持清洁、维护良好,并能提供必要的保护,避免污染食品原料和交叉

management system shall be established, the quality and hygiene of which shall be inspected regularly, and any food raw materials which have deteriorated or exceeded the shelf life shall be cleaned up in time. The shipment order of the warehouse shall follow the principle of first-in first-out, and the shipment order shall be determined according to the characteristics of different food raw materials when necessary.

3. Food additive.

a. When purchasing food additives, the license of the supplier and the certificate of conformity of the product shall be examined. Food additives must be approved before they can be used.

b. The tools and containers for transporting food additives shall be kept clean and well maintained and shall provide the necessary protection against contamination of food additives.

c. The storage of food additives shall be managed by special personnel, the quality and hygiene of food additives shall be inspected regularly, and food additives that have deteriorated or exceeded the shelf life shall be cleaned up in a timely manner. The shipment order of the warehouse shall follow the principle of first-in first-out, and the shipment order shall be determined according to the characteristics of the food additives when necessary.

4. Food-related products.

a. Food-related products, such as food packaging materials, containers, detergents, disinfectants and other shall be subject to inspection of the qualified documents of the products, and the food related products subject to licensing administration shall also examine the licenses of the suppliers. Food packaging materials and other food-related products must pass the acceptance before use.

b. The tools and containers for transporting food-related products shall be kept clean and well maintained and shall provide the necessary

污染。

③食品相关产品的贮藏应有专人管理,定期检查质量和卫生情况,及时清理变质或超过保质期的食品相关产品。仓库出货顺序应遵循先进先出的原则。

(5)其他。盛装食品原料、食品添加剂、直接接触食品的包装材料的包装或容器,其材质应稳定,无毒无害,不易受污染,符合卫生要求。

食品原料、食品添加剂和食品包装材料等进入生产区域时应有一定的缓冲区域或外包装清洁措施,以降低污染风险。

8)生产过程的食品安全控制
(1)产品污染风险控制。
①应通过危害分析方法明确生产过程中的食品安全关键环节,并设立食品安全关键环节的控制措施。在关键环节所在区域,应配备相关的文件以落实控制措施,如配料(投料)表、岗位操作规程等。

②鼓励采用危害分析与关键控制点体系(HACCP)对生产过程进行食品安全控制。
(2)生物污染的控制。
①清洁和消毒。
A. 应根据原料、产品和工艺的特点,针对生产设备和环境制定有效的清洁消毒制度,降低微生物污染的风险。

B. 清洁消毒制度应包括以下内

protection against contamination of food raw materials and cross-contamination.

c. The storage of food-related products shall be managed by special personnel, quality and hygiene shall be inspected regularly, and food related products that have deteriorated or exceeded the shelf life shall be cleaned up in time. The shipment order of the warehouse shall follow the principle of first-in first-out.

5. Other. Packaging or containers containing food raw materials, food additives and packaging materials directly in contact with food should be stable, non-toxic and harmless, not easily contaminated, and meet the hygiene requirements.

When food raw materials, food additives and food packaging materials enter the production area, there should be some buffer area or outer packaging cleaning measures to reduce the risk of pollution.

H. Food Safety Control in the Production Process
1. Risk Control of Product pollution.
a. The key links of food safety in the production process should be defined by hazard analysis method and the control measures for the key links of food safety should be set up. In key areas, relevant documents should be provided to implement control measures, such as burden (feeding) sheet, post operating procedures, etc.

b. Encourage the use of the hazard analysis and critical control point system (HACCP) to control food safety in the production process.
2. Biological pollution control.
a. Cleaning and disinfection.
a) According to the characteristics of raw materials, products and processes, an effective cleaning and disinfection system should be established for production equipment and environment to reduce the risk of microbial contamination.

b) The cleaning and disinfection system shall

容:清洁消毒的区域、设备或器具名称;清洁消毒工作的职责;使用的洗涤、消毒剂;清洁消毒方法和频率;清洁消毒效果的验证及不符合的处理;清洁消毒工作及监控记录。

C. 应确保实施清洁消毒制度,如实记录;及时验证消毒效果,发现问题及时纠正。

②食品加工过程的微生物监控。

A. 根据产品特点确定关键控制环节进行微生物监控;必要时应建立食品加工过程的微生物监控程序,包括生产环境的微生物监控和过程产品的微生物监控。

B. 食品加工过程的微生物监控程序应包括微生物监控指标、取样点、监控频率、取样和检测方法、评判原则和整改措施等,应结合生产工艺及产品特点制定。

C. 微生物监控应包括致病菌监控和指示菌监控,食品加工过程的微生物监控结果应能反映食品加工过程中对微生物污染的控制水平。

(3)化学污染的控制。
①应建立防止化学污染的管理制度,分析可能的污染源和污染途径,制定适当的控制计划和控制程序。

include the following: the name of the area, equipment or appliance for cleaning and disinfection; the duties of the cleaning and disinfection work; the detergent and disinfectant used; and the method and frequency of cleaning and disinfecting; validation of the efficacy of cleaning and disinfection and treatment of non-conformance; cleaning and disinfection work and monitoring records.

c) It is necessary to ensure the implementation of a cleaning and disinfection system and record truthfully, verify the efficacy of disinfection in time, and correct problems in time.

b. Microbiological monitoring of food processing.

a) According to the characteristics of the product, the key control links should be determined to carry out microbial monitoring; if necessary, the microbial monitoring procedures for the food processing should be established, including the microbiological monitoring of the production environment and the microbiological monitoring of the process products.

b) The microbiological monitoring procedures for food processing shall include microbial monitoring indexes, sampling points, monitoring frequencies, sampling and testing methods, evaluation principles and corrective measures, etc., which can combine with the production process and product characteristics.

c) Microbiological monitoring should include pathogenic bacteria monitoring and indicator bacteria monitoring. The microbiological monitoring results of food processing should reflect the control level of microbial contamination in food processing.

3. Chemical pollution control.

a. Management systems should be established to prevent chemical pollution, possible sources of pollution and pollution routes should be analysed, and appropriate control plans and procedures should be developed.

②应当建立食品添加剂和食品工业用加工助剂的使用制度,按照 GB 2760 的要求使用食品添加剂。

③不得在食品加工中添加食品添加剂以外的非食用化学物质和其他可能危害人体健康的物质。

④生产设备上可能直接或间接接触食品的活动部件若需润滑,应当使用食用油脂或能保证食品安全要求的其他油脂。

⑤建立清洁剂、消毒剂等化学品的使用制度。除清洁消毒必需和工艺需要,不应在生产场所使用和存放可能污染食品的化学制剂。

⑥食品添加剂、清洁剂、消毒剂等均应采用适宜的容器妥善保存,且应明显标识、分类贮存;领用时应准确计量、作好使用记录。

⑦应当关注食品在加工过程中可能产生有害物质的情况,鼓励采取有效措施减低其风险。

(4)物理污染的控制。
①应建立防止异物污染的管理制度,分析可能的污染源和污染途径,并制定相应的控制计划和控制程序。

②应通过采取设备维护、卫生管理、现场管理、外来人员管理及加工过程监督等措施,最大程度地降低食品受到玻璃、金属、塑胶等异物污染的风险。

③应采取设置筛网、捕集器、磁铁、金属检查器等有效措施降低金属

b. A system for the use of food additives and processing aids for the food industry shall be established, and food additives shall be used in accordance with the requirements of GB 2760.

c. Non-edible chemicals and other substances that may endanger human health other than food additives shall not be added to food processing.

d. Where lubrication is required for moving parts that may have direct or indirect contact with food on the production equipment, the edible oils or other oils that ensure food safety requirements should be used.

e. Establish a system for the use of chemicals such as detergents and disinfectants. In addition to cleaning and disinfection necessary and technological needs, chemical preparations that may contaminate food shall not be used and stored in the place of production.

f. Food additives, detergents, disinfectants and so on should be properly preserved, and they should be marked, classified and stored. When they are used, they should be accurately measured and recorded.

g. Attention should be paid to the possible production of harmful substances in food processing and effective measures to reduce their risk should be encouraged.

4. Physical pollution control.

a. It is necessary to establish a management system to prevent foreign body pollution, analyze possible sources of pollution and ways of pollution, and formulate corresponding control plans and control procedures.

b. The risk of food contamination by foreign bodies such as glass, metal and plastics should be minimized by measures such as equipment maintenance, hygiene management, site management, external personnel management and process supervision.

c. It is necessary to take effective measures to reduce the risk of contaminated food by metal or

或其他异物污染食品的风险。

④当进行现场维修、维护及施工等工作时,应采取适当措施避免异物、异味、碎屑等污染食品。

(5)包装。

①食品包装应能在正常的贮存、运输、销售条件下最大限度地保护食品的安全性和食品品质。

②使用包装材料时应核对标识,避免误用;应如实记录包装材料的使用情况。

9)检验

(1)应通过自行检验或委托具备相应资质的食品检验机构对原料和产品进行检验,建立食品出厂检验记录制度。

(2)自行检验应具备与所检项目适应的检验室和检验能力;由具有相应资质的检验人员按规定的检验方法检验;检验仪器设备应按期检定。

(3)检验室应有完善的管理制度,妥善保存各项检验的原始记录和检验报告。应建立产品留样制度,及时保留样品。

(4)应综合考虑产品特性、工艺特点、原料控制情况等因素合理确定检验项目和检验频次,以有效验证生产过程中的控制措施。净含量、感官要求以及其他容易受生产过程影响而变化的检验项目的检验频次应大于其他检验项目。

other foreign objects, such as a screen mesh, trap, magnet and metal detector.

d. When carrying out site repair, maintenance and construction, appropriate measures should be taken to avoid contaminated food such as foreign bodies, peculiar smell, debris and so on.

5. Packaging.

a. Food packaging should be able to protect the safety and quality of food to the maximum extent under normal storage, transportation and marketing conditions.

b. The use of packaging materials should be checked to avoid the misuse of labels; the use of packaging materials should be recorded truthfully.

I. Test

1. Food inspection records system shall be established through self-inspection or entrusting food inspection agencies with corresponding qualifications to inspect raw materials and products.

2. The self-inspection shall have the examination room and the inspection ability suitable for the items inspected; the inspection personnel with the corresponding qualifications shall inspect according to the prescribed inspection methods; the inspection instruments and equipment shall be inspected on time.

3. The inspection room should have a perfect management system and properly maintain the original records and inspection reports of each inspection. Product retention systems should be established and samples should be kept in time.

4. Factors such as product characteristics, process characteristics and raw material control should be taken into account to determine the inspection items and inspection frequency so as to effectively verify the control measures in the production process. The inspection frequency of the net content, sensory requirements and other items subject to changes in the production process shall be greater than that of other inspection items.

(5)同一品种不同包装的产品,不受包装规格和包装形式影响的检验项目可以一并检验。

10)食品的贮存和运输

(1)根据食品的特点和卫生需要选择适宜的贮存和运输条件,必要时应配备保温、冷藏、保鲜等设施。不得将食品与有毒、有害或有异味的物品一同贮存运输。

(2)应建立和执行适当的仓储制度,发现异常应及时处理。

(3)贮存、运输和装卸食品的容器、工器具和设备应当安全、无害,保持清洁,降低食品污染的风险。

(4)贮存和运输过程中应避免日光直射、雨淋、显著的温湿度变化和剧烈撞击等,防止食品受到不良影响。

11)产品召回管理

(1)应根据国家有关规定建立产品召回制度。

(2)当发现生产的食品不符合食品安全标准或存在其他不适于食用的情况时,应当立即停止生产,召回已经上市销售的食品,通知相关生产经营者和消费者,并记录召回和通知情况。

(3)对被召回的食品,应当进行无害化处理或者予以销毁,防止其再次流入市场。对因标签、标识或者说明书不符合食品安全标准而被召回的食品,应采取能保证食品安全、且便于重

5. Products of the same variety and different packaging may be inspected together without the influence of packaging specifications and packaging forms.

J. Storage and Transport of Food

1. According to the characteristics of food and hygiene needs to choose the appropriate storage and transportation conditions, if necessary, it should be equipped with insulation, refrigeration, preservation and other facilities. Food shall not be stored and transported with toxic, harmful, or odorous items.

2. Appropriate warehousing systems should be established and implemented, and anomalies should be handled in a timely manner.

3. Containers, utensils and equipment for storing, transporting and loading and unloading food shall be safe and harmless, kept clean and reduce the risk of food contamination.

4. During storage and transportation, direct sunlight, rain, significant changes in temperature and humidity and severe impact should be avoided to prevent food from being adversely affected.

K. Product Recall Management

1. Product recall system should be established in accordance with the relevant regulations of the State.

2. When it is found that the food produced does not meet the food safety standards or that there are other unfit for consumption, the production shall be stopped immediately, the food already sold on the market shall be recalled, and the relevant producers and consumers shall be notified, then record the recall and notification.

3. The recalled food shall be treated or destroyed in a harmless manner so as to prevent its re-entry into the market. For food recalled from food safety standards that are not in conformity with labels, identifications or instructions,

新销售时向消费者明示的补救措施。

(4)应合理划分记录生产批次,采用产品批号等方式进行标识,便于产品追溯。

12)培训

(1)应建立食品生产相关岗位的培训制度,对食品加工人员以及相关岗位的从业人员进行相应的食品安全知识培训。

(2)应通过培训促进各岗位从业人员遵守食品安全相关法律法规标准和执行各项食品安全管理制度的意识和责任,提高相应的知识水平。

(3)应根据食品生产不同岗位的实际需求,制定和实施食品安全年度培训计划并进行考核,作好培训记录。

(4)当食品安全相关的法律法规标准更新时,应及时开展培训。

(5)应定期审核和修订培训计划,评估培训效果,并进行常规检查,以确保培训计划的有效实施。

13)管理制度和人员

(1)应配备食品安全专业技术人员、管理人员,并建立保障食品安全的管理制度。

(2)食品安全管理制度应与生产规模、工艺技术水平和食品的种类特性相适应,应根据生产实际和实施经验不断完善食品安全管理制度。

remedial measures shall be taken to ensure food safety and facilitate re-sale to consumers.

4. Production batches should be reasonably divided and marked by product lot number, so as to facilitate the traceability of the product.

L. Training

1. The training system for the relevant posts in food production should be established to train the food processing personnel and the employees in the relevant posts to carry out the corresponding knowledge of food safety knowledge.

2. It is necessary to promote the awareness and responsibility of the employees in various positions to abide by the relevant laws and regulations of food safety and to implement the food safety management system through training so as to improve the corresponding level of knowledge.

3. According to the actual needs of different positions in food production, an annual training plan for food safety should be formulated and carried out, and the training records should be made.

4. When food safety related laws and regulations are updated, training should be carried out in a timely manner.

5. Training plans should be regularly reviewed and revised to assess the effectiveness of the training and conduct regular inspections to ensure the effective implementation of the training plans.

M. Management System and Personnel

1. Food safety should be provided with professional technical personnel, management personnel, and establish a management system to ensure food safety.

2. The food safety management system should be adapted to the scale of production, the level of technology and the variety of food, and the food safety management system should be continuously improved according to the production practice and

(3)管理人员应了解食品安全的基本原则和操作规范,能够判断潜在的危险,采取适当的预防和纠正措施,确保有效管理。

14)记录和文件管理

(1)记录管理。

①应建立记录制度,对食品生产中采购、加工、贮存、检验、销售等环节详细记录。记录内容应完整、真实,确保对产品从原料采购到产品销售的所有环节都可进行有效追溯。

A. 应如实记录食品原料、食品添加剂和食品包装材料等食品相关产品的名称、规格、数量、供货者名称及联系方式、进货日期等内容。

B. 应如实记录食品的加工过程(包括工艺参数、环境监测等)、产品贮存情况及产品的检验批号、检验日期、检验人员、检验方法、检验结果等内容。

C. 应如实记录出厂产品的名称、规格、数量、生产日期、生产批号、购货者名称及联系方式、检验合格单、销售日期等内容。

D. 应如实记录发生召回的食品名称、批次、规格、数量、发生召回的原因及后续整改方案等内容。

②食品原料、食品添加剂和食品包装材料等食品相关产品进货查验记录,食品出厂检验记录应由记录和审核人员复核签名,记录内容应完整。保存期限不得少于2年。

③应建立客户投诉处理机制。对

implementation experience.

3. Managers should be aware of the basic principles and operational norms of food safety, be able to identify potential hazards and take appropriate preventive and corrective action to ensure effective management.

N. Records and Document Management

1. Record management.

a. A record system should be established to record the procurement, processing, storage, inspection and sale of food in detail. The records should be complete and true to ensure that all links from raw material procurement to product sales are effectively traceable.

a) The name, specification, quantity, supplier name and contact information, date of purchase of food raw materials, food additives and food packaging materials shall be recorded truthfully.

b) The food processing process (including process parameters, environmental monitoring, etc.), product storage, inspection lot number, inspection date, inspection personnel, inspection methods and inspection results shall be recorded truthfully.

c) The name, specification, quantity, production date, production lot number, buyer's name and contact information, inspection certificate, sales date and so on shall be recorded truthfully.

d) The names, batches, specifications, quantities, reasons for the recall and the plans for subsequent rectification shall be recorded truthfully.

b. Inspection records of food related products, such as food raw materials, food additives and food packaging materials, etc. and factory inspection record of food shall be checked and signed by the records and examiners, and the contents of the records shall be complete. The period of preservation shall not be less than 2 years.

c. Customer complaint handling mechanism

客户提出的书面或口头意见、投诉,企业相关管理部门应作记录并查找原因,妥善处理。

(2)应建立文件的管理制度,对文件进行有效管理,确保各相关场所使用的文件均为有效版本。

(3)鼓励采用先进技术手段(如电子计算机信息系统),进行记录和文件管理。

should be established. The relevant management departments of the enterprise shall record and find out the reasons for the written or oral opinions and complaints made by the customers and handle them properly.

2. A document management system should be established to effectively manage the documents to ensure that the documents used in all relevant locations are valid versions.

3. Encourage the use of advanced technical means (such as computerized information systems) for records and document management.

任务3 卫生标准操作程序(SSOP)

1. 概述

卫生标准操作程序(Sanitation Standard Operation Procedure,简称SSOP),是食品加工企业为了保证达到良好操作规范所规定的要求,确保加工过程中消除不良的人为因素,使其加工的食品符合卫生要求而制定的指导食品生产加工过程中如何实施清洗、消毒和卫生保持的作业指导文件。

1)卫生标准操作程序的一般要求

(1)加工企业必须建立和实施卫生标准操作程序,以强调加工前、加工中和加工后的卫生状况和卫生行为。

(2)卫生标准操作程序应该描述加工者如何保证某一个关键的卫生条件和操作得到满足。

(3)卫生标准操作程序应该描述加工企业的操作如何受到监控来保证达到良好操作规范规定的条件和要求。

(4)须保持卫生标准操作程序记

Ⅰ. Overview

The abbreviation of Sanitation Standard Operation Procedure is SSOP. It is the operational guidance document on how to carry out cleaning, disinfection and hygiene maintenance in the course of food production and processing in order to ensure that food processing enterprises to meet the requirements stipulated by the GMP, to ensure that the processing process eliminates undesirable human factors, and to make the food processed in accordance with the hygienic requirements.

A. General Requirements for SSOP

1. Processing enterprises must establish and implement SSOP to emphasize hygiene status and hygiene behavior before, during and after processing.

2. SSOP should describe how the processor ensures that a critical sanitary condition and operation is met.

3. SSOP should describe how the operations of the processing enterprise are monitored to ensure that the conditions and requirements set by GMP are met.

4. SSOP records shall be maintained, and at a

录,至少应记录相关的关键卫生条件和操作受到监控和纠偏的结果。

(5)官方执法部门或第三方认证机构应鼓励和督促企业建立书面的卫生标准操作程序。

2)卫生标准操作程序的主要内容

(1)用于接触食品或食品接触面的水、冰的安全。

(2)食品接触表面的卫生状况和清洁程度,包括工器具、设备、手套和工作服。

(3)防止发生食品与不洁物、食品与包装材料、人流和物流、高清洁区的食品与低清洁区的食品、生食与熟食之间的交叉污染。

(4)手的清洗消毒设施及卫生间设施的维护。

(5)保护食品、食品包装材料和食品接触面免受润滑剂、燃油、杀虫剂、清洗剂、冷凝水、涂料、铁锈和其他化学、物理和生物性外来杂质的污染。

(6)有毒化学物质的正确标识、贮存和使用。

(7)直接或间接接触食品的职工健康状况的控制。

(8)害虫的控制及去除(防虫、灭虫、防鼠、灭鼠)。

卫生标准操作程序应由食品生产加工企业根据卫生规范及企业实际情况编制,尤其应充分考虑到其实用性和可操作性。卫生标准操作程序文件一般应包含监控对象、监控方法、监控频率、监控人员、纠偏措施及监控、纠偏结果的记录要求等内容。

minimum the results of monitoring and correction of relevant key hygiene conditions and operations shall be recorded.

5. Official law enforcement agencies or third-party certification bodies shall encourage and urge enterprises to establish written SSOP.

B. Main Contents of SSOP

1. Safety of water and ice used in contact with food or food contact surfaces.

2. Hygienic conditions and cleanliness on the surface of food, including appliances, equipment, gloves and work clothes.

3. Prevent cross-contamination between food and unclean materials, food and packaging materials, stream of people and logistics, food in high clean areas and food in low clean areas, raw food and cooked food.

4. Maintenance of hand cleaning and disinfection facilities as well as toilet facilities.

5. Protect the food, food packaging materials and food contact surfaces from lubricants, fuel, pesticides, cleaning agents, water condensate, paints, rust and other chemical, physical and biological foreign impurities.

6. Correct marking, storage and use of toxic chemicals.

7. Control of health status of workers who have direct or indirect contact with food.

8. Pest control and removal (insect control, pest control, rodent prevent, rodent control).

SSOP should be compiled by the food production and processing enterprises according to the hygienic norms and the actual conditions of the enterprises, especially considering its practicability and operability. SSOP files should generally include monitoring object, monitoring methods, monitoring frequency, monitoring personnel, corrective measures and monitoring, correction results of the record requirements and so on.

2.卫生标准操作程序的基本内容和要求

1)水(冰)的安全

生产用水(冰)的卫生质量是影响食品卫生的关键因素。对于任何食品生产加工企业,首要的一点就是要保证水的安全。

(1)水源。食品加工厂的水源一般由城市供水、自供水和海水构成。

①城市供水,又称公共供水或城乡生活饮用水,是由自来水厂供应的饮用水。使用城市供水具有许多优点,如它具有良好的化学和微生物标准;经过了净化或处理,在决定使用前经过了检验,符合国家饮用标准等。但是它的费用较其他种类的水源高。城市供水是各种水源中最常用的。

②自供水,由自备水井供水。相比较而言自供水供水费用较低,但比城市供水易污染。由于井水中含有大量的可溶性矿物质、不溶性固体、有机物质、可溶性气体及微生物,因此使用井水需进行水处理。

③海水。海水也是食品生产企业经常使用的一种水源。使用海水时应考虑水源周围环境、季节变化、污水排放等因素对海水的污染。

(2)水的贮存和处理。
①水的贮存方式,包括水塔、蓄水池、贮水罐等。
②水的处理方式,包括物理处理(沉淀、过滤)、化学处理(离子交换)。

II. Basic Content and Requirements of SSOP

A. Safety of Water (Ice)

The sanitary quality of production water (ice) is a key factor affecting food hygiene. For any food production and processing enterprises, the first point is to ensure the safety of water.

1. Water source. The source of water in food processing plants is generally composed of urban water supply, self-supplied water and sea water.

a. Urban water supply. Also known as public water supply or urban and rural drinking water, which is the drinking water supplied by the waterworks. The use of urban water has many advantages, such as its good chemical and microbial standards; it has been purified or treated, and tested to meet the national drinking standards before deciding to use it. But it is also more expensive than other types of water. Urban water supply is the most commonly used of all kinds of water sources.

b. Self-supply water. Water is supplied by a self-contained well. Self-supply water is less expensive than urban water supply, but is more prone to pollution than urban water supply. Well water contains a large number of soluble minerals, insoluble solids, organic substances, soluble gases and microorganisms, so the use of well water needs water treatment.

c. Sea water. Seawater is also a common source of water used by food producers. When using seawater, the pollution of seawater should be taken into account, such as the environment around the water source, seasonal change, sewage discharge and so on.

2. Storage and treatment of water.

a. The way of storage of water. Including water tower, reservoir, storage tank and so on.

b. The treatment of water. It includes physical treatment (precipitation, filtration), chemical

水的消毒处理有加氯处理（自动加氯系统）、臭氧处理、紫外线消毒等几种方法。

treatment (ion exchange). The disinfection treatment of water has several methods, such as chlorination treatment (automatic chlorination system), ozone treatment, ultraviolet disinfection and so on.

除了对水进行处理外，还必须对水塔、蓄水池、贮水罐等水的贮存环境进行定期的清洗消毒，清洗消毒的方法和频率必须在卫生标准操作程序中作出规定，清洗消毒的记录应予以保存。

In addition to the treatment of water, the storage environment of water towers, reservoir and storage tanks must be cleaned and disinfected regularly. The method and frequency of cleaning and disinfection must be stipulated in SSOP, and records of cleaning and disinfection must be preserved.

（3）设施。供水设施要完好，一旦损坏后能立即修好。管道的设计要防止冷凝水集聚下滴污染裸露的加工食品。

3. Facilities. The water supply facilities should be in good condition and can be repaired immediately after the damage, the design of the pipeline should prevent condensate to pollute the exposed processed food.

①防虹吸设备：水管离水面距离为2倍水管直径。

a. Anti-siphon equipment: the distance from the water pipe to the water is 2 times the diameter of the pipe.

②防止水倒流：水管管道有一死水区；水管龙头真空阻断。

b. Prevent water back-flow: the water pipe has a slough; the tap of the pipe should be blocked by vacuum.

③洗手消毒水龙头为非手动开关。

c. The disinfectant faucet for hand washing is a non-manual switch.

④加工案台等工具有将废水直接导入下水道的装置。

d. Tools such as the processing desk has the installation that direct the waste water into the sewer.

⑤备有高压水枪。

e. There is a high-pressure water jet.

⑥有蓄水池（塔）的工厂，水池要有完善的防尘、防虫鼠措施，并进行定期清洗消毒。

f. In a factory with a reservoir (tower), the pool should have a sound dust-proof, pest control measures, and regular cleaning and disinfection.

（4）操作。

4. Operation.

①清洗、解冻用流动水，清洗时防止污水溢溅。

a. Liquid water for cleaning and thawing to prevent spattering of sewage.

②软水管颜色要浅，使用不能拖在地面上。

b. The color of the soft water pipe should be shallow and should not be towed on the ground when it is use.

（5）监测。无论城市公用水还是自备水源都必须充分有效地加以监测，有合格的证明后方可使用。

5. Monitoring. Whether urban public water or self-provided water sources must be fully and effectively monitored and certified before they can be used.

①监测频率：

a. Monitoring frequency:

A. 企业对水的余氯每天监测一次，对所有水龙头每年都监测；

B. 企业对微生物至少每周监测一次；

C. 当地卫生部门对城市公用水项目每年至少监测两次，并有报告正本；

D. 对自备水源监测频率要增加。

② 取样计划。每次取样必须包括总出水口；一年内完成所有出水口的取样。

③ 取样方法。先对出水口进行消毒，放水 5 min 后取样。

④ 日常检测采用试纸、比色法、化学滴定方法检测余氯、pH 值及微生物指标。

(6) 污水排放。

① 污水的处理应符合国家环保部门规定；符合防疫的要求；处理池地点的选择应远离生产车间。

② 废水排放设置。

A. 地面处理（坡度）：为便于排水和防止周围的水逆流进入车间，车间整个地面的水平在设计和建造时应该比厂区的地面略高，并在建造时使地面有一定的坡度，一般为 1°～1.5° 斜坡。

B. 加工用水、台案或清洗消毒用水不能直接流到地面，而应直接入沟，以防止地面的污水飞溅，污染产品和工器具。

a) The residual chlorine in water is monitored once a day and all the faucets are monitored in one year;

b) Microorganism monitoring in enterprises at least once a week;

c) The local health department monitors the urban public water project at least two times a year and has the original report;

d) The monitoring frequency of self-provided water sources should be increased.

b. Sampling plan. Each sampling must include a total outlet; all outlets should be completed within one year.

c. Sampling method. First, the outlet was disinfected, then sampling after turning on the water with 5 min.

d. Test papers, colorimetric method, chemical titration method were used to detect the residual chlorine and pH value as well as microbiological indexes in daily test.

6. Sewage discharge.

a. The treatment of sewage should comply with the regulations of the national environmental protection department; meet the requirements of epidemic prevention; and the location of the treatment pool should be far away from the production workshop.

b. Waste-water discharge setting.

a) Ground treatment (slope): in order to facilitate drainage and prevent the flow of surrounding water from entering the workshop, the whole floor level of the workshop should be slightly higher in design and construction than the ground in the factory area, and a slope of 1° to 1.5° is generally applied to the ground at the time of construction.

b) Water used for processing, cleaning and disinfection shall not flow directly to the ground, but shall be entered directly into the ditch to prevent sewage spatter, contamination of products

C. 废水流向应从清洁区向非清洁区。

D. 排水沟应采用表面光滑、不渗水的材料铺砌,施工时不得出现凹凸不平和裂缝。

(7) 纠偏。监控时发现加工用水存在问题,应停止使用这种水源,直到问题得到解决。

监控时发现在硬管道处有交叉连接时,须立即解决。出现问题处若不能被隔离(如用关闭的阀门),加工应终止,直到修好为止。在不合理的情况下生产的产品不能运销,除非其安全性得到验证。

(8) 记录。水的监控、维护及其他问题处理都要记录、保持。

记录一般应包括城市供水水费单、水分析报告、管道交叉污染等日常检查记录、纠偏记录等。

2) 食品接触面表面的清洁度

(1) 与食品接触的表面。与食品接触的表面是指接触人类食品的那些表面,以及在正常加工过程中会将水滴溅在食品或食品接触的表面上的那些表面。根据潜在的食品污染的可能来源途径,通常把食品接触面分成直接与食品接触和间接与食品接触的表面。

① 直接接触的表面有加工设备、工器具、操作台案、传送带、贮冰池、内包装物料、加工人员的工作服、手套等。

② 间接接触的表面有未经清洁消毒的冷库、车间和卫生间的门把手、操作设备的按钮、车间内电灯开关等。

and appliances from the ground.

c) The flow direction of waste water should flow from the clean area to the non-clean area.

d) Drainage ditch should use smooth surface and non-seepage material, when the construction must not appear uneven and crack.

7. Correction of deviation. Problems with processing water found during monitoring should be discontinued until the problem is resolved.

If there is a cross connection found in the hard pipe during monitoring, it should be resolved immediately. If there is a problem that cannot be isolated (e. g. with a closed valve), the processing should be terminated until it is repaired. Products produced under unreasonable conditions cannot be sold unless their safety is verified.

8. Record. Water monitoring, maintenance and other issues must be recorded and maintained.

Records should generally include water bill for urban water supply, water analysis reports, pipeline cross-contamination and other daily inspection records, rectification records, etc.

B. Cleanliness of Food Contact Surfaces

1. Surfaces in contact with food. The surfaces in contact with food are those surfaces that come into contact with human food and those on which water droplets are spattered on the surface of food or food during normal processing. According to potential sources of food contamination, food contact is usually divided into direct and indirect contact with food surfaces.

a. Direct contact with the surface: processing equipment, workmanship, workbench desk, conveyor belt, ice storage tank, inner packaging materials, the work clothes, gloves and so on of the workers.

b. Indirect contact with the surface: the door handles of a cold storage, workshop, and toilet without cleaning and disinfection, the button of the operating equipment, the lamp switch in the

(2)材料要求。食品接触面的材料应采用无毒（无化学物的渗出）、不吸水（不积水和干燥）、抗腐蚀、不生锈，不与清洁剂、消毒剂产生化学反应，表面光滑易清洗的材料。如不用黄铜制品、黑铁或铸铁及含锌、铅材料，竹、木制品，纤维制品等。可采用不锈钢、无毒塑料、混凝土、瓷砖等。

(3)设计安装要求。食品接触面的设计和安装应无粗糙焊缝、破裂、凹陷，要求表面包括缝、角和边在内；无不良的关节连接，已腐蚀部件，暴露的螺丝、螺帽或其他可以藏匿水或污物的地方，真正做到表里如一，始终保持完好的维修状态。安装应满足在加工人员犯错误的情况下不致造成严重后果的要求。

(4)清洗消毒。
①方法。
A.物理方法。
a.臭氧消毒法：一般消毒 1 h，适用于加工车间。
b.紫外线照射消毒法：每 10～15 m² 安装一只 3 W 紫外线灯，消毒时间不少于 30 min；车间低于 20 ℃，高于 40 ℃，湿度大于 60％时，要延长消毒时间，此方法适用于更衣室、厕所等。
c.药物熏蒸法：每平方米用10 mL 过氧乙酸、甲醛，适用于冷库、保温车等。
肉类加工厂应首选 82 ℃ 热水清洗消毒。此外还有电子灭菌消毒法等。

workshop, etc.

2. Material requirements. The materials used for food contact surfaces should be non-toxic (non-chemical seepage), no water absorption (no water retention and drying), corrosion resistant, no rust, have no chemical reaction with cleaners and disinfectants, and smooth and easy cleaning surfaces. If you do not use brass products, black iron or cast iron and containing zinc, lead materials, bamboo, wood products, fiber products, you can use stainless steel, non-toxic plastics, concrete, ceramic tiles and so on.

3. Design and installation requirements. The design and installation of food contact surfaces shall be free of rough welds, cracks and recesses, and shall require surfaces including joints, angles and edges; no defective joint connections, corroded parts, exposed screws, nuts or other places where water or dirt may be hidden, truly to maintain a sound maintenance state, installation should meet the processing personnel in the case of errors will not cause serious consequences.

4. Cleaning and disinfection.
a. Method.
a) Physical methods.
• Ozone disinfection: general disinfection for 1 hour, suitable for processing workshop.
• Ultraviolet irradiation disinfection method: one 3 W ultraviolet lamp should be installed every $10 \sim 15$ m², disinfection time is not less than 30 min, workshop is below 20 ℃, above 40 ℃, humidity is more than 60％, the disinfection time should be prolonged, this method is suitable for changing rooms, toilet and so on.
• Drug fumigation: with peracetic acid, formaldehyde, 10 mL per square meter, suitable for cold storage, heat preservation vehicle, etc.

The meat packing factory should choose 82 ℃ hot water to clean and disinfect. In addition, there are electronic sterilization methods and so on.

B. 化学方法：一般使用含氯消毒剂，如用次氨酸钠 100～150 mg/kg。

②程序。使用化学清洗消毒剂时一般分为 6 个步骤：清除→预冲洗→使用清洁剂→再冲洗→消毒→冲洗。

首先，必须彻底清洗，以除去微生物赖以生长的营养物质。如清除大的残渣，预冲洗去除表面附着的残渣，使用清洁剂清洗顽垢，再冲洗清洁剂并去除顽垢。然后进行消毒，确保消毒效果。接着再进行冲洗，去除残留的化学消毒剂。在清洗过程中应注意清洁剂的使用和浸洗都需要恰当的时间，此外清洁剂的温度也直接影响清洁效果。

清洁剂的类型包括普通清洁剂（GP）、碱、含氯的清洁剂、酸、酶等。

清洁剂的清洁效果与接触时间、温度、物理擦洗及化学等因素有关，应对清洁效果实施监控。

③设备和工器具的清洗消毒及其管理。

A. 清洗消毒频率：大型设备应在每班加工结束之后；清洁区工器具应每 2～4 h 一次；屠宰线上用的刀具每用一次消毒一次（每个岗位至少两把刀，交替使用）；加工设备、器具被污染之后应立即进行清洗消毒。

B. 手和手套。每次进车间前和加工过程中手被污染时，必须洗手消毒。

b) Chemical methods: chlorinated disinfectants, such as sodium methionine (100～150 mg/kg), are commonly used.

b. Procedure. Chemical cleaning disinfectants are generally divided into six steps: clearance→pre-washing→use of detergent→rewash→disinfection→rinse.

First, it must be thoroughly cleaned to remove the nutrients on which microbes grow. Such as the removal of large residues, pre-washing to remove the surface of the residue, using cleaning agents to clean the stubborn dirt, washing cleaning agent and removal of stubborn dirt. Then sterilize to ensure the disinfection effect. Then rinse it out to remove the residual chemical disinfectant. In the process of cleaning should pay attention to the use of cleaning agents and washing need the appropriate time, in addition to the temperature of cleaning agents also directly affect the cleaning effect.

The types of detergents include general cleaning agents (GP), alkali, chlorinated cleaning agents, acids, enzymes, and so on.

The effect of cleaning agents is related to contact time, temperature, physical scrubbing and chemistry, so the cleaning effect should be monitored.

c. Cleaning and disinfection of equipment and apparatus and its management.

a) Frequency of cleaning and disinfection. Large equipment: after each overtime is over. Cleaning area appliances: every 2 to 4 hours. Tools used in slaughtering lines: sterilize once each time (at least two knives per post are convenient to use alternately). Cleaning and disinfection of processing equipment and utensils shall be carried out immediately after contamination.

b) Hands and gloves. Wash your hands and sterilize your hands every time are contaminated

要做到必须在车间的入口处、车间流水线和操作台附近设有足够的洗手消毒设备,在清洁区的车间入口处还应派专人检查手的清洗消毒情况,检查是否戴首饰、是否留过长的指甲等。手套一般在一个班次结束或中间休息时更换。手套不得使用线手套,所用材料应不易破损和脱落。手套清洗消毒后贮存在清洁的密闭容器中送往更衣室。

C. 工作服。工作服应在专用的洗衣房进行集中清洗和消毒。洗衣设备、能力与实际需求相适应。不同清洁要求区域的工作服应分开清洗,不同清洁区的工作服分别清洗消毒。清洁工作服与脏工作服分区域放置,存放工作服的房间应设有臭氧消毒、紫外线等设备,且干净、干燥和清洁。工作服必须每天清洗消毒。一般工人至少配备两套工作服。工人出车间、去卫生间,必须脱下工作服、帽和工作鞋。

D. 工器具清洗消毒的注意事项。要有固定的清洗消毒场所或区域,推荐使用82 ℃的热水;要根据清洗对象的性质选择相应的清洗剂;在使用清洗剂、消毒剂时要考虑接触时间和温度;冲洗时要用流动的水,同时应防止

before entering the workshop and during processing. To be sure, adequate hand washing and disinfection equipment must be installed at the entrance of the workshop, near pipelines and the operating table of the workshop. At the entrance of the workshop in the cleaning area, a special person should be sent to check the cleaning and disinfection of the hands. Check to see if you wearing jewelry, have long nails, etc. Gloves are usually replaced at the end of a shift or during a break. Gloves shall not be used in line gloves, and the materials used should not be easily damaged and shed. After cleaning and disinfection, the gloves are stored in a clean sealed container and sent to the dressing room.

c) Work clothes. The work clothes should be centralized cleaned and disinfected in the special laundry room. Laundry equipment and capacity shall be adapted to actual requirements. The work clothes of different cleaning areas should be cleaned separately, and the work clothes of different cleaning areas should be cleaned and disinfected separately. Clean work clothes and dirty work clothes should be placed in different areas. The rooms where the uniforms are stored should be equipped with ozone disinfection, ultraviolet radiation and other equipment, and they should be tidy, dry and clean. Work clothes must be cleaned and disinfected every day. The general worker is equipped with at least two sets of work clothes. When workers leave the workshop and go to the bathroom, they must take off their work clothes, hats and work shoes.

d) Matters needing attention in cleaning and disinfecting utensils. Have a fixed cleaning and disinfection site or area, it is recommended to use hot water at 82 ℃; appropriate cleaning agents should be selected according to the nature of the cleaning object; contact time and temperature

清洗、消毒水溅到产品上造成污染。设有隔离的工器具洗涤消毒间,不同清洁工器具应分开清洗。

(5)监控。

①监控对象:食品接触面的状况;食品接触面的清洁和消毒;使用消毒剂的类型和浓度;可能接触食品的手套和外衣是否清洁卫生,且状态良好。

②监控方法。

A.感官检查:表面状况良好。表面已清洁和消毒;手套和外衣清洁且保养良好。

B.化学检测:消毒剂的浓度是否符合规定的要求。

C.表面微生物检测:检测方法包括平板法、棉签法和发光法。

③监控频率。

A.感官监控频率:每天加工前、加工过程中以及生产结束后进行。洗手消毒主要在员工进入车间时、从卫生间出来后和加工过程中检查。

B.实验室监控频率:按实验室制定的抽样计划,一般每周1~2次。

(6)纠偏。在检查发现问题时应采取适当的方法及时纠正,如再清洁、消毒、检查消毒剂浓度、培训等。

(7)记录。作卫生监控记录的目

should be taken into account in the use of cleaning agents and disinfectants; and flowing water should be used when rinsing. At the same time, the cleaning and disinfecting water should be prevented from causing pollution on the product. There is a separate equipment washing and disinfection room, different cleaning equipment should be cleaned separately.

5. Monitoring.

a. Monitoring target: state of food contact surface; cleaning and disinfection of food contact surfaces; type and concentration of disinfectants used; sanitary and in good condition for gloves and coats that may be in contact with food.

b. Monitoring method.

a) Sensory examination: the surface is in good condition. The surface has been cleaned and disinfected; gloves and coats are clean and well maintained.

b) Chemical testing: the concentration of disinfectants is in accordance with the prescribed requirements.

c) Surface microbial detection: detection methods include flat, cotton swab and luminescence.

c. Monitoring frequency.

a) Sensory monitoring frequency: daily monitoring is performed before, during and after production. Hand washing and disinfection are mainly checked when the employee enters the workshop, after coming out of the bathroom and during processing.

b) Frequency of laboratory monitoring: usually one to two times a week according to the sampling plan developed by the laboratory.

6. Correction of deviation. When checking and discovering problems, we should take appropriate measures to correct them in time, such as re-cleaning, disinfection, checking the concentration of disinfectant, training and so on.

7. Record. The purpose of the health monitoring

的是提供证据,证实工厂消毒计划充分,并已执行,此外发现问题能及时纠正。记录包括检查食品接触面状况、消毒剂浓度、表面微生物检验结果等。记录的种类包括每日记录、监控记录、检查、纠偏记录等。

3)防止交叉污染

交叉污染是指通过生的食品、食品加工人员和食品加工环境把生物的、化学的污染物转移到食品上去的过程。防止交叉污染的途径包括防止员工操作造成的产品污染;生的和即食食品的隔离;内外包装材料存放的隔离,以及外包装与内包装操作间的隔离;防止工厂设计造成的污染。

(1)污染的来源。交叉污染的来源包括:工厂选址、设计、布局不合理;加工人员个人卫生不良;清洁消毒不当,卫生操作不当;生、熟食品未分开;原料和成品未隔离。

(2)预防。
①工厂选址、设计。

A. 为了使工厂和车间的选址、设计、布局尽量合理,企业应提前与有关政府主管部门取得联系,了解有关规定和要求。

B. 车间的布局既要便于各生产环节的相互联结,又要便于加工过程的卫生控制,防止交叉污染的发生。

C. 加工工艺布局合理,能采取物理隔离的地方尽量采取物理隔离。应

records is to provide evidence that the plant disinfection plan is adequate and implemented, and that problems can be corrected in a timely manner. Records include inspection of food contact status, disinfectant concentration, surface microbial test results, etc. Types of records include daily records, monitoring records, checks, correction records, etc.

C. Prevent the Cross-contamination

Cross-contamination refers to the process of transferring biological and chemical pollutants to food through raw food, food processors and the food processing environment. Ways to prevent cross-contamination include product contamination caused by employee operations; isolation of raw and ready-to-eat foods; isolation of internal and external packaging materials and between outer and inner packaging operations; and prevention of contamination caused by factory designs.

1. Sources of pollution. The sources of cross-contamination include: unreasonable site selection, design and workshop; poor personal hygiene of workers; improper cleaning and disinfection, improper sanitary operation; raw and cooked products are not separated; raw materials and finished products are not isolated.

2. Prevention.

a. Site selection and design.

a) In order to make the location, design and layout of factories and workshops as reasonable as possible, enterprises should contact relevant government departments in advance to understand the relevant regulations and requirements.

b) The layout of workshop is not only convenient for the connection of each production link, but also convenient for the hygienic control of the processing process to prevent the occurrence of cross-contamination.

c) The layout of processing technology is reasonable and physical isolation can be taken as far as possible

遵守如下原则：前后工序，如生熟之间、不同清洁度要求的区域之间应完全隔离；原料库、辅料库、成品库、内包装材料库、外包装材料库、化学品库、杂品库等应专库专用。

where physical isolation can be taken. The following principles shall be observed: complete isolation of the front and back processes, such as between raw and matured, between areas with different cleanliness requirements; the material storehouse, the excipient material storehouse, the finished product storehouse, the inner packing material storehouse, the outer packing material storehouse, the chemical storehouse, the miscellaneous goods storehouse and so on shall be specially reserved.

D. 同一车间不能同时加工不同类别的产品。

d) Different types of products cannot be processed in the same workshop at the same time.

E. 明确人流、物流、水流、气流的方向。人流应从高清洁区到低清洁区；物流应不造成交叉污染，可用时间、空间分隔；水流应从高清洁区到低清洁区；气流应采用进气控制、正压排气、鼓风排气、非抽气等措施控制，注意采用负压排气时需有一个回气孔，以免从下水道抽气。

e) Define the direction of the stream of people, logistics, water flow and airflow. The stream of people should be from the high clean area to the low clean area; the logistics should not cause cross pollution and can be separated in time and space; the flow of water should be from the high clean area to the low clean area; air flow should be controlled by air intake control, positive pressure exhaust, blast exhaust, non-extraction and other measures. Pay attention to the use of negative pressure exhaust should have a return air hole, so as not to exhaust from the sewers.

②卫生操作防止交叉污染：将生食与熟食或即食食品加工活动充分隔离；贮藏中的产品充分隔离或保护；食品处理或加工区域的设备充分清洁和消毒；员工卫生，衣着和手的清洁；员工食品加工操作和技艺；员工在厂区附近的活动。

b. Sanitary operation to prevent cross-contamination: adequate isolation of raw food from cooked or ready-to-eat food processing activities; adequate isolation or protection of products in storage; adequate cleaning and disinfection of equipment in food handling or processing areas; staff hygiene, clothing and hand cleaning; staff food processing operations and workmanship; employee activities near the site.

③隔离生的和即食产品：当接收产品或辅料时；在加工整理操作期间；贮存期间；运输期间。

c. Isolate raw and ready-to-eat products: when receiving products or accessories; during processing and finishing operations; during storage; during transportation.

④防止加工中的交叉污染：指定区域将生的和即食产品的加工区

d. Prevent the cross-contamination in processing: the designated area separate the processing zones of raw

隔;控制设备由一个加工区域向另一个加工区域的移动;控制人员由一个加工区域通往另一个加工区域。

(3)监控:

①在开工时、交接时、餐后进入生产车间;

②采用生产连续监控;

③产品贮存区域(如冷库)每次检查。

(4)纠偏:

①发生交叉污染,采取措施防止再发生,必要时停产直到改进,如有必要需对产品的安全性进行评估;

②必要时对车间布局进行改造,避免不同清洁区人员交叉流动及工器具的交叉使用;

③清除顶棚上的冷凝物,调节空气流通和房间温度以减少水的凝结,安装遮盖物防止冷凝物落到食品、包装材料或食品接触面上;

④清扫地板,清除地面上的积水;

⑤及时清洗消毒被污染的食品接触面;

⑥在非产品区域操作有毒化合物时,设立遮蔽物以保护产品;

⑦增加培训程序,加强对员工的培训,纠正不正确的操作;

⑧转移或丢弃没有标签的化学品。

(5)记录:

①消毒控制记录;

②改正措施记录。

and ready-to-eat products; control the movement of equipment from one processing area to another; and control the movement of personnel from one processing area to another.

3. Monitoring.

a. During the start and handover, the after-meal processing continues to enter the production workshop.

b. Use continuous production monitoring.

c. Product storage area (such as cold storage) is checked every time.

4. Correction of deviation.

a. When cross contamination occurs, take measures to prevent recurrence, if necessary, stop production until improvement, and if necessary, evaluate the safety of the product.

b. When necessary, the layout of the workshop is modified to avoid the cross movement of personnel and the cross use of tools in different cleaning areas.

c. Remove condensation from the ceiling, adjust air circulation and room temperature to reduce water condensation, install cover to prevent condensate from falling on food, packaging material or food contact surface.

d. Clean the floor and remove the water from the floor.

e. Clean and disinfect contaminated food contact surfaces in time.

f. When operating toxic compounds in a non-product area, establish a shield to protect the product.

g. Increase training procedures, strengthen staff training and correct incorrect operations.

h. Transfer or discard unlabeled chemicals.

5. Record:

a. Disinfection control records;

b. Corrective action record.

4)手的清洗与消毒,厕所设备的维护与卫生保护

(1)洗手消毒设施。

①洗手消毒设施应设在车间入口处、车间内加工岗位的附近和卫生间。

②洗手消毒设施包括非手动开关的水龙头、冷热水、皂液器、消毒槽、干手设施、流动消毒车等。此外还应注意温水一般43 ℃为宜;每10~15人设一水龙头为宜。洗手消毒液应保持清洁且有效氯含量至少为100 mg/kg。

(2)厕所设施。

①位置:与车间相连接或不连接;门不能直接朝向车间;卫生间的门应能自动关闭;卫生间最好不在更衣室内,确保在更衣室脱下工作服和工作鞋后方能上厕所。

②数量:与加工人员相适应,每15~20人设一个为宜。

③结构:严禁使用无冲水的厕所;避免使用大通道冲水式厕所,应采用蹲便器或坐便器。

④配套设备,包括冲水装置、手纸和纸篓、洗手消毒设备、干手设施。

⑤卫生要求:通风良好,地面干燥,保持清洁卫生,光照充足,不漏水,有防蝇、防虫设施,进入厕所前要脱下工作服和换鞋,方便之后要洗手和消毒。

以上要求适用于所有的厂区、车间和办公楼厕所。

D. Hand Cleaning and Disinfection, Maintenance and Sanitary Protection of Toilet Equipment

1. Hand washing and disinfection facilities.

a. Hand washing and disinfection facilities should be located at the workshop entrance, near the workshop processing position and bathroom.

b. Hand washing and disinfection facilities include non-manual switch faucet, hot and cold water, soap dispenser, sterilizer, hand drying facilities, mobile disinfection vehicle and so on. In addition, attention should be paid to warm water generally at 43 ℃; every 10 to 15 people per set up a faucet is appropriate. Hand sanitizer should be kept clean and have an effective chlorine content of at least 100 mg/kg.

2. Toilet facilities.

a. Location: connected to or not connected to the workshop; door cannot be directly oriented to workshop; toilet door should be automatically closed; toilet is best not in the dressing room, ensure that in the dressing room after the removal of work clothes and shoes can go to the toilet.

b. Quantity: adapt to processing personnel, one toilet for every 15 to 20 people.

c. Structure: it is strictly forbidden to use the toilet without flushing water; avoid the use of large passage flush toilet, squat toilet or toilet should be used.

d. Supporting equipment: including water washing device, toilet paper and paper basket, hand washing and disinfection equipment, and hand drying facilities.

e. Health requirements: good ventilation, dry ground, clean and sanitary, light enough, no water leakage, anti flies and pest control facilities. Before going to the toilet, we need to take off the work clothes and change shoes. After that, we need to wash hands and disinfect.

The above requirements apply to all factory premises, workshops and office toilets.

(3)设备的维护与卫生保持：

①设备保持正常运转状态；

②卫生保持良好,不造成污染。

(4)监控：

①每天至少检查一次设施的清洁与完好状况；

②卫生监控人员巡回监督；

③化验室定期做表面样品检验；

④检查消毒液的浓度。

(5)纠偏。检查发现不符合时应立即纠正。纠正可以包括：修理或补充厕所和洗手处的洗手用品；若手部消毒液浓度不适宜,则将其倒掉并配新的消毒液；当发现有令人不满意的状况出现时,记录所进行的纠正措施；修理不能正常使用的厕所。

(6)记录。每日卫生监控记录包括厕所或洗手池和厕所设施的状况，如厕所或水槽和厕所设施的状况及其位置；手部消毒间、池或洗手消毒液的状况；洗手消毒液的浓度；当发现有令人不满意的状况出现时所采取的纠正措施。

5)防止食品被污染

防止食品、食品包装材料和食品所有接触表面被微生物、化学品及物理的污染物沾污,如清洁剂、燃料、杀虫剂、废弃物、冷凝物以及各种污物等。

(1)污染物的来源。

①物理性污染物：无保护装置的照明设备的碎片,天花板和墙壁的脱落物；工具上脱落的大漆片、铁锈,竹木器具上脱落的硬质纤维；头发等。

3. Maintenance and hygiene of equipment：

a. The equipment remain in normal working condition；

b. Good hygiene is kept free of pollution.

4. Monitoring：

a. Check the cleanliness and integrity of the facility at least once a day；

b. Itinerant supervision of health monitors；

c. Surface samples are regularly tested by the laboratory；

d. Check the concentration of the disinfectant.

5. Correction of deviation. Any nonconformance found during inspection should be corrected immediately. Corrections may include：repairing or replenishing hand-washing items in toilets and hand washing places；dumping hand disinfectant solutions and adding new disinfectant if the concentration of hand disinfectant is not appropriate；recording corrective actions taken when unsatisfactory conditions are found to arise；repair toilets that don't work properly.

6. Record. Daily health monitoring records include the condition of the toilet or wash basin and toilet facilities, including：the condition and location of the toilet or sink and toilet facility；the condition of the hand disinfection room, pool or hand sanitizer；and the concentration of hand sanitizer；corrective action taken when unsatisfactory conditions are detected.

E. Prevent the Contamination of Food

Prevent contamination of food, food packaging materials and all food contact surfaces with microbial, chemical and physical contaminants such as detergents, fuels, pesticides, wastes, condensates, and various contaminants.

1. Sources of pollutants.

a. Physical contaminants：including fragments of unprotected lighting equipment, ceilings and walls；large lacquer pieces, rust from tools, hard fibers from bamboo and wood utensils；hair, etc.

②化学性污染物：润滑剂、清洁剂、杀虫剂、燃料、消毒剂等。

③微生物污染物：被污染的水滴和冷凝水，空气中的灰尘、颗粒、外来物质，地面污物，不卫生的包装材料，唾液，喷嚏等。

（2）防控。

①水滴和冷凝水的控制。应保持车间通风，避风量要大于排风量，防止空调管道形成冷凝水。在有水蒸气产生的车间，要安装适当的排气装置。此外还应采取控制车间温度，尤其是控制温差；顶棚呈圆弧形；提前降温，尽量缩小温差等措施。

②防止污染的水溅到食品上。及时清扫，保持车间干燥。车间内设有专用工器具清洗消毒间；待加工原料或半成品远离加工线或操作台；车间内没有产品时才冲洗台面、地面；车间内的洗手消毒池旁没有产品；车间台面、池子中的水不能直接排到地面，应排进管道并引入下水道。

③包装物料的控制。包装物料存放库要保持干燥、清洁、通风、防霉，内外包装分别存放，上有盖布下有垫板，并设有防虫鼠设施。每批包装物进雨水后要进行微生物检验（细菌数＜100 个$/cm^2$，致病菌不得检出），必要时进行消毒。

b. Chemical pollutants: lubricants, cleaners, pesticides, fuels, disinfectants, etc.

c. Microbial pollutants: contaminated water droplets and condensed water, dust in the air, particles, foreign substances, ground dirt, unsanitary packaging materials, saliva, sneezing, etc.

2. Prevention and control.

a. Control of droplets and condensed water. The ventilation of the workshop should be maintained, the amount of avoiding air should be greater than the amount of exhaust air, and the air conditioning pipeline should be prevented from forming condensed water. Appropriate exhaust devices should be installed in workshops where steam is produced. In addition, measures should be taken to control the workshop temperature, especially to control the temperature difference; the ceiling is circular arc; reduce the temperature in advance, try to reduce the temperature difference as far as possible and other measures.

b. Prevent contaminated water from spilling onto food. Clean in time, keep workshop dry. The workshop is equipped with special equipment cleaning and disinfection room; raw materials or semi-finished products to be processed are far away from the processing line or operation table; when there are no products in the workshop, the countertop and floor are washed; there are no products next to the washing and disinfection pool in the workshop; the water in the workshop table and pool should not be discharged directly to the ground, it should be drained into the pipe and introduced into the sewer.

c. Control of packaging materials. Packaging materials storage to keep dry, clean, ventilation, mildew prevention, internal and external packaging is stored separately, the gremial on the top and under the backing board, and has pest control facilities. After each batch of packaging was infiltrated into rain water, microbiological

(3)监控。任何可能污染食品或食品接触面的掺杂物,如潜在的有毒化合物、不卫生的水(包括不流动的水)和不卫生的表面所形成的冷凝物,建议在开始生产时及工作时间每 4 h 检查一次。

(4)纠偏:
①除去不卫生表面的冷凝物;
②用遮盖方法防止冷凝物落到食品、包装材料及食品接触面上;
③清除地面积水、污物,清洗化合物残留;
④评估被污染的食品;
⑤培训员工正确使用化合物。
(5)记录:每日卫生控制记录。
6)有毒化学物质的标记、贮存和使用

食品加工企业使用的化学物质包括洗涤剂、消毒剂、杀虫剂、润滑剂、实验室用品、食品添加剂等,它们是工厂正常运转所必需的。但在使用中必须做到按照产品说明书使用,正确标记、安全贮存,否则存在加工的食品被污染的风险。

(1)常用有毒化学物质。食品加工厂有可能使用的有毒化学物质包括清洗剂、消毒剂(如次氯酸钠)、杀虫剂(如 1605)、灭害灵、除虫菊酯、机械润滑剂、实验室用品(如检查化验用的各种试剂)、食品添加剂(如亚硝酸钠)等。

(2)有毒化学物质的贮存和使用。
①有毒化学物质的贮存:

tests were carried out (the number of bacteria was less than $100/cm^2$, the pathogenic bacteria could not be detected) and disinfected if necessary.

3. Monitoring. Any adulteration that may contaminate food or food contact surfaces, such as potentially toxic compounds, unsanitary water (including unflowing water) and condensates formed on unsanitary surfaces, is recommended to be inspected every 4 hours at the start of production and working hours.

4. Correction of deviation:
a. Remove condensates from unsanitary surfaces;
b. A covering method is used to prevent condensate from falling onto food, packaging material and food contact surface;
c. Remove stagnant water, dirt and residue of cleaning compounds from the ground;
d. Assess contaminated food;
e. Train the employees to use compounds correctly.
5. Record:daily health control records.
F. Label, Storage and Use of Toxic Chemicals

The chemicals used by food processing enterprises include detergents, disinfectants, insecticides, lubricants, laboratory supplies, food additives, etc. They are necessary for the normal operation of a factory, but must be used in accordance with the product instructions. Properly marked and stored safely, otherwise there is a risk that food processed by the enterprise will be contaminated.

1. Common toxic chemicals. Toxic chemicals likely to be used in food processing plants include cleaning agents, disinfectants such as sodium hypochlorite, insecticides such as 1605, mirex, pyrethroids, mechanical lubricants, laboratory supplies such as various reagents for laboratory testing and food additives such as sodium nitrite.

2. Storage and use of toxic chemicals.
a. Storage of toxic chemicals:

A. 食品级化学品与非食品级化学品分开存放；

B. 清洗剂、消毒剂与杀虫剂分开存放；

C. 一般化学品与剧毒化学品分开存放；

D. 贮存区域应远离食品加工区域；

E. 化学品仓库应上锁，并有专人保管。

② 有毒化学物质的正确管理和使用。

A. 原包装容器的标签应标明容器中化学品的名称、生产厂名、厂址、生产日期、批准文号、使用说明和注意事项等。

B. 工作容器的标签应标明容器中的化学品名称、浓度、使用说明和注意事项。

C. 建立化学物品台账（入库记录），以有毒化学物质一览表的形式标明库存化学物品的名称、有效期、毒性、用途、进货日期等。

D. 建立化学物品领用、核销记录。

E. 建立化学物品使用登记记录，如配制记录、用途、实际用量、剩余配置液的处理等。

F. 制定化学物品进厂验收制度和标准，建立化学物品进厂验收记录。

G. 制定化学物品包装容器回收、处理制度，严禁将化学物品的容器用来包装或盛放食品。

H. 对化学物品的保管、配制和使用人员进行必要的培训。

a) Food grade chemicals and non-food grade chemicals are stored separately;

b) Cleaners, disinfectants and pesticides are stored separately;

c) General chemicals are stored separately from highly toxic chemicals;

d) Storage area should be away from food processing areas;

e) Chemical warehouse should be locked and kept by special person.

b. Proper management and use of toxic chemicals.

a) The label of the original packaging container shall be marked with the name of the chemical in the container, the name of the production plant, the site of the factory, the date of production, the approval document number, the instructions for use and the points for attention, etc.

b) The label of the working container shall indicate the name of the chemical in the container, the concentration, the instructions for use and the precautions.

c) Set up a chemical standing book (warehousing record), and mark the name, validity period, toxicity, usage and date of purchase of the chemical products in the form of toxic chemicals list.

d) Establish chemical receipt and write-off records.

e) Establish a record of the use of chemicals, such as the preparation record, use, actual dosage, disposal of the remaining configuration liquid, and so on.

f) Establish the acceptance system and standard of chemical products, and establish the acceptance record of chemical products.

g) Develop chemical packaging container recovery, treatment system, chemical containers are strictly prohibited to package or hold food.

h) Provide necessary training to the personnel responsible for the custody, preparation and use of

l. 化学物品应采用单独的区域贮存,使用带锁的柜子,防止随便乱拿。

(3)有毒化学物质的监控:

①监控内容应包括标识、贮藏及使用过程;

②经常检查确保符合要求;

③建议一天至少检查一次;

④全天都应注意。

(4)纠偏:

①转移存放错误的化合物;

②标签不全、标记不清的应退还给供应商;

③对于不能正确辨认内容物的工作容器应重新标记;

④不适合或已损坏的工作容器弃之不用或销毁;

⑤评价不正确使用有毒有害化合物所造成的影响,判断食品是否已遭污染,以确定是否销毁;

⑥加强对保管、使用人员的培训。

(5)记录。应设有进货、领用、配制记录,以及化学物质批准使用证明、产品合格证。

7)雇员的健康卫生控制

食品生产企业的生产人员(包括检验人员)是直接接触食品的人,其身体健康及卫生状况直接影响产品卫生质量。根据食品卫生管理法,凡从事食品生产的人员必须经过体检合格获得健康证方能上岗,并每年进行一次体检。

(1)雇员的健康卫生的日常管理。

chemicals.

i) Chemicals should be stored in separate areas, using locked cabinets to prevent random pick-up.

3. Monitoring of toxic chemicals:

a. Monitoring should include identification, storage and use processes;

b. Check regularly to ensure compliance;

c. It is recommended that they be checked at least once a day;

d. It should be careful all day.

4. Correction of deviation:

a. Transfer the wrong compound;

b. If the label is not complete or clearly marked, it shall be returned to the supplier;

c. Work containers that do not correctly identify the contents should be remarked;

d. Unsuitable or damaged working containers are discarded or destroyed;

e. Evaluate the effects of improper use of toxic and harmful compounds and determine whether the food has been contaminated to determine whether it is destroyed;

f. Strengthen the training of custody and use personnel.

5. Record. There should be purchase, use, preparation records, approval of the use of chemical substances, product certification.

G. Health Control of Employees

The production personnel (including examiners) of food production enterprises are the people who have direct contact with food, and their health and hygiene condition directly affect the quality of product hygiene. According to the law of food hygiene management, all the personnel engaged in food production must pass the physical examination to get a health certificate bcfore they can take up the post, and conduct an annual medical examination.

1. Daily management of employees' health

①食品加工人员不能患有以下疾病,如病毒性肝炎、活动性肺结核、伤寒及其带菌者、细菌性痢疾及其带菌者、化脓性或渗出性脱屑、皮肤病、手外伤未愈等。

②工人上岗前应进行健康检查,发现有患病症状的员工,应立即调离食品工作岗位,并进行治疗。待症状完全消失,并确认不会对食品造成污染后才可恢复正常工作。

③对加工人员应定期进行健康检查,每年进行一次体检,并取得县级以上卫生防疫部门的健康证明。此外食品生产企业应制订体检计划,并设有健康档案。

④生产人员要养成良好的个人卫生习惯,按照卫生规定从事食品加工,进入加工车间更换清洁的工作服、帽、口罩、鞋等,不得化妆、戴首饰及手表等。

⑤食品生产企业应制订卫生培训计划,定期对加工人员进行培训,并记录存档。应教育员工认识到疾病对食品卫生带来的危害,并主动向管理人员汇报自己和他人的健康状况。

(2)监督。监督的目的是控制可能导致的食品、食品包装材料和食品接触面的微生物污染。

①健康检查:员工的上岗前健康检查;定期健康检查,每年进行一次体

and hygiene.

a. Food processors are forbidden to suffer from such diseases as viral hepatitis, active pulmonary tuberculosis, typhoid fever and its carriers, bacillary dysentery and its carriers, suppurative or exudative desquamation, skin diseases, unhealed hand injuries, and so on.

b. Workers should be checked before taking up their jobs, and employees who find symptoms of illness should be immediately transferred from food work and treated, until the symptoms have completely disappeared, and only after they have confirmed that they will not contaminate the food can they resume their normal work.

c. The processing personnel shall be subject to regular health examinations, an annual physical examination and a health certificate from the department of health and epidemic prevention at or above the county level. In addition, food production enterprises should draw up a medical examination plan, and have a health file.

d. Production personnel should develop good personal hygiene habits, engage in food processing in accordance with hygiene regulations, enter processing workshops to change clean work clothes, hats, masks, shoes, etc., and shall not wear makeup, jewelry, watches, etc.

e. Food production enterprises shall draw up health training plans, regularly train processing personnel, and keep records on file. Employees should be educated to recognize the dangers of disease to food hygiene and take the initiative to report to the management the health of themselves and others.

2. Supervision. The aim of the supervision is to control microbial contamination that may lead to food, food packaging materials and food contact.

a. Health examination: physical examination before mount guard; a regular health checkup is

检；每日健康状况检查，观察员工是否患病或有伤口感染的迹象，要注意加工厂员工的一般症状和状况，如发烧伴有咽喉疼痛、黄疸症（眼结膜或皮肤发黄）、手外伤未愈合等。

②员工个人卫生监控：洗手、消毒程序执行情况；工作服是否干净、整齐，是否身上粘有异物，指甲是否过长，手面是否有伤或化脓现象；与生产无关的物品严禁带入车间，员工不得佩戴首饰，不得化妆、涂指甲油等；生产车间严禁吸烟、吃食物、喝饮料；进入卫生间的更衣洗手情况；工作人员不得串岗；工作过程中每个环节按要求定时洗手、消毒的执行情况。

(3) 纠偏。将患病员工调离生产岗位直至痊愈。

(4) 记录：

①健康检查记录；

②每日上岗前及生产线上员工卫生健康检查记录；

③出现不满意状况和相应的纠正措施记录。

8) 虫、鼠害防治

昆虫、鸟、鼠等物会带一定种类的病原菌，还会直接消耗、破坏食品，并在食品中留下令人厌恶的东西，如粪便或毛发。因此，对虫害的防治对食品加工厂来说是至关重要的。

(1) 防治计划。防治范围覆盖全厂，包括生活区甚至工厂周围也在灭

conducted annually; daily health checks to observe whether employees are ill or have signs of wound infection, and to pay attention to the general symptoms and conditions of the workers in the processing plant. Such as fever with throat pain, jaundice (eye conjunctiva or skin yellowish), hand trauma and other phenomena.

b. Personal health monitoring of employees: the execution of the procedures for washing hands and disinfection; whether the work clothes are clean and tidy, whether there are foreign objects on the body, whether nails are too long, whether there is injury or purulent on the surface of the hands; the production of articles unrelated to the production is strictly prohibited from being brought into the workshop, and no jewelry shall be worn by the staff, No makeup, nail polish, etc.; no smoking, food and drink in the production workshop; changing clothes and washing hands into the bathroom; staff members must not randomly join the post; every link in the working process should regularly wash hands and sterilize according to the requirements.

3. Correction of deviation. Transfer sick staff from production until recovery.

4. Record:

a. Health examination records;

b. Daily health examination records of employees before starting duty and on the production line;

c. Records of unsatisfactory situation and corresponding corrective actions.

H. Control of Insect and Rodent Pests

Insects, birds, mice, and so on will carry certain types of pathogens, and will consume, destroy food and leave in food offensive things such as feces or hair. Therefore, pest control is crucial to food processing plants.

1. Prevention plan. The prevention and cure scope includes the whole plant scope, the living area

鼠工作计划之内。应编制灭鼠分布图、清扫消毒执行规定等。

防治计划应考虑厂房和地面管理、结构布局、工厂机械、设备和工器具、原材料、仓库及室内环境的管理、废物处理、杀虫剂的使用和其他控制措施。

(2)重点。虫、鼠害防治的重点包括厕所、下脚料出口、垃圾箱、原料和成品库周围及食堂。

(3)防治措施：
①消除滋生地及周边环境，包装物、原材料防虫、鼠是第一位的；

②采用风幕、水幕、纱窗、门帘、挡鼠板、翻水弯等预防虫、鼠进入车间；

③厂区采用杀虫剂；
④车间入口用灭蝇灯；

⑤防鼠用粘鼠胶、鼠笼，不能用灭鼠药。

(4)纠偏：
①增加设施；
②加强环境卫生控制；
③增加杀灭频率。

even includes the factory surrounding also in the work plan of rodent control. It is necessary to draw up the distribution map of rodent control and the implementing regulations of cleaning and disinfection.

The prevention plan shall take into account the management of plant and ground, structural layout, plant machinery, equipment and appliances, raw materials, stores and indoor environment, waste disposal and use of pesticides and other control measures.

2. Highlight. The focus of pest and rodent control includes latrines, waste material exports, garbage bins, raw materials and finished products around warehouses and canteens.

3. Control measures:

a. Eliminate breeding grounds and surrounding environment, packaging, raw materials to prevent insects, rats are the first;

b. Adopt wind curtain, water curtain, screen window, door curtain, rat block board, turn water bend and so on to prevent insects and rats from entering the workshop;

c. Insecticides are used in the plant;

d. A fly-proof lamp is used at the entrance of the workshop;

e. Mouse stick glue can used to prevent mice, but rat cages can not use rodenticide.

4. Correction of deviation:

a. Additional facilities;

b. Strengthen environmental health control;

c. Increase kill frequency.

任务4　依据卫生标准操作程序要求编写卫生标准操作规范

6.4.1　基础知识

"卫生标准操作规范"由一系列文件化的程序组成,通过对操作性前提方案的执行,达到食品安全管理的要求。食品生产组织应按照良好操作规范的规定,编写本组织的卫生标准操作程序(SSOP)。

6.4.2　实训任务

编写速冻蔬菜产品的"卫生标准操作规范"。

实训组织:对学生进行分组,每个组参照"基础知识"中的内容并利用网络资源,编写一份速冻蔬菜产品的"卫生标准操作规范"。

实训成果:速冻蔬菜产品的"卫生标准操作规范"。

实训评价:由速冻蔬菜生产企业质量负责人或主讲教师进行评价。

思考题

1. 简述食品加工中安全危害的来源,并分别举例。
2. 简述良好操作规范的主要内容。
3. 简述卫生标准操作程序的基本内容和要求。
4. 防止交叉污染的主要措施有哪些?
5. 卫生标准操作规范有什么作用和意义?
6. 卫生标准操作规范编写的注意事项有哪些?

拓展学习网站

1. 国家市场监督管理总局(http://samr.saic.gov.cn)
2. 各省市场监督管理局
3. 食品伙伴网(http://www.foodmate.net)

项目 7 危害分析与关键控制点(HACCP)体系的建立与认证

项目概述

通过学习,掌握建立危害分析与关键控制点体系的法律依据和方法,以及认证准备和申报技能,并有效运行。

任务 1 危害分析与关键控制点体系

7.1.1 基础知识

1. 认证基本知识

1)认证发展历史

20 世纪初,在工业化国家率先开展了一种由不受产销双方经济利益所支配的第三方用科学公正的方法对上市商品进行评价、监督,以正确指导产品生产和公众购买,保证消费者基本利益的方式。这种第三方的评价行为逐渐演化形成了认证制度。认证制度是市场和社会需求催生的产物,它独立于供需双方,不受供需双方的经济利益支配,因此可保证认证结果的客观性和公正性。

我国的认证认可制度从 1981 年建立了第一个产品认证机构——中国电子元器件认证委员会开始,到 2003 年《中华人民共和国认证认可条例》颁布,我国的认证认可工作已经进入国家统一管理,全面规范化、法治化

Ⅰ. **Basic Knowledge of Certification**

A. The Development History of Certification

At the beginning of the 20th century, the third party, which was not dominated by the economic interests of both producers and sellers, took the lead in the industrialized countries to evaluate and supervise the listed commodities in a scientific and impartial way, in order to correctly guide the production of products and the purchase of public goods and ensure the basic interests of consumers. This kind of third party's appraisal behavior gradually evolves to form the certification system. Certification system is the product of the market and social demand, it is independent of the supply and demand, not subject to the economic interests of the supply and demand, so it can ensure the objectivity and impartiality of the certification results.

The certification and accreditation system established the first product certification body in 1981. From the beginning of China Electronic Components Certification Commission to the promulgation of the *Regulations of the People's Republic of China on Certification and Accreditation*

阶段。

2)认证相关定义

(1)合格评定:对产品、过程、体系、人员或机构符合规定要求的验证。

注1 合格评定的专业领域包括检验、检查和认证,以及对合格评定机构的认可。

(2)合格评定机构:从事合格评定服务的机构。

(3)认证:与产品、过程、体系或人员有关的第三方证明。

注1 管理体系认证有时也被称为注册。

注2 认证适用于除合格评定机构自身以外的所有合格评定对象,认可适用于合格评定机构。

(4)认可:正式表明合格评定机构具备实施特定合格评定工作的能力的第三方证明。

(5)审核:获取记录、事实陈述或其他相关信息并对其进行客观评定,以确定规定要求的满足程度的系统的、独立的和形成文件的过程。

in 2003, the certification and accreditation work in China has entered the stage of national unified management, comprehensive standardization and rule of law.

B. Definition of Certification

1. Conformity assessment: verification of compliance with specified requirements relating to products, processes, systems, personnel or institutions.

NOTE 1 Professional areas of conformity assessment include test, inspection and certification, as well as recognition of conformity assessment bodies.

2. Conformity assessment institutions: institutions engaged in conformity assessment services.

3. Certification: third party certification relating to a product, process, system, or person.

NOTE 1 Management system certification is sometimes referred to as registration;

NOTE 2 The certification is applicable to all the conformity assessment objects outside the conformity assessment organization itself, and the certification applies to the conformity assessment organization.

4. Accreditation: third party certification that formally indicates that the conformity assessment body has the ability to perform a specific conformity assessment.

5. Audit: the systematic, independent, and documented process of obtaining and objectively evaluating records, statements of facts, or other relevant information to determine the level of satisfaction required.

2. 认证认可监管机构

图7-1为认证认可监管机构。我国建立了以认证认可监督管理委员会(CNCA)为主管部门的包括产品认证、体系认证、实验室认可、认证认可人员注册、认证认可咨询过程在内的全面监管体制。其中,危害分析与关键控制点体系认证属于体系认证的一部分。

项目 7 危害分析与关键控制点(HACCP)体系的建立与认证

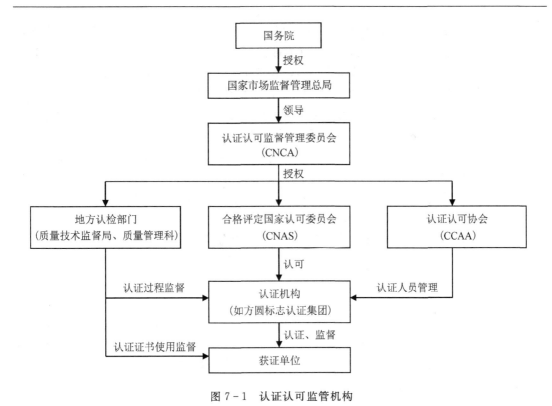

图 7-1 认证认可监管机构

3. 《危害分析与关键控制点(HACCP)体系认证实施规则》简介和制定目的

《危害分析与关键控制点(HACCP)体系认证实施规则》是中国认监委于 2011 年 12 月 31 日发布,2012 年 5 月 1 日实施的,是用于规范危害分析与关键控制点体系认证的纲领性规则,其编号为 CNCA-N-008:2011。

(1)为规范食品行业危害分析与关键控制点体系认证工作,根据《中华人民共和国食品安全法》《中华人民共和国认证认可条例》等有关规定,制定本规则。

该规则是依据《食品安全法》和《认证认可条例》制定的。其中,《认证认可条例》是我国认证行业最高层次法规,是认证行业进入法律层次的标志。我国在加入世界贸易组织时已经承诺,对重要的进口产品质量安全许可制度和我国产品安全认证制度将实行"四个统一"(即统一产品目录,统一技术规范的强制性要求、标准和合格评定程序,统一标志,统一收费标准),要使我国的认证认可工作符合世界贸易组织规则。为了履行我国政府加入世界贸易组织的承诺,2003 年 8 月 20 日国务院第 18 次常务会议审议通过《中华人民共和国认证认可条例》,并于 2003 年 11 月 1 日正式实施。它的颁布为整顿和规范认证认可市场秩序,适应社会生产力发展需要,提高我国产品、服务质量和管理水平提供了有力的法律保障。

(2)本规则规定了从事危害分析与关键控制点体系认证的认证机构(以下简称认证机构)实施危害分析与关键控制点体系认证的程序与管理的基本要求,是认证机构从事危害分析与关键控制点体系认证活动的基本依据。

(3)本规则适用于无专项危害分析与关键控制点体系认证实施规则的危害分析与关键控制点体系认证。有专项规则的危害分析与关键控制点体系认证应按照相应认证实施规则实施。

所谓"专项"是指某些食品行业为了更严格地规范危害分析与关键控制点,设有专项审核标准。例如乳制品企业认证危害分析与关键控制点体系,其认证依据为 GB/T 27342—2009《危害分析与关键控制点(HACCP)体系 乳制品生产企业要求》,而不是 GB/T 27341—2009《危害分析与关键控制点(HACCP)体系 食品生产企业通用要求》。其背景是在 2008 年三聚氰胺事件之后,为了对乳制品企业进行更严格的监管。也就是说,乳制品企业建立危害分析与关键控制点体系需认证专项的乳制品危害分析与关键控制点体系。

(4)在中华人民共和国境内从事危害分析与关键控制点体系认证的认证机构和认证人员应遵守本规则的规定,遵守本规则的规定,并不意味着可免除其所承担的法律责任。

只要在我国境内从事危害分析与关键控制点认证,都必须遵循本规则。这里主要指外商独资或合资的认证机构,其在我国境内从事危害分析与关键控制点认证也必须遵守本规则。另一方面,企业遵守本规则并不意味着可以免除其他法律责任。例如,企业按本规则要求获得危害分析与关键控制点体系认证证书后,还必须遵守《食品安全法》第二十八条之规定,禁止生产相关的十一条食品。

4. 认证依据与认证范围

认证依据是指管理体系认证或产品认证时所遵守的标准。例如,管理体系认证包括质量管理体系、危害分析与关键控制点体系等,产品认证包括绿色食品、有机食品等。质量管理体系认证的认证依据是 GB/T 19001 和 ISO9001,危害分析与关键控制点体系认证的认证依据是 GB/T 27341—2009 和 GB 14881—2013。

认证范围是管理体系(例如危害分析与关键控制点体系)所覆盖的产品、过程、活动、场所的概述。例如,某酒类生产企业有两个生产车间,一个生产葡萄酒,一个生产白酒(浓香型)。白酒生产按照 GB/T 27341—2009 建立危害分析与关键控制点体系,其过程包括白酒生产、销售等。那么它的危害分析与关键控制点体系认证范围为"白酒(浓香型)的生产和销售",而不包括葡萄酒的生产销售,在审核时也只看有关白酒的部分。一般认证范围可参考企业生产许可证上的"产品名称"。同样,危害分析与关键控制点体系认证也有其规定的认证范围,也就是什么样的企业可以认证危害分析与关键控制点体系。

认监委发布的危害分析与关键控制点体系认证依据是《危害分析与关键控制点(HACCP)体系 食品生产企业通用要求》(GB/T 27341—2009)和《食品企业通用卫生规范》(GB 14881—2013)。认证机构可在上述认证依据基础上,增加符合《认证技术规范管理办法》规定的技术规范作为认证审核补充依据。认监委发布的危害分析与关键控制点体系认证范围如表 7-1 所示。

也就是说,上述食品生产企业可以认证危害分析与关键控制点体系,获得认证证书。

7.1.2 实训任务

实训组织:到认监委网站查询各部门分工及组织机构图。
实训成果:综述类文章,包括网络截图或组织机构描述。
实训评价:5 分制,组织机构图 2 分,各部门分工描述 3 分。

表 7-1 认监委发布的危害分析与关键控制点体系认证范围

代码	行业类别	种类示例
C	加工1(易腐烂的动物产品),包括农业生产后的各种加工,如屠宰	C1 畜禽屠宰及肉制品加工 C2 蛋及蛋制品加工 C4 水产品的加工 C5 蜂产品的加工 C6 速冻食品制造
D	加工2(易腐烂的植物产品)	D1 果蔬类产品加工 D2 豆制品加工 D3 凉粉加工
E	加工3(常温下保存期长的产品)	E1 谷物加工 E2 坚果加工 E3 罐头加工 E4 饮用水、饮料的制造 E5 酒精、酒的制造 E6 焙烤类食品的制造 E7 糖果类食品的制造 E8 食用油脂的制造 E9 方便食品(含休闲食品)的加工 E10 制糖 E11 盐加工 E12 制茶 E13 调味品、发酵制品的制造 E14 营养、保健品制造
G	餐饮业	G1 餐饮及服务

任务2 危害分析与关键控制点体系认证对食品企业的价值

7.2.1 基础知识

1. 危害分析与关键控制点体系的特点

以危害分析与关键控制点理论为基础的危害分析与关键控制点体系作为一种科学、简便和实用的预防性食品安全管理手段,被国际权威机构认可为控制由食品引起的疾病的最有效方法,一些国家和国际组织已制定或

Ⅰ. The Characteristics of HACCP System

As a scientific, simple and practical preventive food safety management method, the HACCP system based on HACCP theory has been recognized by international authorities as the most effective method to control food borne diseases. A number of countries and international organizations

正在着手制定以危害分析与关键控制点为基础的相关技术法规和标准,作为对食品的强制性管理措施或实施指南。该体系强调组织本身的作用,而不是依靠对最终产品的检测或政府部门的取样分析来确定产品的质量。危害分析与关键控制点是评估危害并建立控制体系的手段,其重点在于预防。与一般传统的监督方法相比较,它具有较高的经济效益和社会效益。

危害分析与关键控制点从生产角度来说是安全控制系统,是使产品从投料开始至成品保证质量安全的体系。如果使用了危害分析与关键控制点的管理系统最突出的特点是:

(1)使食品生产对最终产品的检验(即检验是否有不合格产品)转化为控制生产环节中潜在的危害(即预防不合格产品);

(2)应用最少的资源,做最有效的事情。

危害分析与关键控制点是决定产品安全性的基础,食品生产者利用危害分析与关键控制点控制产品的安全性比利用传统的最终产品检验法要可靠,实施时也可作为谨慎防御的一部分。

have developed or are in the process of developing relevant technical regulations and standards based on HACCP as mandatory management measures or implementation guidelines for food. The system emphasizes the role of the organization itself rather than relying on the testing of the final product or the sampling analysis of the government to determine the quality of the product. HACCP is a means of assessing harm and establishing a control system, which focuses on prevention. Compared with the traditional methods of supervision, it has higher economic and social benefits.

HACCP is a safety control system from the point of view of production. It is a system to ensure the quality and safety of a product from the start of feeding to the finished product. The most prominent feature of the management system using HACCP is:

1. Turn the inspection of the final product (that is, whether there is an unqualified product) in food production into controlling the potential harm in the production link (that is to prevent the unqualified product);

2. Use the least amount of resources to do the most effective things.

HACCP is the basis of determining product safety. Food producers use HACCP to control the safety of products more accurately than traditional products. Final product testing should be reliable and implemented as part of a cautious defense.

2. 食品企业实施危害分析与关键控制点体系认证的意义

(1)适应我国加入世界贸易组织后形势的需要。实施危害分析与关键控制点体系认证,可使企业的管理体系与国际接轨,当市场把认证作为准入要求时,增加出口和进入市场的机会。

(2)有利于卫生注册。企业建立和实施危害分析与关键控制点体系并获得认证,也表明良好的操作规范已通过认证,有利于一般食品企业(餐饮类)卫生许可证的取得;对于出口企业,则表明已获得出口食品生产企业卫生注册的基本条件,才可申请出口卫生注册登记。

(3)提高企业形象。危害分析与关键控制点体系是目前国际上公认的最有效的食品安全管理体系,企业通过寻求并获取危害分析与关键控制点体系认证,可以向外界表明已具备可靠生产安全卫生食品的能力,进而取得更大的竞争优势,增强客户对产品的信心,扩大消费者满意度。

(4)降低投资风险。当今食品生产已日趋规模化,只有将食品危害控制在最安全的范围内,投资风险才能降低。危害分析与关键控制点的预防机制使因食品问题的投诉和索赔受到控制,避免发生重大危害事件造成的损失。危害分析与关键控制点体系认证能作为公司敬业的依据,降低负债倾向。

(5)节约管理成本。危害分析与关键控制点体系是预防性的食品安全控制体系,重在预防危害发生,从而可减少企业和监督机构人力、物力和财力的支出。而且危害分析与关键控制点体系认证能通过定期审核来维持体系运行,防止系统崩溃。

7.2.2 实训任务

实训组织:论述我国危害分析与关键控制点体系的现状和发展趋势。
实训成果:综述类文章,总字数 1000 字以上。
实训评价:5 分制,现状 2 分,发展趋势 3 分。要求内容全面,条理清楚。

任务3 危害分析与关键控制点体系认证标准

7.3.1 基础知识

危害分析与关键控制点体系是一种质量保证体系、一种预防性策略,它是简便、易行、合理、有效的食品安全保证系统,其为政府机构实行食品安全管理提供了实际内容和程序。危害分析与关键控制点体系是确定、评估和控制重要的食品安全危害的一个系统。

1. 危害分析与关键控制点体系的基本术语

食品法典委员会(CAC)在《危害分析与关键控制点体系及其应用准则》中规定的危害分析与关键控制点基本术语有以下几种。

(1)控制(control,动词)指采取一切必要行动,以保证和保持符合危害分析与关键控制点计划所制定的指标。

(2)控制(control,名词)指遵循正确的方法和达到安全指标时的状态。

HACCP is a quality assurance system, it is a preventive strategy, and is a simple, easy, reasonable and effective food safety assurance system, which provides the actual content and procedure for government agencies to implement food safety management. HACCP is a system to identify, evaluate and control important food safety hazards.

Ⅰ. **Basic Terminology of the HACCP System**

The basic terms of HACCP as defined by the Codex Alimentarius Commission in the *Guidelines for the HACCP System and Their Application* are:

1. Control (verb) refers to taking all necessary actions to ensure and maintain compliance with the targets set out in the HACCP plan.

2. Control (noun) refers to the state of being in compliance with the correct method and when a safety indicator is met.

(3)控制措施(control measure)指用以防止或消除食品安全危害或将其降到可接受的水平所采取的任何行动和活动。

(4)纠偏行动(corrective action)指监测结果表明关键控制点(CCP)失控时,在关键控制点上所采取的措施。

(5)关键控制点(critical control point)指可进行控制,并能有效防止或消除食品安全危害,或将其降低到可接受水平的必需的步骤。

(6)关键限值(critical limit)指区分可接受与不可接受水平的指标。

(7)偏离(deviation)指不符合关键限值。

(8)流程图(flow diagram)指生产或制造特定食品所用操作顺序的系统表达。

(9)危害分析与关键控制点(hazard analysis critical control point)指对食品安全显著危害加以识别、评估以及控制的体系。

(10)危害分析与关键控制点计划(hazard analysis critical control point plan)指危害分析与关键控制点原理所制定的用以确保所考虑食品链的各环节中对食品有显著意义的危害予以控制的文件。

(11)危害(hazard)指食品中产生的潜在的对人体健康有危害的生物、化学或物理因子或状态。

(12)危害分析(hazard analysis)指收集信息和评估危害及导致其存在的条件的过程,以便决定哪些对食品安全有显著意义,从而应被列入危害分析与关键控制点计划中。

(13)监控(monitor)指为了评估关键控制点是否处于控制之中,对被控制参数所做的有计划的、连续的观

3. Control Measure refers to any action or activity taken to prevent, eliminate or reduce food safety hazards to acceptable levels.

4. Corrective Action refers to the measures taken on CCP by monitoring results indicating that CCP is out of pressing.

5. Critical Control Point refers to the necessary steps to control, effectively prevent or eliminate food safety hazards, or reduce them to acceptable levels.

6. Critical Limit refers to an indicator that distinguishes between acceptable and unacceptable levels.

7. Deviation refers to a failure to meet the critical limit.

8. Flow Diagram is a systematic representation of the sequence of operations used to produce or manufacture a particular food.

9. The Hazard Analysis Critical Control Point is a system for identifying, evaluating, and controlling significant food safety hazards.

10. The HACCP program (Hazard Analysis Critical Control Point Plan) refers to a document developed by the HACC principle to ensure that significant food hazards are controlled in all parts of the food chain under consideration.

11. Hazard refers to biological, chemical, or physical factors or states that are potentially harmful to human health in foods.

12. Hazard Analysis refers to the process of gathering information and assessing the conditions that cause the hazard and its existence in order to determine what is significant for food safety and should be included in the HACCP plan.

13. Monitor refers to the planned, continuous observation or measurement of the controlled parameters to assess whether the CCP is under

察或测量活动。

（14）步骤（step）指包括原材料及从初级生产到最终消费的食品链中的某个点、程序、操作或阶段。

（15）确认（validation）指获得证据，证明危害分析与关键控制点计划的各要素是有效的过程。

（16）验证（verification）指除监控外，用以确定是否符合危害分析与关键控制点计划所采用的方法、程序、测试和其他评估方法的应用。

2. 危害分析与关键控制点的基本原理

1999年，食品法典委员会（CAC）在《食品卫生总则》附录《危害分析和关键控制点（HACCP）体系应用准则》中，确定了危害分析与关键控制点的七个原理。

1）原理一：进行危害分析

危害分析是危害分析与关键控制点体系七个原理的基础，是危害分析与关键控制点体系的核心之一。所谓危害分析是通过以往资料分析、现场实地观测、实验采样检测等方法，对食品生产全过程各个环节中可能发生的危害及危害的严重性进行科学、客观、全面的分析和评估，以判断危害的性质、程度和对人体健康的潜在影响，从而确定哪些危害对食品安全是重要的，应被列入危害分析与关键控制点计划中并制定相应的预防控制措施。主要包括以下三个部分的内容。

（1）危害识别。危害指食品中可能影响人体健康的生物性、化学性和物理性因素或状态，尤以生物性危害（特别是微生物危害）最为严重，也最易发生，具体如下。

control.

14. Step refers to a point, process, operation, or stage in the food chain that includes raw materials and from primary production to final consumption.

15. Validation refers to the process of obtaining evidence that the elements of a HACCP plan are valid.

16. Verification refers to the application of methods, procedures, tests and other evaluation methods used to determine compliance with the HACCP plan in addition to monitoring.

II. The Basic Principles of HACCP

In 1999, the Codex Alimentarius Commission established the 7 principles of HACCP in the appendix of *Hazard Analysis Critical Control Points (HACCP) System Application Guidelines* in the *General Provisions of Food Safety*.

A. Principle 1: Hazard Analysis

Hazard analysis is the basis of seven principles of HACCP system and one of the core of HACCP system. The so-called hazard analysis is a scientific, objective and comprehensive analysis and assessment of the possible hazards and the severity of hazards in the whole process of food production by means of previous data analysis, field observation, experimental sampling detection and other methods, in order to judge the nature and extent of the hazards and the potential impact on human health, so as to determine which hazards are important to food safety, it should be included in the HACCP plan and formulate the corresponding preventive and control measures. Mainly includes the following three parts.

1. Hazard identification. Hazard refers to biological, chemical and physical factors or states in food that may affect human health, especially biological hazards (especially microbial hazards), which are most serious and prone to occurrence, the details are as follows.

①生物危害。生物危害包括病原性微生物、病毒和寄生虫。

病原性微生物一般会导致食源性疾病的发生,且发病率较高。病原性微生物对人体健康造成的伤害包括食源性感染和食源性中毒。食源性感染会造成腹泻、呕吐等症状;食源性中毒,即食物中毒,对人体造成的危害更加严重。病原性微生物主要的来源是,在适宜的环境如营养成分、pH 值、温度、水活度、气体(氧气)等条件下,微生物会快速繁殖,从而引起食物腐败变质。

病毒比细菌更小,食品携带上病毒后,可以通过感染人体细胞从而引起疾病。病毒污染食品的途径一般如下:一是动植物原料环境感染了病毒,如上海甲肝病流行就是人们食用的毛蚶生长水域感染了甲肝病毒;二是原料动物携带病毒,如牛患狂犬病或口蹄疫;三是食品加工人员带有病毒,如乙肝患者。

寄生虫通常寄生在宿主体表或体内,通过食用携带寄生虫的食品而感染人体,可能出现淋巴结肿大、脑膜炎、心肌炎、肝炎、肺炎等症状。比如人们比较熟悉的猪囊虫病,就是人们食用了未煮熟的囊虫病猪肉而被感染。寄生虫污染食品的途径有以下几种:一是原料动物患有寄生虫病;二是食品原料遭到寄生虫卵的污染;三是粪便污染,食品生熟不分。

a. Biological hazards. Biological hazards include pathogenic microorganisms, viruses and parasites.

Pathogenic microorganisms generally lead to food-borne diseases, and the incidence is high. The harm of pathogenic microorganism to human health includes food-borne infection and food-borne poisoning. Food-borne infections can cause diarrhea, vomiting and other symptoms; food-borne poisoning, that is food poisoning, causing more serious harm to the human body. The main source of pathogenic microorganisms is that under suitable conditions such as nutrition, pH value, temperature, water activity, gas (oxygen) and other conditions, microorganisms will multiply rapidly, resulting in food spoilage.

Viruses are smaller than bacteria, and food can cause disease by infecting human cells with the virus. Virus contamination of food is generally as follows: first, animal and plant raw materials environment infected with the virus, such as the prevalence of hepatitis A in Shanghai is the infection of hepatitis A virus in the growing waters of the edible clam; second, raw animals carry viruses, such as rabies or foot-and-mouth disease in cattle; third, food processing personnel with the virus, such as hepatitis B patients.

Parasites usually parasite in the host body surface or body, through eating food carrying parasites and infected human body, it may appear lymphadenopathy, meningitis, myocarditis, hepatitis, pneumonia and other symptoms. For example, people are more familiar with cysticercosis, that is, people eat undercooked pork infected with cysticercosis. Parasites contaminate food in the following ways: first, raw animals suffer from parasitic diseases; second, food raw materials are contaminated by parasite eggs; third, fecal contamination, food raw and cooked are not distinguished.

②化学危害。化学危害一般可分为天然的化学危害、添加的化学危害和外来的化学危害。

天然的化学危害来自化学物质，这些化学物质在动物、植物自然生长过程中产生，如人们常说的毒蘑菇，某些生长在谷物上的霉菌可以生成毒素（比如黄曲霉毒素可以致癌），河鲀中含有的毒素，某些贝类因食用一些微生物和浮游植物而产生贝毒素。

添加的化学危害是人们在食品加工、包装、运输过程中加入的食品色素、防腐剂、发色剂、漂白剂等，如果超过安全水平使用就会产生危害。

外来的化学危害主要来源于以下几种途径：一是农用化学药品，如杀虫剂、除草剂、化肥等的使用；二是兽用药品，如兽医治疗用药、饲料添加用药在动物体内的残留；三是工业污染如铅、砷、汞等化学物质进入动植物及水产品体内，食品加工企业使用的润滑剂、清洁剂、灭鼠药等化学物质污染食品。化学危害对人体可能造成急性中毒、慢性中毒，影响人体发育，致畸、致癌，甚至致死等后果。

③物理危害。物理危害是指在食品中发现的不正常有害异物，当人们误食后可能造成身体外伤、窒息或其他健康问题。比如食品中常见的金属、玻璃、碎骨等异物对人体的伤害。物理危害主要来源于以下几种途径：植物收获过程中掺进玻璃、铁丝铁钉、石头等；水产品捕捞过程中掺杂鱼钩、铅块等；食品加工设备上脱落的金属

b. Chemical hazards. Chemical hazards can be divided into natural chemical hazards, chemical hazards added and foreign chemical hazards.

The natural chemical hazards come from chemicals, which are produced in the natural production of animals and plants, such as the death cup, the certain mildew that grow on grains can produce toxins (such as aflatoxin that can cause cancer), toxins found in puffer fish, and shellfish that produce shellfish toxins from eating microorganisms and phytoplankton.

The chemical hazards added are food pigments, preservatives, colorants, bleach and so on, which are added in food processing, packaging and transportation. If they exceed the safety level, they will become harmful.

The foreign chemical hazards mainly come from the following ways: first, the use of agrochemicals, such as pesticides, herbicides, fertilizers and so on; second, veterinary drugs, such as veterinary treatment drugs, feed additive drug residues in the animal body; third, industrial pollution, such as lead, arsenic, mercury and other chemicals into the body of animals and plants and aquatic products, food processing enterprises used lubricants, detergents, rodenticide and other chemicals to contaminate food. Chemical hazards may cause acute poisoning, chronic poisoning, affect human development, teratogenesis, carcinogenesis and even death and other consequences.

c. Physical hazards. Physical hazard refers to abnormal harmful foreign matter found in food, which may cause physical injury, asphyxia or other health problems after people eat it by mistake. For example, metal, glass, broken bone and other foreign bodies in food are harmful to human body. The physical harm mainly comes from the following ways: mixing glass, iron wire nail, stone and so on in the process of harvesting plants;

碎片,灯具及玻璃容器破碎造成的玻璃碎片等;畜禽在饲养过程中误食铁丝,畜禽肉和鱼剔骨时遗留骨头碎片或鱼刺。

(2)危害评估。通过危害评估可以判断已识别的危害是否为显著危害。作为显著危害有两个必要条件——可能性和严重性,缺少一项就不能成为显著危害。所谓显著危害是指极有可能发生,如不加以控制就有可能导致消费者不可接受的健康或安全风险的危害。危害分析与关键控制点体系中的危害分析主要针对显著危害。

(3)建立预防措施。危害分析完成后,还要制定出所有危害尤其是显著危害的控制措施和方法,以消除或减少危害发生,确保食品质量与卫生安全。对于生物性危害中的微生物危害,原辅料、半成品可采用无害化生产;加工过程可采用调 pH 值与控制水分活度,并辅以其他方法进行处理。昆虫、寄生虫等可采用加热、冷冻、辐射等处理方法。对于化学性危害,应严格控制产品原辅料的卫生,防止重金属污染和农药残留,不添加人工合成色素和有害添加剂,防止储藏过程中有毒化学成分的产生。对于物理性危害,可采用原料严格检测、提供质量保证证书、避光、去杂、加抗氧化剂、用金属检测器(如磁铁等)检查金属碎片等方法,并用卫生标准操作程序控制一般危害。

危害分析表可用来确定食品安全

mixing fish hooks and lead blocks in the fishing process of aquatic products, and so on; shedding metal fragments on food processing equipment, lamps and the glass fragments caused by the breakage of a glass container; bone fragments or spines left over from the evisceration of livestock and poultry meat and fish in the course of feeding by mistake.

2. Hazard assessment. Through hazard assessment, it can be judged whether the identified hazard is a significant hazard. As a significant hazard, there are two necessary conditions-possibility and severity, and the absence of one cannot be considered as a significant hazard. The so-called significant hazards are those that are highly likely to occur and that, if left unchecked, may lead to unacceptable health or safety risks for consumers. Hazard analysis in HACCP system is mainly aimed at significant hazards.

3. Establishment of preventive measures. After hazard analysis, the control measures and methods of all hazards, especially the significant hazards, should be worked out to eliminate or reduce the occurrence of hazards and ensure food quality and health safety. For the biological hazards of microbiological hazards, raw materials, semi-finished products can be produced innocuously, pH value and water activity can be adjusted, and other methods are used to deal with it. Insects, parasites and so on can be heated, frozen, radiation and other treatment. For chemical hazards, we should strictly control the hygiene of raw and auxiliary materials, prevent heavy metal pollution and pesticide residues, do not add synthetic pigments and harmful additives, and prevent the production of toxic chemical components in the storage process. For physical hazards, raw materials can be strictly inspected, provide the quality assurance certificates, avoid light, remove impurities, add antioxidants, check metal fragments with metal detectors (such as magnets, etc.), together with control general harm with SSOP.

Hazard analysis table can be used to determine

危害。加工流程图的每一步被列在第(1)栏中,危害分析的结果被记录在第(2)栏中,显著危害的判定结果记录在第(3)栏中,在第(4)栏中对第(3)栏的判断提出了依据。表7-2是危害分析表的一种格式。

food safety hazards. Each step of the process flow chart is listed in column(1). The results of the hazard analysis are recorded in column(2). The results of the determination of significant hazards are recorded in column(3), and the judgment in column(3) is provided in column(4). Table 7-2 is a format for hazard analysis tables.

表7-2 危害分析表

配料、加工步骤	确定本步骤中引入的、受控制的或增加的潜在危害	潜在的食品安全危害是显著的吗（是/否）	对第(3)栏的判断提出依据	能用于显著危害的预防措施是什么	该步骤是关键控制点吗（是/否）
(1)	(2)	(3)	(4)	(5)	(6)
1					
2					
3					
4					
5					
6					

公司名称：　　　　　　　　　　　产品名称：
公司地址：
贮藏和销售方式：　　　　　　　　预期用途和客户：
签名：
日期：

当危害分析证明没有发生食品安全危害的可能时,可以没有危害分析与关键控制点计划,但危害分析表必须予以记录和保存,它是危害分析与关键控制点计划验证和审核(内审和外审)的依据。

2)原理二：确定关键控制点(CCP)

关键控制点是指对食品加工过程中的某一步骤或工序进行控制后,就可以防止食品安全危害或使其减少到可接受水平。这里所说的食品安全危害是显著危害,需要通过危害分析与关键控制点来控制,也就是每个显著危害都必须通过一个或多个关键控制

When the hazard analysis proves that there is no food safety hazard, there can be no HACCP plan, but the hazard analysis table must be recorded and kept. It is the basis of HACCP plan verification and audit (internal audit and external audit).

B. Principle 2：Determination of Critical Control Points (CCP)

CCP (critical control points) refer to a certain point, steps or processes in the food processing process can be controlled to prevent, eliminate food safety hazards or reduce them to an acceptable level. The food safety hazards mentioned above are significant hazards, which need HACCP control, that is, each significant hazard must be controlled

点来控制。另外,一个关键控制点可以控制多个危害,如加热可以消灭致病性细菌及寄生虫,冷冻、冷藏可以防止致病性微生物生长和组胺的生成。反过来,有些危害则需多个关键控制点来控制,如鲭鱼罐头通过原料收购、缓化、切台三个关键控制点来控制组胺的形成。

确定关键控制点的原则:如果分析的显著危害在某一步骤可以被控制、预防、消除或降低到可接受水平,那么这一步骤就是关键控制点。包括三种情况。

(1)当危害能被预防时,该点可以被认为是关键控制点。如通过控制原料接受来预防病原体或药物残留,包括供应商的证明;改变食品中的 pH 值到 4.6 以下,可使致病菌不能生长;添加防腐剂、冷藏或冷冻能防止细菌生长;改进食品的原料配方,防止不当或过量食品添加剂危害的发生。

(2)能将危害消除的点可以被认为是关键控制点。如金属碎片能通过金属探测器检出;加热能杀死所有的致病性细菌;冷冻到 -38 ℃以下可以杀死寄生虫。

(3)能将危害降低到可接受水平的点可以被认为是关键控制点。如通过过滤装置或自动收集使外来杂质减少到可接受水平;灯检或肉眼挑拣可使明显可见的杂质减少到可接受水平。

关键控制点的准确和完整的识别是控制食品安全危害的基础。在进行危害分析过程中产生的资料,对于危

by one or more CCP. In addition, a CCP may be able to control multiple hazards, such as heating to kill pathogenic bacteria and parasites, or freezing, which prevents the growth of pathogenic microorganisms and histamine production. In turn, some hazards need more than one CCP to control, such as canned mackerel, in raw materials acquisition, retarding and cutting table, three CCP to control the formation of histamine.

The principle of identifying critical control points: this step is the critical control point if the significant hazard of analysis can be controlled, prevented, eliminated or reduced to acceptable levels at this step. This includes three situations.

1. The point can be considered a critical control point when the hazard can be prevented, such as preventing pathogen or drug residues by controlling raw material acceptance, such as supplier certification, changing pH value in food to below 4.6, and preventing pathogenic bacteria from growing; adding preservatives, refrigeration or freezing can prevent the growth of bacteria; improving the formula of food raw materials to prevent improper or excessive food additives harm.

2. The hazard elimination point can be considered as the critical control point. For example, metal fragments can be detected by metal detectors; heating can kill all pathogenic bacteria; freezing to below -38 ℃ can kill parasites.

3. Points that reduced hazards to acceptable levels can be considered as the critical control points. For example, external impurities can be reduced to acceptable levels by filtering devices or automatic collection, and visible impurities can be reduced to acceptable levels by light detection or naked eye sorting.

The accurate and complete identification of CCP is the basis of controlling food safety hazards. The information generated in the process of hazard analysis is

项目 7 危害分析与关键控制点(HACCP)体系的建立与认证

害分析与关键控制点小组识别加工工序中的关键控制点是非常重要的。由于工厂的布局、设施设备、原辅材料的选择、加工过程的不同,生产同样食品的不同工厂可能在各类危害和关键控制点的确定上各不相同。国际上一般推荐采用"关键控制点判断树"的逻辑推理法来确定关键控制点。该法通过回答一系列逻辑连贯的问题来完成对关键控制点的判定。在危害分析的基础上应用判断树原则确定关键控制点,一个危害分析与关键控制点体系的关键控制点数量一般应控制在 6 个以内。一个危害可由一个或多个关键控制点控制到可接受水平;同样,一个关键控制点可以控制一个或多个危害。判断树如图 7-2 所示。

very important for HACCP teams to identify the CCP in the processing process. Due to the different layout of the plant, facilities and equipment, the choice of raw and auxiliary materials, processing process, different factories producing the same food may have different kinds of hazards and CCP determination. It is generally recommended to use the logical reasoning method of "CCP judgment tree" to determine CCP in the world. This method completes CCP judgment by answering a series of logically coherent questions. On the basis of hazard analysis, the number of critical control points of a HACCP system should be limited to less than 6 by using the principle of judgment tree. A hazard can be controlled to an acceptable level by one or more critical control points; likewise, a critical control point can control one or more hazards. The judgment tree as shown in Figure 7-2.

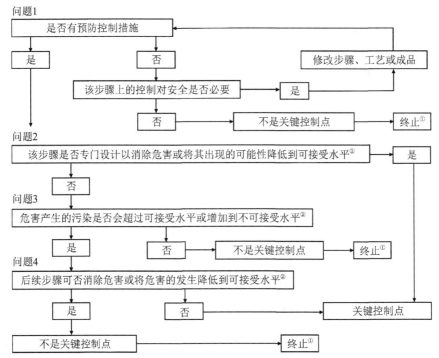

① 按描述的过程进行至下一个危害。
② 在识别危害分析与关键控制点计划中的关键控制点时,需要在总体目标范围内对可接受水平和不可接受水平作出规定。

图 7-2 确定关键控制点的判断树

判断树由四个连续问题组成。

问题1:在加工过程中存在并确定的显著危害,是否在这一步或后面的工序中有预防措施? 如果有,则回答问题2。如果无,则回答是否有必要在这一步控制食品安全危害。如果回答"否",则不是关键控制点。如果回答"是",则说明加工工艺、原料或原因不能控制以保证必要的食品安全,应重新改进产品设计,包括预防措施。另外,只有显著危害而没有预防措施,则不是关键控制点,需改进。

问题2:这一加工步骤是否能消除可能发生的显著危害或降低其到一定水平(可接受水平)? 如果回答"是",还应考虑这步是否最佳,如果是,则是关键控制点。如果答案是"否",则回答问题3。

问题3:已确定的危害是否能影响对产品可接受水平的判定,或者这些危害会增加到使产品不可接受? 如果回答"否",则不是关键控制点。主要考虑污染或危险干预,即是否存在、是否要发生或是否会增加? 如果是,回答问题4。

问题4:下边的工序是否能消除已确定的危害或使其减少到可接受的水平? 如回答"否",则这一步是关键控制点。如回答"是",则这一步不是关键控制点,而下一道工序才是关键控制点。

3) 原理三:确定关键限值(CL)

关键限值是指为确保各关键控制点处于控制之下以防止显著危害发生的预防性措施,必须达到能将可接受

The judgment tree consists of four continuous problems.

Question 1: Are there preventive measures in this step or in subsequent processes for identified significant hazards in the process? If the answer is "YES", answer question 2. If the answer is "NO", then answer whether it is necessary to control food safety hazards at this step. If the answer is "NO", it is not CCP. If "YES" is answered, indicate that the processing process, raw materials or reasons are not controlled to ensure the necessary food safety, and that the design of the product, including preventive measures, should be improved. In addition, there are only significant hazards, and no prevention measures. It is not CCP, and it needs to be improved.

Question 2: Can this process eliminate significant hazards that may occur or reduce them to a certain level (acceptable level)? If the answer is "YES", you should also consider whether this step is the best, and if so, it is CCP. If you answer "NO", answer question 3.

Question 3: Are the hazards identified to affect the determination of acceptable levels of the product, or will these hazards increase to the unacceptable level of the product? If "NO" is answer, it is not CCP. The main consideration is pollution or intervention of hazards, that is, does it exist or is it going to happen or is it going to increase? If "YES", answer question 4.

Question 4: Are subsequent processes capable of eliminating identified hazards or reducing them to acceptable levels? If answer "NO", this step is CCP. If answer "YES", this step is not CCP, and the next process is CCP.

C. Principle 3: Determine the Critical Limits (CL)

Critical limit is a preventive measure to ensure that each CCP is under control to prevent the occurrence of significant hazards, it is a judgement

水平与不可接受水平区分开的判断指标、安全目标水平或极限,是确保食品安全的界限。关键限值在某一关键控制点上将物理的、生物的、化学的参数控制到最大或最小水平,从而可防止或消除所确定的食品安全危害发生,或将其降低到可接受水平。

关键限值的选择和设置非常重要。应在大量收集资料并充分考虑被加工产品的内在因素和外部加工工序的基础上,合理地确定关键限值,使其具有科学性和可操作性,并且经过证实。它可以来自强制性标准、指南、文献、实验结果和专家的建议,如科学刊物、学术刊物、食品科学教科书、法规性指南、国家地方指南、美国食品药物管理局指南、标准、专家、学术权威、设备制造商、大学附设机构、实验研究、实验室、试生产等。

一般选择快速、准确、方便且可操作性强的指标作为关键限值。如果关键限值过严,即使没有发生影响到食品安全的危害,就要求去采取纠正措施,造成资源浪费、成本上升;如果过松,又会造成不安全的产品到了顾客手中。实际操作中,常采用直观、可连续监测的物理指标(如时间、温度等)和化学指标(如pH值、水分活度、盐度等)作为关键限值,而尽量少用微生物学指标。好的关键限值应该直观、易监测,通常采用的指标包括对温度、时间、水分含量、湿度、水活度、pH值、余氯浓度等的测量以及感官参数。另外应考虑实际可操作性,如测虾片的

index, safety target level, or limit that must be reached to distinguish acceptable levels from unacceptable levels, and it is the boundaries of ensuring food safety. At a critical control point, critical limits control physical, biological, and chemical parameters to maximum or minimum levels to prevent or eliminate the occurrence of certain food safety hazards, or reduce identified food safety hazards to acceptable levels.

The selection and setting of critical limits is important. On the basis of collecting a large amount of data and taking fully into account the internal factors and external processing procedures of the products to be processed, critical limit is reasonably determined to make it scientific and operable, and has been verified. It can come from mandatory standards, guidelines, literature, experimental results, and expert advice. Such as scientific journals, academic journals, textbooks on food science, regulatory guidelines, national and local guidelines, FDA guidelines, standards, experts, academic authorities, equipment manufacturers, university-affiliated institutions, experimental research, laboratories, trial-produce, etc.

Generally choose fast, accurate, convenient and operable indexes as critical limits. If the selection of critical limits is too strict, even if there is no impact on food safety hazards and the need to take corrective measures, resulting in waste of resources, cost increases; if too loose, it will cause unsafe products into the hands of users. Physical indexes (such as time, temperature, etc.) and chemical indexes (such as pH value, water activity, salinity, etc.), which can be monitored continuously and intuitively, are often used as critical limits in practical operation, but microbiological indexes are seldom used as far as possible. Good critical limits should be intuitive and easily monitored. Commonly used indicators

中心温度操作性不强,而测油的温度、传送带速度以及虾片厚度可达到相应的要求。

确定关键限值的三项原则：
(1)有效。在此限值内,显著危害能被防止、消除或降低至可接受水平。

(2)简单。简便快捷,易于操作,可在生产线不停顿的情况下快速监控。

(3)经济。只需较少的人力、物力、财力的投入。

4)原理四:关键控制点的监控

监控是指对每个关键控制点对应的关键限值的定期测量或观察,以评估一个关键控制点是否受控,并且为将来验证时提供准确的记录。监控需要形成文件的监控程序,其目的是跟踪加工过程,查明和注意可能偏离关键限值的趋势,并及时采取措施进行加工调整,使整个加工过程在关键限值发生偏离前恢复到控制状态。同时,当一个关键控制点发生偏离时,可以很快查明何时失控,以便及时采取纠偏行动。另外,监控记录可以为将来的验证提供必需的资料。通常情况下,每个监控程序必须包括四个要素,即监控什么、怎样监控、何时监控、谁来监控。

(1)监控什么是指通过观察和测量加工过程的特征,来评估一个关键控制点是否在关键限值内进行操作。

include the measurement of temperature, time, moisture content, humidity, water activity, pH value, residual chlorine concentration, and sensory parameters. In addition, practical maneuverability should be taken into account, for example, the maneuverability of the central temperature of the Prawn Crackers is not strong, while the temperature of oil measurement, the speed of conveyor belt and the thickness of Prawn Crackers can meet the corresponding requirements.

Three principles for determining critical limit:

1. Effective. Within this limit, significant hazards can be prevented, eliminated or reduced to acceptable levels.

2. Simple. The utility model is simple, quick and easy to operate, and can quickly monitor and control the production line without pausing.

3. Economy. Only a small amount of human, material and financial input is needed.

D. Principle 4: Monitoring of Critical Control Point

Monitoring refers to the periodic measurement or observation of the critical limit corresponding to each CCP to assess whether a CCP is controlled and to provide accurate records for future validation. Monitoring procedures that need to be documented in order to track the processing process, identify and monitor trends that may deviate from the critical limits, and take timely steps to make processing adjustments. At the same time, when a CCP deviates, it can quickly find out when out of control, in order to take corrective action in time. In addition, monitoring records can provide the necessary information for future validation. Normally, each monitor must include four elements: what to monitor, how to monitor, when to monitor, and who to monitor.

1. What to monitor is meant to evaluate whether a CCP operates within the critical limit by observing and measuring the characteristics of the

监控对象也可以包括检查一个关键控制点的预防措施是否实施。例如,检查原料供应商的许可证;检查原料肉表面或包装上的屠宰场注册证号,以保证其是自己注册的屠宰场。

(2)怎样监控是指对定量的关键限值通过物理或化学的检测方法,对定性的关键限值采用检查的方法来进行监控。由于生产中没有时间等待分析实验结果,而且关键限值的偏离要快速判定,必须在产品销售之前采取适当的纠偏行动。通常物理和化学的测量手段快速、方便,是较理想的监控方法。

(3)监控时间可以是连续的,也可以是间断的,如果有可能的话要尽量采取连续监控。但是一个能连续记录监控值的监控仪器本身并不能控制危害,还需要定期观察连续的监控记录,必要时采取适当的措施,这也是监控的一个组成部分。当出现关键限值偏离时,检查间隔的时间长短将直接影响到返工和产品损失的数量。在所有情况下,检查必须及时进行以确保不正常产品在出厂前被分离出来。当不可能连续监控一个关键控制点时,也可以实施非连续监控(间断性监控),但应尽量缩短监控的时间间隔,以便及时发现可能的偏离。

(4)制订危害分析与关键控制点

process. Monitoring objects can also include checking whether a CCP's preventive measures are implemented. For example, check the license of the raw material supplier; inspect the surface of the raw meat or the registration number of the slaughterhouse on the package to ensure that it is the slaughterhouse of its own registration.

2. How to monitor is to monitor the quantitative critical limit value by physical or chemical method, and to monitor the qualitative critical limit value by means of inspection method. Because there is no time to wait for the results of the analysis experiment for a long time in production, and the deviation of the critical limit value is to be judged quickly, it is necessary to take appropriate corrective action before the product is sold. Usually physical and chemical means of measurement are fast, convenient and ideal monitoring methods.

3. The monitoring time can be continuous or intermittent. If possible, continuous monitoring should be taken as far as possible. However, a monitoring instrument that can record the monitoring value continuously can not control the harm by itself. It also needs to observe the continuous monitoring records regularly and take appropriate measures if necessary, and this is also a part of the monitoring. When critical limit deviates, the length of the check interval will directly affect the amount of rework and product loss. In all cases, inspections must be carried out in a timely manner to ensure that abnormal products are separated before they leave the factory. When it is not possible to continuously monitor a CCP, discontinuous monitoring (intermittent monitoring) may also be implemented, but the time interval for monitoring should be shortened as much as possible in order to detect possible deviations in time.

4. When making the HACCP plan, it should be

计划时,应该明确由谁来监控。从事关键控制点监控的人员可以是流水线上的人员、设备操作者、监督员、维修人员或质量保证人员。作业的现场人员进行监控是比较合适的,因为这些人能比较容易地发现异常情况的发生。负责关键控制点监控的人员必须方便工作,能够对监控活动提供准确的报告,能够及时报告关键限值偏离情况,以便迅速采取纠正措施。监控人员的责任是及时报告异常事件和关键限值偏离情况,以便在加工过程中采取调整。所有关键控制点的有关记录必须有监控人员的签名。

made clear who should monitor. The personnel engaged in CCP monitoring can be pipeline personnel, equipment operators, supervisors, maintenance personnel or quality assurance personnel. It is more appropriate for field workers to monitor because it is easier for them to detect anomalies. The person in charge of CCP monitoring must be convenient to work on the job; be able to provide accurate reports on monitoring activities; be able to report critical limit deviation in a timely manner so that corrective action can be taken quickly. It is the responsibility of the supervisor to report the abnormal events and the deviation of the critical limit in time in order to make adjustments during the processing process. All CCP records must be signed by the monitor.

表 7-3 是一个危害分析与关键控制点计划,其中监控程序的内容填写在计划表的第(4)至(7)栏中。

Table 7-3 is an example of an HACCP schedule, in which the contents of the monitoring program are filled in columns(4) to (7) of the schedule.

表 7-3 危害分析与关键控制点计划表

关键控制点	显著危害	关键限值	监控				纠偏措施	验证	记录
			对象	方法	频率	人员			
(1)	(2)	(3)	(4)	(5)	(6)	(7)	(8)	(9)	(10)
蒸制	致病菌残存	蒸制温度≥105 ℃,蒸制时间≥15 min	蒸制时间和温度	观察数字式温度计、计时器	连续观察,每3 min记录一次,发现异常随时记录	蒸制时间和人员	调整温度和时间,确认偏离的产品,隔离待评估,延长蒸制时间	每日审核记录,每周用标准温度计对数字式温度计校正一次,每年检定标准温度计,每周抽取蒸制后的产品进行微生物化验	蒸制记录

企业名称:××食品有限公司
企业地址:××省××市××路××号
产品种类:速冻蒸熟猪肉包子,塑料袋包装后装纸箱
销售和贮存方法:-18 ℃以下冷藏
预期用途和消费者:解冻后加热食品,一般公众
签署: 日期:

5) 原理五：纠偏措施

纠偏措施是指在关键控制点上，监控结果表明失控时所采取的任何措施。在食品生产过程中，任何关键控制点的关键限值即使是在建立完善的关键控制点监控程序后，不发生偏离是几乎不可能的。因此，为了使监控到的失控关键控制点或发生偏离的关键限值得以恢复正常并处于控制之下，必须建立相应的纠偏行动或措施以确保关键控制点再次处于控制之下。纠偏措施的目的是使关键控制点重新受控。可以通过以下四个步骤进行处理：①确定产品是否存在安全方面的危害；②如果产品不存在危害，可以解除隔离，放行出厂；③如产品存在潜在的危害，则需要确定产品可否再加工、再杀菌，或改作其他用途安全使用；④如果不能按第③步进行处理，产品必须予以销毁。这样做付出的代价最高，通常到最后才选择该处理方法。各个关键控制点纠偏程序应事先制定并包括在危害分析与关键控制点计划内，将纠偏措施的详细情况记录下来是非常重要的。

纠偏行动由两部分组成。

(1) 查出原因并予以消除，使生产过程恢复控制。纠偏措施必须把关键控制点尽可能短地恢复到控制状态。为了避免继续生产不良产品和将不合格的产品剔除，有时需要停止生产，查出原因并予以消除，防止以后再次发生。对于没有预料到的关键限值，或再次发生的偏差，应该调整加工工艺（改变温度、时间，调整 pH 值，改变原料配比等）或重新评估危害分析与关

E. Principle 5: Corrective Measures

Corrective measures are any action taken at a critical control point where the monitoring results indicate that it is out of control. In the food production process, even after the establishment of a sound CCP monitoring procedures, it is almost impossible for any critical limit of CCP not to deviate. Therefore, in order to make the uncontrolled CCP or critical limit deviated from the monitoring to return to normal and under control, it is necessary to establish corresponding corrective actions or measures to ensure that CCP is under control again. The aim of the corrective measures is to bring the CCP back under control. The following 4 steps can be done: a. determine whether there is a safety hazard in the product; b. if the product is not hazardous, you can release the isolation and release from the factory; c. if there is a potential hazard to the product, it is necessary to determine whether the product can be reprocessed, resterilized, or converted to safe use for other purposes; d. if the product cannot be disposed of in step c, the product must be destroyed. This is the most expensive thing to do, and it's usually the last thing you choose to do. Each CCP corrective procedure should be prepared in advance and included in the HACCP plan, and it is important to document the details of corrective action.

Corrective action consists of two parts.

1. Identify the causes and eliminate them to restore control of the production process. Corrective action must restore critical control points to the control state for as short a time as possible. Sometimes it is necessary to stop production in order to avoid the continued production of defective products and the removal of unqualified products. Identify causes and eliminate them to prevent recurrence in the future. For unexpected critical limits or recurring deviations, the

键控制点计划。

(2)确定、隔离并存放偏离期间生产的产品,评估后采取适当的处理方式(如分选、特采、返工、销毁)。应记录偏离和产品的处置方法。进行评估的人员应经过专门的培训或有这方面的经验。

通常情况下,纠偏措施应在制订危害分析与关键控制点计划时预先制定,并将其填写在危害分析与关键控制点计划表(表7-3)的第(8)栏里。纠偏措施应由对过程、产品和危害分析与关键控制点计划有全面理解,并有权作出决定的人来负责实施。如果有可能的话,在现场纠正问题,会带来满意的结果。有效的纠偏措施依赖于充分的监控程序。

危害分析与关键控制点计划应包含一份独立的文件,其中所有的偏离和相应的纠偏措施要以一定的格式记录进去。这些记录可以帮助企业确认再发生的问题和危害分析与关键控制点计划被修改的必要性。表7-4是一份纠偏措施报告。

process should be adjusted (change in temperature, time, pH value, ratio of raw materials, etc.) or the HACCP plan should be reassessed.

2. Identify, isolate and store the products produced during the deviation period, and assess and adopt appropriate disposal methods (e. g. separation, mining, rework, destruction). Deviations and disposal methods of the product shall be recorded. The personnel conducting the assessment should be specially trained or experienced in this regard.

In general, corrective measures should be prepared in advance of the preparation of the HACCP plan and included in column(8) of the HACCP schedule (Table 7-3). Corrective measures should be implemented by a person who has a comprehensive understanding of the process, product and HACCP plan and has the right to make a decision. If possible, correcting the problem on the spot will lead to satisfactory results. Effective corrective measures depend on adequate monitoring procedures.

The HACCP plan should contain a separate document in which all deviations and corresponding corrective measure are recorded in a format. These records can help businesses identify recurring problems and the need for HACCP plans to be modified. Table 7-4 is a corrective measure report.

表7-4 纠偏措施报告

公司名称:		编　　号:	
地　　址:		日　　期:	
加工步骤:		关键限值:	
监控人员		发生时间		报告时间	
问题及其发生的描述					
采取的措施					
问题的解决及现状					
危害分析与关键控制点小组意见					
审核人:			日期:		

6) 原理六:建立有效的验证程序

(1)危害分析与关键控制点计划的确认。危害分析与关键控制点计划使用前应进行确认,以确定所有危害已被识别并有效控制。如果原料及其来源、产品配方、加工方法或体系、销售体系或预期用途、计算机及软件发生可能影响以前所作危害分析结果的变化时,加工者应重新评估危害分析的适应性。

确认危害分析与关键控制点计划的信息通常包括专家的意见和科学研究成果,生产现场的观察、测量和评价。例如,加热过程的确认应包括杀灭致病微生物所需加热时间和温度的科学证据,以及加热设备的热分布研究。

(2)危害分析与关键控制点计划的验证。企业应定期审查危害分析与关键控制点计划的有效性,验证危害分析与关键控制点计划是否正确执行,审查关键控制点监视记录和纠偏行动记录。验证内容包括:①复查收到的消费者投诉,以确定它们是否与危害分析与关键控制点计划的实施有关,或发现存在未确定的关键控制点;②监控仪器的校准;③定期的成品、半成品的检测,对产品进行有关指标菌(如大肠杆菌生物Ⅰ型)检测以验证杀菌处理的有效性;④复核记录的完整性,以及是否按照计划进行了适当控制。由经过危害分析与关键控制点培训的人员在一周内完成复核。需要复核的记录至少包括关键控制点的监控记录、纠偏行动记录、关键控制点控制仪器的校准记录,以及定期对成品和

F. Principle 6: Establish Effective Validation Procedures

1. Confirmation of the HACCP plan. The HACCP plan should be validated before use to ensure that all hazards have been identified and effectively controlled. If changes have taken place in the raw materials and their sources, product formulations, processing methods or systems, marketing systems or intended uses, including computers and software, and may affect the results of previous hazard analyses, processors should reassess the suitability of hazard analysis.

The information that confirms the HACCP plan usually includes expert opinion and scientific research results, observation, measurement and evaluation of the production site. For example, the identification of the heating process should include scientific evidence of the heating time and temperature required to kill pathogenic microorganisms, as well as studies of the heat distribution of the heating equipment.

2. Verification of HACCP plan. The enterprise shall regularly review the effectiveness of the HACCP plan, verify that the HACCP plan is executed correctly, and review the CCP monitoring records and corrective action records. Verification includes: a. review customer complaints received to determine if they are related to the implementation of the HACCP plan, or to identify undetermined critical control points; b. the calibration of the monitoring instrument; c. regular inspection of finished and semi-finished products, in order to verify the efficacy of bactericidal treatment, the relative index bacteria (e. g. coli biotype Ⅰ) were tested; d. review the integrity of the records and whether appropriate controls are in place in accordance with the plan. The review shall be completed within one week by the trained personnel of the HACCP. The records to be reviewed include at least monitoring records of

加工过程中产品检验的记录。

critical control points, records of corrective actions, calibration records of critical control point control instruments, and periodic inspection records of finished products and products in process.

表 7-5 是一个验证计划的实例。

表 7-5　验证计划表

活　动	频　率	负责人	审查人
验证活动的计划	每年一次或当危害分析与关键控制点体系变化时	危害分析与关键控制点负责人	工厂负责人
危害分析与关键控制点计划的首次确认	在计划首次实施前和实施中	独立专家	危害分析与关键控制点小组
危害分析与关键控制点计划的随后确认	当关键限值变化时；当加工过程有明显变更时；当设备改变时；当体系失效后等	独立专家	危害分析与关键控制点小组
按计划对关键控制点监控的验证	危害分析与关键控制点计划确定的频率（如每班一次）	危害分析与关键控制点计划确定的负责人（如生产线监督员）	危害分析与关键控制点计划确定的审查人（如质量控制人员）
监控、纠偏行动记录的审查，以确定是否与计划相符	每月一次	质保部门	危害分析与关键控制点小组
综合性危害分析与关键控制点体系验证	每年一次	独立专家	工厂负责人

（3）危害分析与关键控制点管理体系执行后，每年至少进行一次内部验证。当加工过程出现任何改变影响危害分析或危害分析与关键控制点计划时，应及时进行危害分析与关键控制点管理体系的验证。危害分析与关键控制点管理体系的内部验证应包括危害分析与关键控制点计划及卫生标准操作程序的验证。企业的管理层可指定危害分析与关键控制点小组进行危害分析与关键控制点管理体系的内部验证，也可以委托第三方对危害分析与关键控制点管理体系进行审核。

3. HACCP management system is carried out at least once a year after implementation, and HACCP management system verification should be carried out in time when any change in machining process affects hazard analysis or HACCP plan. The internal verification of HACCP management system should include the verification of HACCP plan and SSOP. The management of the enterprise may designate the HACCP team to carry out the internal verification of the HACCP management system, or may entrust the third party to carry on the audit to the HACCP management system.

7) 原理七:建立记录保存程序

文件和记录保存程序应事先建立并认真实施,危害分析与关键控制点管理体系应有效,过程应文件化,并准确地保存记录。文件和记录的保存应与实际情况相适应。危害分析与关键控制点管理体系的文件和记录应包括但不限于如下内容。

(1)危害分析与关键控制点计划和支持性文件,包括危害分析与关键控制点计划表、危害分析工作单、危害分析与关键控制点小组名单和各自的责任、食品特性的描述、销售方法、预期用途和消费人群、流程图、计划确认记录等。

(2)监控记录。危害分析与关键控制点的监控记录将反映所监控的值是否超过关键限值。这些记录必须与各关键控制点所设的关键限值相对应,监控记录可以为审核员判断是否遵守危害分析与关键控制点计划提供证据。通过监控记录,操作人员和管理人员可以对加工进行必要调整和控制。

(3)纠偏措施记录。当超过关键限值并采取了纠偏措施,必须予以记录。纠偏措施记录应包括以下内容:产品描述、涉及产品的数量、对偏离情况的描述、采取的纠偏措施、执行纠偏措施的责任者和评估结果。

(4)验证记录。验证记录应包括因原料、配方、加工、包装及销售等改

G. Principle 7: Establish Record Keeping Procedures

Document and record keeping procedures should be established in advance and carefully implemented, the HACCP management system should be effective, the process should be documented, and records should be kept accurately. The preservation of documents and records should be adapted to the actual situation. The documentation and records of the HACCP management system shall include, but are not limited to, the following:

1. HACCP plan and supporting document including HACCP plan table, hazard analysis worksheet, HACCP team list and their respective responsibilities, description of food characteristics, sales methods, intended use and consumer population, flow chart, plan confirmation record, etc.

2. Monitoring record. The monitoring record of the HACCP will reflect whether the monitored value exceeds the critical limit. These records must correspond to the critical limits set at each critical control point, and the monitoring records can provide evidence for the auditor to determine whether or not to comply with the HACCP plan. Through the monitoring records, operators and managers can make necessary adjustments and control of the processing.

3. Record of corrective action. When the critical limit is exceeded and corrective action is taken, it must be recorded. The corrective action record shall include the following: product description, number of products involved, description of deviations, corrective actions taken, responsible persons and evaluation results for the implementation of corrective actions.

4. Verification record. Verification records shall include records of modified in the HACCP

变导致危害分析与关键控制点计划修改的记录,为确保供应商证明的有效性进行的审核记录,监测设备的校准记录。

(5)执行卫生标准操作程序记录。所有记录都必须至少包括加工者或供应商的名称和地址,记录所反映的工作日期和时间,操作者的签字或署名,适当的时间,包括产品的特性和代码。加工过程或其他信息资料也应包括在记录中。

记录的保存期限:对于冷藏产品,一般至少保存一年;对于冷冻或货架期稳定的商品应至少保存两年;对于其他说明加工设备、加工工艺等方面的研究报告、科学评估的结果应至少保存两年。可以采用计算机保存记录,但要求保证数据的完整和一致。

危害分析与关键控制点体系建立和实施过程中有大量的技术文件和各种日常工作监测记录,而完整准确地记录和妥善保存这些资料是成功建立和实施危害分析与关键控制点体系的关键之一。因此,在建立和实施危害分析与关键控制点体系过程中,所有程序、记录必须文件化,所有文件必须妥善保存且符合操作特性和规范。记录保存在危害分析与关键控制点计划表(表7-3)的第(10)栏中。

以上七个原理中,原理一至五是环环相扣的步骤,显示了危害分析与关键控制点体系极强的科学性、逻辑

plan resulted from changes in raw materials, formulation, processing, packaging and sales, records of audits to ensure the validity of supplier certification, and records of calibration of monitoring equipment.

5. Record of execution of Sanitation Standard Operation Procedure. All records must or at least include the following: the name and address of the processor or supplier, the date and time of work reflected in the record, the signature or attribution of the operator, the appropriate time, including the product characteristics and the code, as well as the process or other information, shall also be included in the record.

Record retention period: for refrigerated products, generally at least one year; for frozen or stable goods should be kept for at least two years; the results of scientific evaluation shall be kept for at least two years for other research reports describing the processing equipment, processing technology, etc. You can use a computer to keep records, but you need to ensure that the data is complete and consistent.

There are a large number of technical documents and various daily work monitoring records in the process of HACCP system establishment and implementation, the complete and accurate records as well as proper preservation of these data is one of the keys to the successful establishment and implementation of HACCP system. Therefore, in the process of establishing and implementing the HACCP system, all procedures and records must be documented, and all documents must be properly preserved and kept in accordance with operational characteristics and specifications. Record keeping is completed in column (10) of the HACCP plan table (Table 7-3).

Among the above 7 principles, principle 1~5 is an interlocking step, which shows that the HACCP system is extremely scientific and logical,

性;而原理六和七哪一个在前都可以,显示了危害分析与关键控制点体系的灵活性。这七个原理中,危害分析是基础,关键控制点及其关键限值的确定是根本,监控程序、纠偏行动、验证程序以及科学完整的记录及其保存程序是关键。

while the order of principles 6 and 7 has no influence and shows the flexibility of the HACCP system. Among the 7 principles, hazard analysis is the basis, the determination of CCP and critical limits is fundamental, monitoring procedures, corrective action, verification procedures and scientific integrity of the record and keeping procedure are the key.

3. 危害分析与关键控制点体系食品生产企业通用要求的内容

《危害分析与关键控制点(HACCP)体系 食品生产企业通用要求》(GB/T 27341—2009)是 ISO9001 和危害分析与关键控制点原理的结合体,所以结构与质量管理体系类似,核心部分包含了危害分析与关键控制点原理,详见表 7-6 和表 7-7。

表 7-6 ISO9001:2015 与 GB/T 27341—2009 结构对照表

ISO9001:2015			GB/T 27341—2009
范围	1	1	范围
规范性引用文件	2	2	规范性引用文件
术语和定义	3	3	术语和定义
文件要求	4.4.2	4.2	文件要求
质量手册	4.2.2	4.2.2	HACCP 手册
领导作用	5	5	管理职责
领导作用和承诺	5.1	5.1	管理承诺
方针	5.2	5.2	食品安全方针
组织的岗位、职责和权限	5.3	5.3	职责权限与沟通
沟通	7.4	5.3.2	沟通
管理评审	5.6	5.5	管理评审
人员	7.1.2	6.2	人力资源保障计划
基础设施	7.1.3	6.3	良好生产规范(部分)
		6.6	维护保养计划
标识和可追溯性	8.5.2	6.7.1	标识和追溯计划

表 7-7 危害分析与关键控制点体系计划的十二个步骤与 GB/T 27341—2009 结构对照表

危害分析与关键控制点的十二个步骤		GB/T 27341—2009	
建立 HACCP 小组	步骤 1	7.2.1	HACCP 小组的组成
产品描述	步骤 2	7.2.2	产品描述
识别预期用途	步骤 3	7.2.3	预期用途的确定
制作流程图	步骤 4	7.2.4	流程图的制定
现场确认流程图	步骤 5	7.2.5	流程图的确认

续表 7-7

危害分析与关键控制点的十二个步骤			GB/T 27341—2009
危害分析	步骤 6（原理一）	7.3	危害分析和制定控制措施
确定关键控制点（CCP）	步骤 7（原理二）	7.4	关键控制点（CCP）的确定
对每个关键控制点（CCP）确定关键限值	步骤 8（原理三）	7.5	关键限值的确定
对每个关键控制点（CCP）建立监视系统	步骤 9（原理四）	7.6	CCP 的监控
建立纠偏措施	步骤 10（原理五）	7.7	建立关键限值偏离时的纠偏措施
建立验证程序	步骤 11（原理六）	7.8	HACCP 计划的确认和验证
建立文件和记录保存程序	步骤 12（原理七）	7.9	HACCP 计划记录的保存

表 7-6 所列内容在项目 5 中进行了详细阐述，表 7-7 所列内容在项目 7 中进行了详细阐述。

1）良好的操作规范（GMP）

良好的操作规范在我国一方面表现为符合 GB 14881 及其专项标准。也就是说，食品企业除满足 GB 14881 的要求外，还应满足相应生产产品的专项标准，例如生产白酒还要满足 GB 8951—2016《食品安全国家标准 蒸馏酒及其配制酒生产卫生规范》，生产糕点应满足 GB 8957—2016《食品安全国家标准 糕点、面包卫生规范》。参见本项目任务 1。另一方面体现为企业通过食品生产许可审查。同样，食品生产许可审查也有《食品生产许可审查通则》和各专项细则。例如饼干生产应满足《饼干生产许可证审查细则》，茶饮料生产应满足《茶饮料类生产许可证审查细则》。

2）卫生标准操作程序（SSOP）

卫生标准操作程序是为控制食品企业加工卫生所总结归纳出的八个控制方面，是实施危害分析与关键控制点的重要前提条件。该部分内容在项目 6 中已详细介绍，在此不作过多说明。

3）产品召回计划

我国《食品安全法》第六十三条规定："国家建立食品召回制度。食品生产者发现其生产的食品不符合食品安全标准或者有证据证明可能危害人体健康的，应当立即停止生产，召回已经上市销售的食品，通知相关生产经营者和消费者，并记录召回和通知情况。"因此，企业应建立食品召回制度。召回程序一般包括：组成召回小组分配任务→通知经销商、超市等销售终端下架产品→隔离库房中未销售产品→产品运输回企业→评价产品→制定处理措施（重新加工、改作他用、销毁）→实施措施→总结经验，防止再次发生。如召回原因与食品安全相关还应向当地监管机构报告。在正常生产销售状态，无召回情况发生时，可定期进行召回演练，以验证召回程序有效、全面。

4）应急预案

制定应急预案的目的一方面是为了防止紧急事件（如突然停水、停电，自然灾害等）可能造成的有安全隐患的产品流入市场，对消费者造成伤害；另一方面最大限度地降低企业的经济损失。紧急事件包括自然灾害，动物疫情，突然的环境污染，有毒有害化学品的泄漏，火灾，高空坠物、机械伤害，突然停电、停水等。

7.3.2 实训任务

实训组织：

1. 分组绘制液奶乳制品企业、蛋糕生产企业、饮料生产企业的车间平面图。

2. 上网查找有关 GB/T 27341—2019"7.6 CCP 的监控"和"7.8 HACCP 计划的确认和验证"的案例各 3 个,分析、判断是否符合标准及其原因。

实训成果:

1. 食品企业车间平面图。
2. 案例综述类文章,总字数 500 字以上。

实训评价:5 分制,平面图 2 分,案例分析 3 分。要求内容全面,条理清楚。

任务 4 危害分析与关键控制点体系建立方法

7.4.1 基础知识

危害分析与关键控制点体系建立是一个循序渐进的过程,不能要求一次成形,也不会立竿见影。在建立过程中,通常应伴随着培训、贯标,使企业的员工通过文件编写和相关标准的学习,逐渐理解体系的作用,真正体会危害分析与关键控制点体系的精髓。危害分析与关键控制点体系的建立不是无章可循的,而是按一定的步骤进行的。常规的实施危害分析与关键控制点体系的计划由十二个步骤组成,如图 7-3 所示。

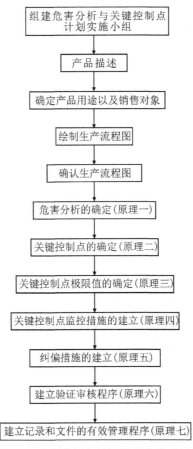

图 7-3 危害分析与关键控制点体系计划

1. 危害分析与关键控制点体系实施的必备程序与条件

要实施危害分析与关键控制点体系计划,必须具备一些程序和基本条件,必备程序有良好的操作规范(GMP)和卫生标准操作程序(SSOP)、管理层的支持、人员的素质要求和培训、校准程序、产品标志的可追溯性、产品回收计划的实施。除此之外,还应该在生产设备、过程方法等方面具备一定条件。

2. 组建危害分析与关键控制点计划实施小组

危害分析与关键控制点体系涉及的学科内容有食品方面的生产、技术、管理、储运、采购、营销、环境、统计等,因而危害分析与关键控制点计划实施小组应由多个成员组成。组建一支相互支持、相互鼓励、团结协作、专业素质好、业务能力强、技术水平高的危害分析与关键控制点计划实施小组,是有效实施危害分析与关键控制点体系的核心保障。

3. 产品描述

产品描述是危害分析与关键控制点体系实施小组对产品的名称、成分、重要性能等进行说明。描述包括食品安全有关的特性(含盐量、酸度、水分活度等)、加工方式(热处理、冷冻、盐渍、烟熏等)、计划用途(主要消费对象、分销方法)、食用方法、包装形式、保质期、销售点、标签说明、特殊储运要求(环境湿度、温度)、装运方式等,尤其对某些产品应该有警示声明,如"本产品未经巴氏杀菌,可能含有导致儿童、老人和免疫力差人群疾病的有害细菌"。产品描述实例如表 7-8 所示。

I. The Necessary Procedures and Conditions for the Implementation of the HACCP System

To implement the HACCP system plan, we must have some necessary procedures and basic conditions, such as GMP and SSOP, management support, personnel quality requirements and training, calibration procedures, traceability of product signs, and implementation of product recovery plans. In addition, we should also have certain conditions in production equipment, process methods and so on.

II. Set up a Team to Implement the HACCP Plan

The HACCP system involves food production, technology, management, storage and transportation, procurement, marketing, environment, statistics and so on. Therefore, the HACCP program implementation team should be composed of many members. It is the core guarantee for the effective implementation of HACCP system to set up a HACCP plan implementation team with mutual support, mutual encouragement, unity and cooperation, good professional quality, strong business ability and high technical level.

III. Product Description

The product description is a description of the name, composition and important properties of the product by the HACCP system implementation team. The description includes characteristics related to food safety (salinity, acidity, moisture activity, etc.), processing (heat treatment, freezing, salting, fumigation, etc.), planned uses (main consumers, distribution methods), eating methods, packaging form, shelf life, point of sale, label description, special storage and transportation requirements (environmental humidity, temperature), method of shipment, etc. In particular, there should be a warning for certain products, such as "this product is not pasteurized and may contain harmful bacteria that cause diseases in children, the elderly and people with

表 7-8 桑果浓缩汁的产品描述

项目	内容
产品名称	桑果浓缩汁
重要特征 (含水量、pH 值、矿物质、主要维生素量)	固形物:50°Bx±1°Bx;总酸:11~16 g/100 g; 维生素 C;有机酸;pH<4.6
食用方式	即时用水调配(13°Bx)饮用或与其他饮料调配饮用
包装方式	复合袋密封罐装
货架寿命	18 个月
销售地点及对象	批发、零售;销售对象无特殊规定
标签说明	开封后,请冷藏保存
特殊分销控制	贮藏温度:−18 ℃

poor immunity". Examples of product descriptions are shown in Table 7-8.

对产品进行必要的表述,可以帮助消费者或后续的加工者识别产品在形成过程中以及包装材料中可能存在的危害,便于考虑易感人群是否接受该产品。

The necessary description of the product can help consumers or subsequent processors to identify possible hazards in the formation of the product and in the packaging materials, thus facilitating the consideration of whether the susceptible population accepts the product.

4. 确定产品用途以及销售对象

确定产品用途以及销售对象是确定产品的预期消费者和消费者如何消费产品(如该产品是直接食用,还是加热后食用,或者再加工后才能食用等)、产品的销售方法等。对于不同的用途和不同的消费者,食品的安全保证程度不同。尤其是婴儿、老人、体弱者、免疫功能不全者等社会弱势群体以及对该产品进行再加工的食品企业,更要充分了解和把握产品的特性。

Ⅳ. Determine the Use of the Product and the Target for Sale

Determining the use of the product and the object of sale is to determine the expected consumer of the product and how the consumer will consume the product (such as whether the product is eaten directly or heated or reprocessed before being edible), the sales method of the product, etc. For different uses and different consumers, the degree of food safety assurance varies. Especially infants, the elderly, the frail, the immune dysfunction and other social vulnerable groups as well as the reprocessing of the product food enterprises should fully understand and grasp the characteristics of the product.

5. 绘制生产流程图

生产流程图由危害分析与关键控制点计划实施小组制定，是对从原辅料购入到产品贮存的全过程所作的简单明了且全面的情况说明。它概括了整个生产、产品贮存过程的所有要素和细节，准确地反映了从原辅料到产品贮存全过程中的每一个步骤。流程图表明了产品形成过程的起点、加工步骤、终点，确定了危害分析和制定危害分析与关键控制点计划的范围，是建立和实施危害分析与关键控制点体系计划的起点和焦点。

一张完整的实用型流程图，要有以下一些必要的技术性资料作支持：

（1）原辅料及包装材料的物理、化学、微生物学方面的数据；

（2）加工工艺步骤及顺序；

（3）所有工艺参数；

（4）生产中的温度-时间对应图；

（5）产品的循环或再利用线路；

（6）设备类型和设计特征，有无卫生或清洗死角存在；

（7）高、低危害区的分隔；

（8）产品贮存条件。

生产流程图无统一格式要求，以简明扼要、易懂、实用、无遗漏、清晰、准确为原则，形式可以多样化，通常是由简洁的文字表述配以方框图和若干的箭头按顺序组成。

生产流程图是危害分析的基础，要能反映出每一个技术环节。流程图中对应的加工步骤，应有适当的文字

Ⅴ. Draw the Process of the Flow Sheet

The process of flow sheet developed by the HACCP program implementation team, which is a simple, clear and comprehensive description of the entire process from raw and auxiliary materials to product storage. It summarizes all the elements and details of the whole production and product storage process and accurately reflects every step from raw and auxiliary materials to the whole process of product storage. The flow sheet shows the starting point, the processing step and the end point of the product formation process, and determines the scope of hazard analysis and HACCP plan, which is the starting point and focus of establishing and implementing the HACCP system plan.

A complete practical flow sheet should be supported by the following necessary technical information:

1. Physical, chemical and microbiological data of raw materials and packaging materials;

2. Process steps and sequence;

3. All process parameters;

4. Temperature-time map in production;

5. The circulation or reutilization of a product;

6. Equipment types and design features, with or without sanitary or cleaning dead-corners;

7. Separation of high and low hazard areas;

8. Product storage conditions.

The process of flow sheet has no uniform format requirement, and the principle is to be concise, easy to understand, practical, without omission, clear and accurate, and the form can be diversified. The common text is composed of simple words with block diagrams and a number of arrows in sequence.

The process of the flow sheet is the basis of hazard analysis, which should reflect every technical link. The corresponding processing steps

性工艺表述,这样有利于对危害的识别。对于一些用流程图描述不太清楚的技术内容,如环境或加工过程中出现的其他危害(冰、水、清洗、消毒过程、工作人员、厂房结构、设备特点等),要以文字性的形式附在流程图后面,作为流程图的补充内容列出。

6. 生产流程图的现场确认

危害分析与关键控制点计划实施小组对于已制作的流程图进行生产现场确认,以验证流程图中表达的各个步骤与实际是否一致。发现有不一致或有遗漏,就应对流程图作相应的修改和补充。

现场确认可分为四个阶段。

(1)对比阶段。将拟定的生产流程图与实际操作过程作对比,在不同的操作时间查对工艺过程与工艺参数、生产流程图中的有关内容,检验生产流程图对生产全过程的实效性、指导性、权威性。

(2)查证阶段。查证与实际生产不吻合部分,对生产流程图作适当修改。

(3)调整阶段。在出现配方变动或设备更换时,也要适时调整生产流程图,以确保生产流程图的准确性和完整性,使之更具可操作性和科学性。

(4)确认阶段。通过前面三个阶

in the flowchart should be properly described in writing, which is beneficial to the identification of hazards. For some technical content that is not clearly described in flowcharts, for example, environmental or other hazards occurring during processing (ice, water, cleaning, disinfection process, staff, plant structure, equipment characteristics, etc.) should be appended to the flowchart in a written form and listed as a supplement to the flow sheet.

Ⅵ. Site Confirmation of the Purpose of the Production Process

The HACCP plan implementation team is confirmed on the production site to verify that the steps expressed in the flowchart are consistent with the practice. If any inconsistencies or omissions are found, the flowchart shall be modified and supplemented accordingly.

Site confirmation can be divided into four phases.

1. Contrast phase. Compare the process of flow sheet drawn up with the actual operation process, check the process engineering, the process parameters and the relevant contents of the process of flow sheet at different operating times. Check the effectiveness, guidance and authority of the process of the flow sheet for the whole production process.

2. Verification phase. Check the parts that do not match with the actual production, and make appropriate modifications to the process of flow sheet.

3. Adjustment phase. In order to ensure the accuracy and completeness of the process of flow sheet, it is necessary to adjust the process of the flow sheet in time when the formula changes or equipment changes, making it more operable and scientific.

4. Confirmation phase. Through the first

段的工作,对生产流程图作出客观的确认与定夺,作为生产中的执行规范下发企业各个部门和所有人员,并监督执行。

7. 危害分析的确定(原理一)

1)危害分析与危害程度判别

危害是指一切使食品变得不安全的因素,一半来自于生物、化学、物理三个方面。危害分析与关键控制点体系计划实施小组进行的危害分析,就是要确定食品中每一种潜在的危害及其可能出现的点,尤其要注意危害具有变动性的特征;还应对危害达到什么样的程度作出评价。

一般来说,食品中的危害通常来自于下面几个方面:

(1)原辅材料,如食品生产所用动植物原辅材料的生长环境会带来物理性(土块、石屑、杂草、玻璃、金属等异物)、化学性(农药、抗生素、杀虫剂)、生物性(微生物、致病菌)污染;

(2)加工引起的食品成分理化特性变化,如微生物或制品中酶类的存在及加工条件等使食品特性发生变化,造成毒素的生成、颜色的改变、酸度的增加等;

(3)车间设施及设备,如设备、仪器、仪表运行不正常时出现机油渗漏、碎玻璃、金属碎片等,以及设备消毒不彻底,卫生未达标;

(4)人员健康状况,如个人卫生不符合要求、操作不符合卫生规范等;

three stages of the work, the process of the flow sheet should be made objective confirmation and decision, as the implementation of the norms of production under the enterprise departments and all personnel, and supervise the implementation.

Ⅶ. Determination of Hazard Analysis(Principle 1)

A. Hazard Analysis and Hazard Degree Discrimination

Harm refers to all the factors that make food unsafe, half of which come from three aspects: biology, chemistry and physics. The hazard analysis carried out by the HACCP system implementation team is to identify each potential hazard in food and where it may be born, paying particular attention to the variable nature of the hazard; and the extent of the harm should also be assessed.

Generally speaking, the hazards in food usually come from the following aspects:

1. Raw and auxiliary materials, for example, the growth environment of animal and plant raw materials used in food production will bring physical (soil, stone debris, weeds, glass, metal and other foreign bodies), chemical (pesticides, antibiotics, insecticide), and biological (microorganisms, pathogenic bacteria) pollution;

2. The changes of physical and chemical properties of food components caused by processing, such as the existence and processing conditions of enzymes in microbes or products to change the food characteristics, which causes the formation of the toxin, the change of color, the increase of acidity, and so on;

3. Workshop facilities and equipment, such as equipment, instruments and apparatuses not normal operation, there will be oil leakage, broken glass, metal debris, and equipment disinfection is not complete, hygiene is not up to standard;

4. Health status of personnel, for example, the personal hygiene does not meet the

(5)包装方面,包装材料及包装方式不卫生,包装标签内容含糊不清;

(6)食品的储运与销售,如光照、不密封等,不适当的储运条件往往导致或加重危害程度;

(7)消费者对食品不正确的消费行为也会导致或加重危害的产生;

(8)消费对象的身体健康状况和体质特异性或体质差异,同样会导致或显现出危害。

2)危害分析的顺序

危害分析一般遵循以下顺序。

(1)确定产品品种和加工地点。

(2)根据流程图,确认加工工序的数量。当存在两个以上不同加工工序时,应分别进行危害分析。

(3)复查每一个加工工序对应的流程图是否准确,对存在偏差的,要作出调整。

(4)列出污染源。对照加工工序,从生物性、化学性、物理性污染三个方面考虑并确定在每一个加工步骤上可能引入的、增加的或受到限制的食品危害,属于卫生标准操作程序范畴的潜在危害也应一并列出。

(5)明显危害的判定。判定原则为潜在危害风险性和严重性的大小。属于卫生标准操作程序范畴的潜在危害若能由卫生标准操作程序计划消除的,就不属于明显危害,否则将对其进

requirements, operations does not comply with the health standards, and so on;

5. Packaging, packaging materials and packaging methods are not sanitary, packaging labels are ambiguous;

6. Food storage and marketing, such as light, unsealed, improper storage and transportation conditions often lead to or aggravate the degree of harm;

7. Consumer's incorrect consumption behavior to food will also lead to or aggravate the production of harm;

8. The physical health of the consumer and physical specificity or physical differences will also lead to or show harm.

B. Sequence of Hazard Analysis

Hazard analysis generally follows the following order.

1. Determine the product variety and processing location.

2. Confirm the number of processing procedures according to the flow chart. Hazard analysis should be carried out separately when there are more than two different processing procedures.

3. Review the flow chart of each processing procedure is accurate or not, for the existence of deviation to make adjustments.

4. List sources of pollution. Refer to processing procedures, consider and identify possible, increased or restricted food hazards in each of the processing steps from three aspects of biological, chemical and physical contamination, potential hazards falling within the category of SSOP should also be listed.

5. Determination of obvious harm. The principle of determination is the magnitude of potential hazard risk and severity. If the potential harm in the category of SSOP can be eliminated by SSOP plan, it is not obvious harm; otherwise, it

行判定。判定的依据应科学、正确、充分,应针对每一个工序和每一个步骤进行。

(6)预防措施的建立。对已确定的每一种明显危害,要制定相应的预防控制措施,要求是列出控制组合、描述控制原理、确认控制的有效性。

危害分析的确定是一个危害分析与关键控制点计划实施小组广泛讨论、广泛发表科学见解、广泛听取正确观点、广泛达成共识的集思广益、经历思维风暴的必然过程。

按照危害分析的顺序完成分析过程后,形成危害分析结果。经过确定后,可以以危害分析工作单的形式记录下来。表7-9是美国食品药物管理局推荐的一份表格式危害分析工作单。

will be judged. The basis for the determination shall be scientific, correct and adequate, and shall be carried out for each process and every step.

6. The establishment of preventive measures. For each defined obvious harm, the corresponding preventive control measures should be formulated, the requirements are to list the control combination, describe the control principle, and confirm the effectiveness of the control.

The determination of hazard analysis is an inevitable process of benefit by mutual discussion and experiencing the storm of thinking for HACCP plan implementation group to discuss widely, widely publish scientific opinions, listen to the correct views and reach consensus widely.

According to the sequence of hazard analysis, the result of hazard analysis is formed after the analysis process is completed. Once determined, it can be recorded in the form of the hazard analysis work sheet. Table 7-9 is a tabular hazard analysis worksheet recommended by FDA in the United States.

表7-9 危害分析工作单

企业名称:　　　　　　　　　　　　　　企业地址:

加工步骤	食品安全危害	危害显著(是/否)	判断依据	预防措施	关键控制点(是/否)
	生物性				
	化学性				
	物理性				
	生物性				
	化学性				
	物理性				
	生物性				
	化学性				
	物理性				

危害分析报告单形成后,纳入危害分析与关键控制点记录。

The hazard analysis report form was included in the HACCP record.

8. 关键控制点(CCP)的确定(原理二)

控制点是指食品加工过程中那些能防止物理性、化学性、生物性危害产生的任意一个步骤或工艺,它也包括对食品的风味、色泽等非安全危害要素的控制。

关键控制点判定的一般原则:

(1)在某点或某个步骤中存在卫生标准操作程序无法消除的明显危害;

(2)在某点或某个步骤中存在能够防止、消除或降低明显危害到允许水平以下的控制措施;

(3)在某点或某个步骤中存在的明显危害,通过本步骤中采取的控制措施,将不会再现于后续的步骤中,或者在以后的步骤中没有有效的控制措施;

(4)在某点或某个步骤中存在的明显危害,必须通过本步骤与后序步骤中控制措施的联动才能被有效遏制。

只有符合上述判断原则中的某几条或全部四条的点或加工步骤,才能判断为关键控制点。

根据关键控制点的概念,通常将其分为一类关键控制点(CCP1)和二类关键控制点(CCP2)两种。CCP1是指可以消除或预防危害的控制点;CCP2是指可以将危害最大限度减少或降低到能够接受的水平以下的控制点。

9. 关键控制点极限值的确定(原理三)

关键控制点的极限值又称为关键限值,是指所采用的措施达到使危害消除、预防或降低到允许水平以下的

Ⅷ. Determination of Critical Control Points (CCP) (Principle 2)

Control point refers to any steps or process that can prevent physical, chemical and biological hazards in the whole process of food. It also includes the control of food flavor, color and other non-safety hazards.

General principles for the determination of critical control points:

1. At a point or step, there is an obvious harm that SSOP cannot eliminate;

2. Control measures exist at a point or step that can prevent, eliminate, or reduce apparent harm below the allowable level;

3. The implementation of the control measures taken through this step will not be repeated in subsequent steps, or there will be no effective control measures in subsequent steps if there is a clear hazard at one point or one step;

4. Any apparent harm at a certain point or step must be effectively contained by the linkage of control measures in this step with the subsequent step.

It can be judged as the critical control point only if it conforms to some of the above judgment principles, as well as the points or processing steps that meet the above four principles.

According to the concept of critical control point, it is usually divided into two kinds: one is class A critical control point (CCP1) and the other is class B critical control point (CCP2). CCP1 is a control point that can eliminate or prevent hazards. CCP2 is a control point that can minimize or reduce the harm to an acceptable level.

Ⅸ. Determination of Critical Control Point Limit (Principle 3)

The limit value of the critical control point, also known as the critical limit, refers to the maximum or minimum parameter value of the measures used to

最大或最小参数值,即食品安全生产、销售全过程中的最大或最小参数值。

关键限值确定的原则是尽可能地有效、简单、经济。有效是指此限值确实能预防、消除或降低危害到允许水平以下。便于操作,可以在不停产的情况下快速监控,这就是简单。投入较少的人力、物力、财力即为经济。

关键限值的确认步骤是:

(1)确认在本关键控制点上需要控制的明显危害与相应措施的对应关系。

(2)分析明确此项措施对明显危害的控制原理。

(3)根据原理,确定实现关键限值的最佳载体和种类,如温度、纯度、酸度、水分活度、厚度、残留农药限量等。

(4)确定关键限值的数值。关键限值可以是法规、法典和权威组织公布的数据,如残留农药限量,也可以是科学文献和科技书籍的记载,还可以根据现场实验的准确结论得出。

完成关键限值的确定后,应紧接着进行关键限值技术报告的编制,并把它纳入危害分析与关键控制点体系的支持文件中。

10. 关键控制点监控措施的建立(原理四)

监控就是针对关键控制点实施有效的监督与调控的过程,通过监控了解关键控制点是否处于控制当中。

eliminate, prevent or reduce the hazard below the allowable level. That is, the maximum or minimum parameter value in the whole process of production and sale of food safety without harm.

The critical limit is determined by the principle that it is as efficient, simple and economical as possible. Validity means that this limit does prevent, eliminate, or reduce the hazard below the allowable level. Easy to operate refers to the ability to quickly monitor without stopping production, which is simple. Less investment in human, material and financial resources is the economical.

The confirmation steps of the critical limit are:

1. Confirm the corresponding relationship between the obvious hazards to be controlled on this CCP and the corresponding measures.

2. The control principle of this measure to obvious harm is analyzed.

3. According to the principle, the best carrier and type to realize the critical limit value were determined, such as temperature, purity, acidity, water activity, thickness, pesticide residue limit and so on.

4. Determine the value of the critical limit. The critical limits can be based on data published by laws, codes and authoritative organizations, such as pesticide residues limits; on the basis of scientific literature and scientific and technical books; and on the precise conclusions of field experiments.

After the determination of the critical limit is completed, the technical report of the critical limit should be followed up. And put it in the HACCP support file.

Ⅹ. Establishment of Monitoring Measures for Critical Control Points (Principle 4)

Monitoring is the process of implementing effective supervision and control for critical control points, and through monitoring to understand

监控措施应起到这样的作用,即跟踪各项操作,及时发现有偏离关键限值的趋势,迅速进行调整;查明关键控制点出现失控的时刻和操作点;提交异常情况的书面文件。

监控对象常常是关键控制点的某一个或某几个可测量或可观察的特征,如酸度是关键控制点,pH值就是监控对象;温度是关键控制点,监控对象就是加工或储运的温度;蒸煮或加热、杀菌是关键控制点,温度与时间就是监控对象。

监控过程受限于每一个具体的关键控制点的关键限值、监控设备、监测方法。监测方法一般有在线(生产线上)检测和不在线(离线)检测两种。在线检测可以连续地随时提供检测情况,如温度、时间的检测;离线检测是离开生产线的某些检测,可以是间歇的,如 pH 值、水分活度等的检测。与在线检测比较,离线检测稍显滞后,不如在线检测那么及时。

11. 纠偏措施的建立(原理五)

纠偏措施是当发现关键控制点出现失控(关键限值发生偏离)时,找到原因并为了让关键控制点重新恢复到控制状态所采取的行动。

纠偏措施包括:

(1)列出每个关键控制点对应的关键限值。

(2)寻找偏离的原因、途径。

(3)采取措施纠正和消除偏离的原因和方法,以防止再次出现偏离。当生产参数接近或刚超过操作限值不多时,立即采取纠偏措施。例如在牛

whether CCP is in control.

Monitoring measures should play such a role, that is to track all operations, detect deviations of critical limits and make quick adjustments, find out the runaway time and operation points of CCP, and submit written documents for abnormal situations.

Typically, the monitoring object is one or more measurable or observable features of the CCP, such as the acidity is the CCP, pH value is the monitoring object; the temperature is CCP, the monitoring object is the temperature of processing or storage; cooking or heating, sterilization is CCP, temperature and time is the monitoring object.

The monitoring process is limited by the critical limits of each specific CCP, monitoring equipment, and monitoring methods. There are two kinds of monitoring methods: on-line (on the production line) detection and off-line (not on the production line) detection. On-line detection can be continuously provided at any time, such as temperature, time detection; off-line detection is some detection off the production line, it can be intermittent, such as pH value, water activity detection. Compared with on-line detection, off-line detection appears to be slightly lagging, and not as timely as on-line detection.

XI. Establishment of Correction Measures (Principle 5)

The corrective measures are the action taken to bring the CCP back to control when it is found to be out of control (critical limit deviation) and the cause is found.

Corrective measures include:

1. List the critical limits for each critical control point.

2. Find out the reasons and ways of deviation.

3. The measures used to correct and eliminate the causes and ways of deviation to prevent the recurrence of deviation. When the production parameters are close to or have just exceeded the

奶的巴氏杀菌中,没有达到杀菌温度的牛奶,通过开启的自动转向阀,重新进入杀菌程序。

(4)启用备用的工艺或设备,如生产线某处出现故障后,启用备用的工艺或设备继续进行生产。

(5)对有缺陷的产品(关键控制点出现失控时的产品)应及时处理,如缺陷产品的返工或销毁。对经过返工程序的食品的安全性要有正确评估,无危害性的才可以流入市场。

必须预先制定每一个关键控制点偏离关键限值的书面纠偏措施,形成"纠偏措施技术报告"。纠偏工作要紧紧围绕关键控制点恢复受控进行,关键控制点实施小组应研究纠偏措施的具体步骤,建立适当的纠偏程序,并记录下来。在"纠偏措施技术报告"中明确指定防止偏离和纠正偏离的具体负责人,以减少或避免纠偏行动中可能出现的混乱和争论,影响纠偏的效果。

应当引起重视的是,当在某个关键控制点上,纠偏措施已被正确实施却仍反复发生偏离关键限值的情况,就需要重新评价危害分析与关键控制点计划,并对整个危害分析与关键控制点计划作出必要的调整和修改。

"纠偏措施技术报告"要纳入危害分析与关键控制点体系的支持文件。

operating limit of a few, immediately take corrective measures. For example, in pasteurization of milk, milk that does not reach the sterilization temperature is re-entered into the sterilization process through an open automatic steering valve.

4. Enable a standby process or equipment, if a fault occurs somewhere in the production line, the standby process or equipment will be enabled to continue the production.

5. The defective products (product when CCP is out of control) should be disposed of in a timely manner, such as rework or destruction of defective products. For the rework process of food, its safety must have a correct assessment, no harmful food can enter the market.

Written corrective measures for each critical control point deviating from the critical limit must be developed in advance to form a "Technical Report on Correction Measures". The corrective work should be controlled tightly around the CCP recovery, the implementation group of CCP should study the concrete steps of rectifying measures, establish appropriate correction procedures, and record them. Specific responsibility for preventing and correcting deviations is clearly identified in the "Technical Report on Correction Measures" in order to reduce or avoid possible confusion and controversy in corrective actions, thereby affecting the effect of correction.

What should be paid attention to is that when the corrective measures have been implemented correctly and deviate from the critical limit value repeatedly, it is necessary to re-evaluate the HACCP plan and make necessary adjustments and modifications to the whole HACCP plan.

The "Technical Report on Correction Measures" should be included in the HACCP support file.

12. 建立验证审核程序(原理六)

验证审核是指通过严谨科学的方法,确认危害分析与关键控制点体系是否需要修正,是否得到切实可行的落实,是否有效的过程。验证审核的对象是危害分析与关键控制点体系的计划。

1)确认危害分析与关键控制点体系

确认危害分析与关键控制点体系就是复查消费者投诉,确定是否与危害分析与关键控制点计划的实施有关,是否存在未确定的关键控制点;确认危害分析与关键控制点体系建立的充分性和必要性,危害分析与关键控制点体系是否能有效控制危害因素对食品安全性的侵袭。由危害分析与关键控制点体系实施小组或受过适当培训、有丰富经验的人员,针对危害分析与关键控制点体系中的每一个环节(确认的对象),结合基本的科学原理,应用实际生产中检测的数据和生产全过程中获得的观察检测结果,进行有效性评估,得出危害分析与关键控制点体系运行是否正确的结论。

2)危害分析的确认

危害分析的确认是对危害分析的可靠性进行确认。当企业有内外因素变化波及到危害分析时,要重新进行危害分析确认。

3)关键控制点的验证审核

关键控制点的验证审核有三个过程。

(1)校准及校准记录的复查。要对监控设备进行校准,确保设备灵敏

XII. Establish Verification and Audit Procedures (Principle 6)

Verification audit refers to the process of confirming whether the HACCP system needs to be revised, whether it is feasible to implement, and whether it is effective through rigorous and scientific methods. The object of validation audit is the plan of the HACCP system.

A. Confirm HACCP System

Confirming that the HACCP system is a review of consumer complaints to determine whether it is related to the implementation of the HACCP plan and whether there are critical control points that are not identified, confirm the adequacy and necessity of establishing HACCP system and whether it can effectively control the invasion of harmful factors to food safety. Through the HACCP implementation team or appropriately trained and experienced personnel, for every aspect (confirmed object) of the HACCP system, combining the basic scientific principles, the application of the actual data in production and the observation and detection results obtained in the whole production process, the validity evaluation is carried out, and the conclusion is drawn whether the HACCP system is running correctly or not.

B. Confirmation of Hazard Analysis

The confirmation of the hazard analysis is to confirm the reliability of the hazard analysis. When the enterprise has the internal and external factors changing to hazard analysis, it is necessary to reconfirm the hazard analysis.

C. Verification Audit of CCP

There are three processes to verify and audit the CCP.

1. Calibration and review of calibration records. To calibrate the monitoring equipment to

度符合要求。对设备校准记录(校准日期、校准方法、校准结果、校准结论)进行复查,确定设备灵敏度是否有效。

(2)有针对性的样品检测。对有怀疑的样品的中间产品、成品进行抽样检测,查看实际结果与标准的吻合程度。

(3)关键控制点的记录复查。着重复查关键控制点的记录和纠偏记录,如监控仪器的校准记录、监控记录、纠偏措施记录、产品大肠杆菌等的微生物检验记录等。查看关键控制点是否始终在安全参数范围内运行,发生与操作限值偏离的情形时是否进行了纠偏行动。

4)危害分析与关键控制点体系的验证审核

验证审核是为了检验危害分析与关键控制点体系计划与实际操作之间的符合率,以及危害分析与关键控制点体系的有效性。收集验证活动所需的所有信息,对危害分析与关键控制点体系及记录进行现场观察和复核,以完成对危害分析与关键控制点体系的验证审核工作。

审核危害分析与关键控制点体系的验证活动应包括以下内容:

(1)检查产品说明和生产流程图的准确性;

(2)检查生产中是否按照危害分析与关键控制点体系计划监控了关键控制点;

ensure that the equipment sensitivity meets the requirements, the calibration records (calibration date, calibration methods, calibration results, calibration conclusions) are reviewed to determine whether the equipment sensitivity is effective.

2. Targeted sample testing. Sampling and testing the intermediate products and finished products of suspected samples to check the degree of conformity between the actual results and the standard.

3. The review records of the CCP. Recheck the records of critical control points and correct errors, such as calibration records of monitoring instruments, monitoring records, records of corrective measures, microbiological inspection records of Escherichia coli and so on. See if the CCP is always running within the security parameter range, and whether corrective measures have been taken in the event of a deviation from the operating limit.

D. Verification and Auditing of HACCP System

Validation audits are designed to verify the compliance rate between the HACCP system plan and actual operations and the effectiveness of the HACCP system. Collect all the information required for verification activities, and conduct on-site observation and review of the HACCP system and records to complete the verification and audit of the HACCP system.

Validation activities for auditing the HACCP system shall include the following:

1. Check the accuracy of product description and process of flow sheet;

2. Check that CCP is monitored in production according to the HACCP system plan;

(3)检查所有参数是否在关键限值以内；

(4)记录结果是否在规定时间间隔内完成和是否如实记录；

(5)监控活动是否按照危害分析与关键控制点体系计划规定的频率执行；

(6)当出现关键控制点偏离时，是否有纠偏措施；

(7)设备仪器是否按照危害分析与关键控制点体系计划进行校准。

5)执法机构对危害分析与关键控制点体系的审核验证

执法机构对危害分析与关键控制点体系的审核验证通常分为内部验证和外部验证两类。内部验证由企业内危害分析与关键控制点实施小组进行，又称为内审；外部验证由政府检验机构或有资格的有关人士进行，又称为审核。

执法机构的验证内容有：对危害分析与关键控制点体系计划及其修改的复核；对关键控制点监控记录的复查；对纠偏记录的复查；对验证记录的复查；现场检查危害分析与关键控制点体系计划实施状况；复核危害分析与关键控制点体系计划的记录保存情况；随机抽样分析复核。

危害分析与关键控制点体系计划的确认每年至少一次，当出现影响危害分析与关键控制点体系计划的因素时，应及时进行确认。若确认结论表明危害分析与关键控制点体系计划有效性不符合要求时，应对原来的危害分析与关键控制点体系计划立即进行修订，使之符合要求。

3. Check that all parameters are within the critical limit;

4. Whether the recorded results are completed at specified intervals and whether the records are true;

5. Whether the monitoring activities are performed in accordance with the frequency specified by the HACCP system plan;

6. When the CCP deviation occurs, whether there are corrective measures;

7. Whether the equipment is calibrated according to the HACCP system plan.

E. Verification Audit of HACCP System by Law Enforcement Agencies

The verification audit of HACCP system by law enforcement agencies are usually divided into two types: internal verification and external verification. Internal validation is carried out by the HACCP implementation team in the enterprise, also known as internal audit; external verification is conducted by the government inspection structure or qualified persons, also known as audit.

The contents of the verification of the law enforcement agencies include: the review of the HACCP system plan and its modification; the review of the CCP monitoring record; the review of the rectification record; the review of the verification record; the on-site inspection of the implementation status of the HACCP system plan; review the record keeping of HACCP system plan; random sampling analysis review.

Validation of the HACCP system plan should be made at least once a year and should be confirmed in a timely manner when factors affecting the HACCP system plan arise. If it is confirmed that the validity of the HACCP system plan does not meet the requirements, the original HACCP system plan shall be revised immediately to make it meet the requirements.

13. 建立记录和文件的有效管理程序（原理七）

1）危害分析与关键控制点体系记录

企业是否有效执行了危害分析与关键控制点体系计划，危害分析与关键控制点体系计划的实施对食品安全性是否有效，最具有说服力的就是危害分析与关键控制点体系计划的记录和文件等书面证据。所以，危害分析与关键控制点体系计划的每一个步骤和与危害分析与关键控制点体系计划相关的每一个行为都要求有详尽、翔实的记录，并有效地保存下来。

（1）危害分析与关键控制点体系记录编制的原则：

①题目与内容，题目应简洁明了，内容能体现记录活动的关键特征，内容应完整、准确、简洁；

②形式统一，一般采用表格式，表格各项目之间逻辑正确；

③容易识别，便于企业和部门的识别，应有操作人员的签字和记录日期。

（2）危害分析与关键控制点体系记录包括：

①执行卫生标准操作程序的记录；

②执行危害分析与关键控制点体系计划的记录，包括监控记录、纠偏记录、危害分析与关键控制点体系验证记录、危害分析与关键控制点计划确认记录、危害分析记录、危害分析与关键控制点计划表等；

③书面危害分析和危害分析与关键控制点计划的批准，由企业最高管理者或其代表签署批准。当发生修改、验证、确认时，由企业最高管理者

XIII. Establishment of Effective Procedures for the Management of Records and Documents (Principle 7)

A. HACCP System Records

Whether the enterprise has effectively carried out the HACCP system plan and whether the implementation of the HACCP system plan is effective for food safety is the most convincing documentary evidence such as the records and documents of the HACCP system plan. Therefore, every step of the HACCP system plan and every behavior associated with the HACCP system plan requires detailed records and effective preservation.

1. Principles for the compilation of HACCP system records:

a. Title and content. The title should be concise and clear, and the content should reflect the key features of the recorded activities; the content should be complete, accurate and concise.

b. The form is unified. In general, table format is used, and the logic of each item in the table is correct.

c. Easy to identify. To facilitate the identification of enterprises and departments, the signature and record date of the operator shall be indicated.

2. HACCP system records include:

a. Records that execute SSOP.

b. Records of execution of HACCP system plan, including monitoring record, correction record, HACCP system verification record, HACCP plan confirmation record, hazard analysis record, HACCP schedule, etc.

c. The approval of written hazard analysis and HACCP plan: signed and approved by the supreme management or his representative; and re-signed and approved by the top manager or his

或其代表重新签署批准。

保存的记录应涵盖这样一些项目:说明危害分析与关键控制点体系的各种措施;危害分析采用的所有数据;危害分析与关键控制点体系实施小组会议报告和决议;监控方法和数据、记录;偏差及纠偏记录;验证记录;验证审核报告;危害分析工作表和危害分析与关键控制点体系计划表等各类表格。

记录中应反映的内容有:产品名称与生产地址;记录产生的日期和时间;操作者签字或署名;产品全过程监控情况的实际数据、观测资料和其他信息资料。

(3)重要的记录有:

①危害分析与关键控制点体系计划及支持性材料,包括危害分析与关键控制点体系实施小组成员及其职责,建立危害分析与关键控制点体系的基础工作,如有关科学研究、实验报告和实施危害分析与关键控制点体系的先决程序(良好的操作规范、卫生标准操作程序)等;

②确定关键限值的依据和验证关键限值的记录;

③关键控制点的监控记录;

④纠偏措施的记录;

⑤验证记录,包括监控设备的检查记录、半成品与产品检验记录、验证活动的结果记录等;

⑥修改危害分析与关键控制点体系计划(原辅料、配方、工艺、设备、包装、储运)后的确认记录;

representative in case of modification, validation and confirmation.

The records should cover the following items: description of the various measures of the HACCP system; all data used in hazard analysis; reports and resolutions of the implementation team of the HACCP system; monitoring methods and data, records; deviation and correction records; verification records; verification audit reports; hazard analysis worksheets and HACCP system schedules and other tables.

The contents to be reflected in the record are: product name and production address; date and time when the record was generated; sign or signature of the operator; actual data, observation data and other information data of the whole process of monitoring the product.

3. Important records include:

a. HACCP system planning and supporting materials, including HACCP implementation team members and their responsibilities, the basic work for the establishment of the HACCP system, such as scientific research, experimental reports and the implementation of the HACCP system procedures (GMP, SSOP)and so on;

b. The basis for determining the critical limit and the record for validating the critical limit;

c. CCP surveillance records;

d. The record of corrective measures;

e. Verification records, including inspection records of monitoring equipment, inspection records of semi-finished products and products, records of results of verification activities, etc.;

f. The confirmation record after modifying the HACCP system plan (raw and auxiliary materials, formulations, processes, equipment, packaging, storage and transportation);

⑦产品回收的记录；
⑧人员培训的记录；
⑨危害分析与关键控制点体系计划的验证审核记录。

记录的方式有表格式、文字式（各种报告）、图形式（生产流程图、监控检测图）等。所有的记录应该完整、准确、真实；每周审核记录一次，由审核人签名，注明日期。

记录的保存期限：冷藏产品，至少保存一年；冷冻或货架期稳定的产品，至少保存两年；其他说明加工设备与加工工艺等方面的研究报告、科学评估结果，至少保存两年。

记录应归档放置在安全、固定的场所，便于查阅。记录应专人保存，有严格的借阅手续。记录的保存一般可采用计算机或档案室。所有记录一律要求采用档案化保存。

2) 危害分析与关键控制点体系文件
(1) 危害分析与关键控制点体系文件编制的原则是：
①采用过程方法编制，明确过程运行的预期结果，分析表达各个过程之间的关系；

②全体员工执行危害分析与关键控制点手册的规定，将危害分析与关键控制点体系转化为具体的执行程序，要求员工的操作和危害分析与关键控制点手册规定保持一致；

③具有针对性和可操作性，要将危害分析与关键控制点体系理论与企业实际相结合；

④与支持性文件和记录保持有机

g. Records of product recycling；
h. Record of personnel training；
i. Validation audit records for the HACCP system plan.

The methods of recording are tabular, textual (various reports), graphic (process of flow sheet, monitoring and testing chart) and so on. All records shall be complete, accurate and true; the records shall be reviewed once a week, signed by the reviewer and dated.

Record retention period: refrigerated products for at least one year; frozen or shelf-life stable products for at least two years; and other studies, scientific assessments and so on describing processing equipment and processes for at least two years.

Records should be archived in a safe, fixed place for easy access. Records should be kept by special persons, with strict borrowing procedures. Record saving tools can generally be used in computers or archives. All records are required to be archived.

B. HACCP System File
1. The principles of HACCP system documentation are:

a. The process method is used to make clear the expected results of the operation of the process; analyze and express the relationship between the various processes.

b. All employees follow the HACCP manual. Translate the HACCP system into specific execution procedures and require employees to operate in accordance with the requirements of the HACCP manual.

c. It has pertinence and maneuverability. The theory of HACCP system should be combined with enterprise practice.

d. Maintain organic and complete contact with

的、完整的联系,要对执行危害分析与关键控制点体系所需要的支持性文件和记录提出具体要求。

(2)危害分析与关键控制点体系文件的组成：

①文件控制程序；

②良好的操作规范与卫生标准操作程序；

③设备维修保养控制程序；

④产品回收控制程序；

⑤产品识别代码控制程序；

⑥危害分析与关键控制点体系计划预备步骤控制程序；

⑦危害分析与关键控制点体系计划所有实施步骤的控制程序。

危害分析与关键控制点体系文件的内容包括目的、范围、职责、程序图（过程描述、相关记录、相关文件）。

危害分析与关键控制点体系的支持性文件：相关的法律和法规；相关的技术规范、标准、指南；相关的研究报告和技术报告（危害分析报告）；加工过程的工艺文件（作业指导书、设备操作规程、监控仪器校准规程、产品验收准则）；人员岗位职责和任职条件；相关管理制度。

危害分析与关键控制点体系的支持性文件是危害分析与关键控制点体系建立和实施的技术资源、技术保证、科学依据,也是进行食品无危害生产、保证食品安全的有力工具、标准及行为准则。

危害分析与关键控制点体系建立步骤如表 7-10 所示。

supporting documents and records. Specific requirements for supporting documents and records required to implement the HACCP system.

2. The composition of the HACCP system files：

a. File control procedures；

b. GMP and SSOP；

c. Equipment maintenance control procedures；

d. Product recovery control procedures；

e. Product identification code control procedures；

f. The preparatory step control procedures for the HACCP system plan；

g. The HACCP system plans control procedures for all implementation steps.

The content of the HACCP system document includes the purpose, scope, responsibility, procedure diagram (process description, related records, related documents).

The supporting document of HACCP system is composed of relevant laws and regulations, relevant technical specifications, standards and guidelines; relevant research reports and technical reports (hazard analysis report); process documents of the processing procedure (operating instructions, equipment operating procedures, calibration procedures for monitoring instruments, product acceptance criteria); post responsibilities and working conditions; related management system.

The supporting document of the HACCP system is the technical resource, technical guarantee and scientific basis for the establishment and implementation of HACCP system. It is also a powerful tool, standard and code of conduct for non-hazardous food production and food safety.

表 7-10 危害分析与关键控制点体系建立步骤

步骤	工作任务	责任人	工作内容和要点	输出
第一步：方针目标制定	制定食品安全方针	最高管理者	内容要求：①体现食品安全的重要性；②体现产品特点	写入质量手册，最高管理者签字下发
	制定食品安全总目标	最高管理者	内容要求：①体现产品安全性；②应可测量，以百分比、数量的形式体现	各部门食品安全分目标，直接领导签字下发
	制定各部门分目标	各部门负责人	要求：①分目标要有具体的测量方式和统计频率；②只可高于总目标，不可低于总目标	
第二步：人员准备	任命危害分析与关键控制点体系实施小组组长	最高管理者	要求：①从中高层领导中选择，该领导应德高望重，能有力支持质量体系的推行；②直接向最高管理者负责	危害分析与关键控制点体系实施小组组长任命书，由最高管理者签字任命
	选择成员组成危害分析与关键控制点体系实施小组	组长	要求：①每个部门选派1~2名内审员组成小组；②所有成员组合在一起使小组具有多学科的知识，全面熟悉企业现状	小组任命书，由组长签字任命
	内审员培训	组长	要求：①组长、所有内审员应参加；②培训内容包括GB/T 27341、GB 14881、ISO19011及其他相关法律法规，培训时间不少于30小时内部审核员培训证书	
第三步：部门职责确定	制定组织机构框架，明确部门职责	最高管理者	要求：①各部门职责应明确，杜绝一事多人管，一事无人管的情况；②将GB/T 27341所有条款要求对应到每个部门，不应缺少	推荐：组织机构图、职能分配表、部门职责描述等

续表 7-10

步骤	工作任务	责任人	工作内容和要点	输出
第四步：文件编写	编写危害分析与关键控制点体系文件	组长/小组	文件内容包括危害分析与关键控制点手册、程序文件、危害分析与关键控制点计划、卫生标准操作程序、各部门使用的作业文件，以及按文件要求工作时填写的工作表格	前述文件
第五步：文件学习	文件下发试运行、文件内容培训	组长/小组	要求：①将文件下发至各部门试运行；②以一定频率走访部门，了解文件适用性，并对文件内容进行培训	文件更改审批记录，文件内容学习培训记录
第六步：正常运作	各部门按文件要求运作	各部门	要求：①严格按照文件规定运行；②做任何事情都要有记录	各部门工作记录
	验证	各部门负责人/内审员	要求：各部门领导或内审员按文件规定的要求进行检查	各种检查记录
第七步：内审和管理评审	内部审核	组长/小组	要求：①按 GB/T 27341 全条款审核，全部门参与；②各部门选派审核员组成审核组，不能自审	内审计划、检查表、不合格报告、内审报告
	管理评审	最高管理者	要求：①内审结束后一周内进行，内审组、各部门直接领导参会；②解决内审出现的不合格情况；③提出改进措施	管理评审报告
第八步：外审	申请外部审核	组长	要求：危害分析与关键控制点体系稳定运行三个月以上可以申请外审	不合格报告、外部审核报告，如通过认证获得认证证书

7.4.2 实训任务

实训组织：
1. 设计某食品厂食品安全的方针、目标。
2. 设计危害分析与关键控制点体系实施小组组长及成员的任命书，包括工作职责描述。
3. 设计食品厂组织机构图、职能分配表。

实训成果：有关食品安全的方针、目标，任命书，组织机构图，职能分配表，为《危害分析与关键控制点手册》前半部分作准备。

实训评价：5分制，上述成果每项各1分。要求内容全面，条理清楚。

任务5 危害分析与关键控制点体系文件编写方法

7.5.1 基础知识

1. 文件编写技巧

原则上文件编写没有格式要求，但作为初学者，按照下述技巧进行编写，可使文件更规范。

(1)文件格式：封面、修订页、正文、附件、附表。
(2)封面：文件名称、文件编号、制定部门、生效日期、制定、审查、批准、密级。
(3)正文：目的、范围、定义、职责、内容、参考文件、记录表单、附件。
(4)页眉：公司名称、文件名称、文件编号、生效日期、版本号、页码、制定部门、密级。
(5)页脚：制定、审查、批准。

2. 危害分析与关键控制点计划编写思路

1) 原料、成品描述

在原料库或成品库中寻找所有产品（原料和成品），包括内包装材料的标签，从中找到需要的内容。描述以表格形式体现，内容如表7-11和表7-12所示。

表7-11 原料描述模板

产品名称：	
化学、生物和物理特性	
包装和交付方式	
产地	
生产方法	
贮存条件和保质期	
使用或生产前的预处理	
产品接收标准	
配制辅料的组成	

表 7-12 成品描述模板

产品名称：	
食用方法	
包装方式	
产品特性	
保存期限	
加工方法	
销售对象	
标签说明	
产品标准	
保存条件	
成品规格	

2）工艺流程图的绘制和确认

注意返工点、循环点、废弃物排放点、产品排出点等。必须到生产现场进行确认，以保证与生产实际相符。工艺流程图（图 7-4）必须附工艺说明，将其中的每一个步骤拿出来，详细说明其加工参数。

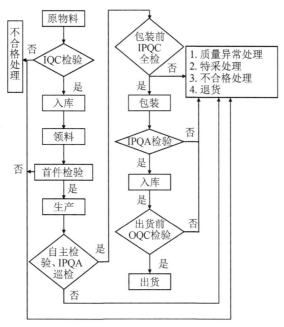

图 7-4 工艺流程图模板

IQC—来料质量控制；IPQC—过程质量控制；IPQA—过程质量保证；OQC—成品出厂检验

3）危害分析

危害分析可采用危害分析工作单的形式体现，如表 7-2 或表 7-9 所示。将上述流程图中的每一个步骤放在"加工步骤"栏中，然后进行危害分析。

4)危害分析与关键控制点计划表

将危害分析工作单中"是否关键控制点"栏判断为"是"的步骤放到危害分析与关键控制点计划表(表7-3)中。

上述内容再加上危害分析与关键控制点实施小组组长的任命书,构成企业危害分析与关键控制点计划的内容。

7.5.2 实训任务

实训组织:分组编写乳品液奶企业、蛋糕生产企业、肉制品生产企业的危害分析与关键控制点计划。

实训成果:"××产品危害分析与关键控制点计划",5000字以上。

实训评价:5分制,原料、成品描述1分,工艺流程图及说明1分,危害分析工作单2分,危害分析与关键控制点计划表1分。要求内容全面,条理清楚。

任务6 内审和管理评审

7.6.1 基础知识

内部审核(简称内审)是企业自我评价的一种手段。危害分析与关键控制点体系内审的步骤和要求与ISO9001质量管理体系内审相同,不同之处在于审核时的侧重点。质量管理体系侧重于顾客满意,危害分析与关键控制点体系侧重于产品是否安全,是否会对消费者造成伤害。本部分只介绍危害分析与关键控制点体系内部审核的重点。

1. 危害分析审核要点

在审核危害分析时,要对与食品生产有关的所有环节进行危害分析,应重点注意危害识别是否全面,包括原材料验收、原材料入库、加工制造、贮存、运输、销售直到与消费有关的全部环节。在审核时要特别关注每个生产步骤引入的、产生的或增加的潜在危害来源及最终产品的可接受水平,必要时审核其有无法律依据、危害发生概率和严重性评估等。

Internal review (abbreviated as internal audit) is a means of enterprise self-evaluation. The steps and requirements of the internal audit of HACCP system are the same as those of ISO9001 quality management system, while the difference lies in the emphasis of audit. The quality management system focuses on customer satisfaction and HACCP system focuses on whether the product is safe or not and whether it will cause harm to consumers. This section only introduces the key points of the internal audit of HACCP system.

Ⅰ. **Audit Points for Hazard Analysis**

When examining the hazard analysis, we should carry out the hazard analysis on all links related to the food production, and should pay attention to whether the hazard identification is comprehensive or not. Acceptance of raw materials, storage of raw materials, processing, manufacturing, storage, transportation, sales and all links related to consumption. In the audit, special attention should be paid to the source of potential hazards introduced, generated or

审核时要注意产品是否包含敏感微生物。例如,肉罐头类重点关注肉毒梭菌,婴幼儿奶粉重点关注阪崎肠杆菌;加工过程中是否有有效消灭微生物的处理步骤;是否存在加工后微生物及其毒素污染的明确危害;在批发和消费过程中是否有由于不良习惯造成危害的可能性;在包装后或家庭食用前是否进行最后的加热处理等。

在审核过程中要明确,通过危害分析,即使不存在显著危害,仍可以制订并实施危害分析与关键控制点计划,以满足产品品质及贸易上的要求。

2. 关键控制点审核要点

在审核过程中,要重点审核影响产品安全的关键控制点。要求审核人员必须懂得生产,熟悉加工工艺。审核企业对关键控制点能否做到有效监控,即监控人员资质能否满足关键点的要求,监控频率是否合理,监控人员是否按照规定的方法实施监控,关键控制点的监控设备是否得到有效校准,监控偏离的情况能否得到有效纠偏,是否定期对监控效果进行验证等。

increased by each production step and the acceptable level of the final product. If necessary to review whether there is a legal basis, the probability of occurrence of harm and severity assessment, and so on.

In the audit, we should pay attention to whether the product contains sensitive microorganisms, such as clostridium botulinum in canned meat and enterobacter sakazakii in infant milk powder. Whether there are effective steps to eliminate microorganisms during processing; whether there is a clear hazard of contamination of processed microorganisms and their toxins; and whether there is a possibility of harm caused by bad habits in wholesale and consumer processes; whether to carry on the final heating treatment after packing or before the domestic food.

During the audit process, we should be clear, through hazard analysis, even if there is no significant harm, we can still formulate and implement the HACCP plan to meet the product quality and trade requirements.

II . Key Points of the CCP Audit

In the process of audit, the critical control points that affect the safety of products should be audited. The auditor must be familiar with production and process technology. Check whether the enterprise can effectively monitor the critical control points, that is, whether the qualification of the monitoring personnel can meet the requirements of the key points, whether the frequency of monitoring is reasonable, and whether the monitoring personnel carry out monitoring in accordance with the prescribed methods, whether the monitoring equipment of the critical control point can be effectively calibrated, whether the monitoring deviation can be effectively corrected, whether the monitoring effect is verified regularly, and so on.

7.6.2 实训任务

实训组织:编制关键控制点的监控检查表。
实训成果:关键控制点的监控检查表,200字以上。
实训评价:5分制。要求内容全面,条理清楚。

思考题

1. 卫生标准操作程序(SSOP)与良好的操作规范(GMP)、危害分析与关键控制点(HACCP)的关系是怎样的?
2. 什么样的点是关键控制点(CCP)?
3. 天然化学物危害及预防措施有哪些?
4. 什么是危害分析与关键控制点体系的审核?

拓展学习网站

1. 食品伙伴网(http://www.foodmate.net)
2. 中国质量认证中心(http://www.cqc.com.cn)
3. 美国食品药物管理局危害分析与关键控制点体系(http://www.fda.gov/Food/GuidanceRegulation/HACCP)

项目 8　ISO22000 食品安全管理体系的建立与认证

项目概述

通过与项目 7 "危害分析与关键控制点(HACCP)体系的建立与认证"的比较,学习 ISO22000 食品安全管理体系的相关内容。

任务 1　ISO22000 食品安全管理体系概况

8.1.1　基础知识

1. ISO22000 的由来

食品安全管理体系(英文简称 FSMS)是以《食品安全管理体系 食品链中各类组织的要求》(ISO22000:2005)为核心的可用于国际标准化组织认证注册的管理体系。它在 ISO9001 质量管理体系的基础之上,将危害分析与关键控制点原理和 ISO9001 框架结合,使危害分析与关键控制点原理进入国际标准化组织的管理体系认证领域,从而促进了危害分析与关键控制点体系在全球的推广。食品安全管理体系的历史和危害分析与关键控制点体系的发展密不可分。

1)危害分析与关键控制点体系的创立阶段

危害分析与关键控制点体系是 20 世纪 60 年代由美国 Pillsbury 公司 H. Bauman 博士等与宇航局和美国陆军 Natick 研究所共同开发的,主要用于航天食品中。1971 年在美国第一次国家食品保护会议上提出了危害分析与关键控制点原理,立即被食品药物管理局(FDA)接受,并决定在低酸罐头食品的良好操作规范中采用。食品药物管理局于 1974 年公布了将危害分析与关键控制点原理引入低酸罐头食品的良好操作规范。1985 年美国科学院(NAS)就食品法规中危害分析与关键控制点的有效性发表了评价结果。随后由美国农业部食品安全检验署(FSIS)、美国陆军 Natick 研究所、食品药物管理局、美国海洋渔业局(NMFS)四家政府机关及大学和民间机构的专家组成的美国食品微生物学基准咨询委员会(NACMCF)于 1992 年采纳了食品生产的危害分析与关键控制点七原则。1993 年食品法典委员会(CAC)批准了《危害分析与关键控制点体系应用准则》,1997 年颁发了新版法典指南《危害分析与关键控制点体系及其应用准则》,该指南已被广泛地接受并得到了国际上的普遍采纳,危害分析与关键控制点概念已被认可为世界范围内生产安全食品的准则。

2)危害分析与关键控制点体系的应用阶段

联合国粮农组织(FAO)和世界卫生组织(WHO)在 20 世纪 80 年代后期就大力推荐危害分析与关键控制点体系,至今不懈。1993 年 6 月食品法典委员会考虑修改《食品卫生的一般性原则》,把危害分析与关键控制点体系纳入该原则内。根据世界贸易组织的协议,食品法典委员会制定的法典规范或准则被视为衡量各国食品是否符合卫生、安全要求的尺度。

另外有关食品卫生的欧共体理事会指令 93/43/EEC 要求食品工厂建立危害分析与关键控制点体系以确保食品安全的要求。在美国,食品药物管理局在 1995 年 12 月颁布了强制性水产品的危害分析与关键控制点体系法规,又宣布自 1997 年 12 月 18 日起所有对美出口的水产品企业都必须建立危害分析与关键控制点体系,否则其产品不得进入美国市场。食品药物管理局鼓励并最终要求所有食品工厂都实行危害分析与关键控制点体系。另一方面,加拿大、澳大利亚、英国、日本等国也都在推广和采纳危害分析与关键控制点体系,并分别颁发了相应的法规,针对不同种类的食品分别提出了危害分析与关键控制点模式。

开展危害分析与关键控制点体系的领域包括饮用牛乳、奶油、发酵乳、乳酸菌饮料、奶酪、冰淇淋、生面条类、豆腐、鱼肉火腿、炸肉、蛋制品、沙拉类、脱水菜、调味品、蛋黄酱、盒饭、冻虾、罐头、牛肉食品、糕点类、清凉饮料、腊肠、机械分割肉、盐干肉、冻蔬菜、蜂蜜、高酸食品、肉禽类、水果汁、蔬菜汁、动物饲料等。

3)我国危害分析与关键控制点体系的应用发展情况

中国食品和水产界较早关注和引进危害分析与关键控制点体系质量保证方法。1991 年农业部渔业局派遣专家参加了美国食品药物管理局等组织的危害分析与关键控制点体系研讨会,1993 年国家水产品质检中心在国内成功举办了首次水产品危害分析与关键控制点体系培训班,介绍了危害分析与关键控制点体系的原则、水产品质量保证技术、水产品危害及监控措施等。1996 年农业部结合水产品出口贸易形势颁布了冻虾等五项水产品行业标准,并进行了宣讲贯彻,开始了较大规模的危害分析与关键控制点体系的培训活动。2002 年 12 月中国认证机构国家认可委员会正式启动对危害分析与关键控制点体系认证机构的认可试点工作,开始受理危害分析与关键控制点体系认可试点申请。

4)《食品安全管理体系 食品链中各类组织的要求》(ISO22000:2005)的产生

随着危害分析与关键控制点原理的广泛应用和推广,监管机构和食品加工企业逐渐看到危害分析与关键控制点体系的优势。但是,包含危害分析与关键控制点原理的法规增多和技术标准的不统一,使食品制造商难以应付。为满足各方面的要求,在丹麦标准协会的倡导下,通过国际标准化组织的协调,将相关的国家标准在国际范围内进行整合,最终形成统一的国际食品安全管理体系。

2001 年初,丹麦标准协会向 ISO/TC34 食品生产秘书处递交了 ISO22000《食品安全管理体系 要求》的提案,TC34 的 14 个成员愿意参加 ISO22000《食品安全管理体系 要求》新标准的制定工作,并建议为该标准的起草成立一个工作组(ISO/TC34/WG)。因此,这一新的国际标准是在国际标准化组织设立在匈牙利标准局(MSZT)负责食品产品的第 34 技术委员会第 8 工作组(ISO/TC34/WG8)直接负责下起草的。除了许多欧洲国家外,阿根廷、澳大利亚、印度尼西亚、坦桑尼亚、泰国、美国和委内瑞拉也提名专家为这一工作组工作。共有 17 个国家和 3 个国际组织(食品法典委员会、食品行业论坛和世界食品安全组织)以及欧盟食品和饮料工业联合会参与此项目的工作。根据工作组 2001 年 11 月份在哥本哈根召开的第一次会议要求,ISO/TC34(ISO 食品技术委员会)/WG8(食品安全管理体系工作组)起草了标准《食品安全管理体系 食品链中各类组织的要求》(ISO22000),该标准经历了工作组草案(WD)、委员会草案(CD)、国际标准草案(DIS)、最终国际标准草案(FDIS)、国际标准(IS)几个阶段,于 2005 年 9 月 1 日正式发布。该标准是认证机构实施食品安全管理体系认证的依据。

5) ISO22000 族标准的组成

(1) ISO22000:2005《食品安全管理体系 食品链中各类组织的要求》于 2005 年 9 月 1 日发布,它提供了全球食品行业产品接受的统一标准。该标准由国际标准化组织来自食品行业的专家,通过与食品法典委员会、联合国粮农组织和世界卫生组织紧密合作产生。ISO22000:2005 是 ISO22000 家族标准中的第一个标准,这一标准可以单独用于认证、内审或合同评审,也可与其他管理体系,如 ISO9001 合并实施。

(2) ISO/TS22004:2005《食品安全管理体系 ISO22000:2005 应用指南》主要是帮助全球的中小企业建立和实施 ISO22000 标准。

(3) ISO/TS22003:2007《食品安全管理体系 对实施食品安全管理体系认证和审核的机构的要求》给出对从事 ISO22000 进行认证审核的机构的认可要求。

(4) ISO22005:2007《饲料和食品链的可追溯性 体系设计和开发的通用原理和指南》为饲料和食品链的可追溯性体系设计和发展提供了总的导则。

2. 食品安全管理体系对食品企业的作用

1) 安全的防护堤

搞好食品安全管理可以防范、减少食物中毒和食源性疾病的发生,有助于保障消费者身体健康。

2) 贸易的通行证

自从我国加入世界贸易组织以来,我国的食品进出口贸易越来越多地受到国际通行准则的影响和限制。西方国家有较为完善的食品安全管理系统,安全管理模式已成为国家通用的标准和进入欧美市场的通行证。我国顺应现代经济潮流,使企业按国际通用标准生产出高质量的安全产品,从而更有利于参与国际竞争,提高经济效益。

3) 发展的动力源

食品企业的管理人员、技术人员和工作人员,都应懂得食品安全管理的基础知识,从整体上把握安全管理的共性,以更好地应用先进、科学的安全管理方法,全面提高企业的安全管理水平。

8.1.2 实训任务

实训组织:用图表的形式总结 ISO22000 的来龙去脉,每一步骤描述字数不超过 30 字。

总结图表模板

实训成果:ISO22000 由来总结图表。

实训评价:5 分制,每个步骤 1 分。

任务 2　ISO22000:2005 标准内容

8.2.1 基础知识

实际上,ISO22000:2005 与 GB/T 27341—2009 在内容上有很多共性部分。因此,本部分通过二者对比的形式介绍前者的内容。

表 8-1　ISO22000:2005 与 GB/T 27341—2009 的对照

ISO22000:2005				GB/T 27341—2009
引言			引言	
范围	1	1	范围	
规范性引用文件	2	2	规范性引用文件	
术语和定义	3	3	术语和定义	
食品安全管理体系	4	4	企业 HACCP 体系	
总要求	4.1	4.1	总要求	
文件要求 总则	4.2 4.2.1	4.2 4.2.1 4.2.2	文件要求 总则 HACCP 手册	
文件控制 记录控制	4.2.2 4.2.3	4.2.3 4.2.4 7.9	文件控制 记录控制 HACCP 计划记录的保持	
管理职责	5	5	管理职责	
管理承诺	5.1	5.1	管理承诺	
食品安全方针	5.2	5.2	食品安全方针	
食品安全管理体系策划	5.3			
职责和权限	5.4	5.3	职责权限与沟通	
食品安全小组组长	5.5	5.3.1	职责、权限	
沟通 外部沟通 内部沟通	5.6 5.6.1 5.6.2	5.3.2	沟通	
应急准备和响应	5.7	6.8	应急预案	
管理评审 总则 评审输入 评审输出	5.8 5.8.1 5.8.2 5.8.3	5.5	管理评审	
资源管理	6			
资源提供	6.1			
人力资源 总则 能力、意识和培训	6.2 6.2.1 6.2.2	6.2	人力资源保障计划	

续表 8-1

ISO22000:2005		GB/T 27341—2009	
基础设施	6.3	6.3	良好生产规范(GMP)
工作环境	6.4		
安全产品的策划和实现	7	7	HACCP计划的建立和实施
总则	7.1	7.1	总则
前提方案(PRP(s))	7.2	6	前提计划
	7.2.1	6.1	总则
	7.2.2	6.3	良好生产规范(GMP)
	7.2.3	6.4	卫生标准操作程序(SSOP)
		6.5	原辅料、食品包装材料安全卫生保障制度
		6.6	维护保养计划
实施危害分析的预备步骤	7.3	7.2	预备步骤
总则	7.3.1		
食品安全小组	7.3.2	7.2.1	HACCP 小组的组成
产品特性	7.3.3	7.2.2	产品描述
预期用途	7.3.4	7.2.3	预期用途的确定
流程图、过程步骤和控制措施	7.3.5	7.2.4	流程图的制定
		7.2.5	流程图的确认
危害分析	7.4	7.3	危害分析和制定控制措施
总则	7.4.1		
危害识别和可接受水平的确定	7.4.2	7.3.1	危害识别
危害评价	7.4.3	7.3.2	危害评估
控制措施的选择和评价	7.4.4	7.3.3	控制措施的制定
		7.3.4	危害分析工作单
操作性前提方案的建立	7.5		
HACCP 计划的建立	7.6	7	HACCP 计划的建立和实施
HACCP 计划	7.6.1	7.1	总则
关键控制点(CCPs)的确定	7.6.2	7.4	关键控制点(CCP)的确定
关键控制点的关键限值的确定	7.6.3	7.5	关键限值的确定
关键控制点的监视系统	7.6.4	7.6	CCP 的监控
监视结果超出关键限值时采取的措施	7.6.5	7.7	建立关键限值偏离时的纠偏措施
预备信息的更新、描述前提方案和HACCP计划的文件的更新	7.7	7.1	总则
验证的策划	7.8	7.8	HACCP 计划的确认和验证
可追溯性系统	7.9	6.7.1	标识和追溯计划

续表 8-1

ISO22000:2005		GB/T 27341—2009	
不符合控制	7.10		
纠正	7.10.1		
纠正措施	7.10.2		
潜在不安全产品的处置	7.10.3		
撤回	7.10.4	6.7.2	产品召回计划
食品安全管理体系的确认、验证和改进	8		
总则	8.1		
控制措施组合的确认	8.2	7.8	HACCP 计划的确认和验证
监视和测量的控制	8.3	7.8	"监控设备校准记录的审核,必要时,应通过有资格的检验机构,对所需的控制设备和方法进行技术验证,并提供形成文件的技术验证报告。"
食品安全管理体系的验证	8.4		
内部审核	8.4.1	5.4	内部审核
单项验证结果的评价	8.4.2	7.8	HACCP 计划的确认和验证
验证活动结果的分析	8.4.3	7.8	HACCP 计划的确认和验证
改进	8.5		
持续改进	8.5.1		
食品安全管理体系的更新	8.5.2		

从表 8-1 可以看出 GB/T 27341—2009 与 ISO22000:2005 内容基本相同,但存在以下不同点。

(1)ISO22000:2005 没有要求必须要编写"HACCP 手册",而 GB/T 27341—2009 要求(4.2.2)。但在实际运作过程中,绝大部分企业还是会编写"HACCP 手册",尤其是食品安全管理体系和质量管理体系同时运行时。

(2)ISO22000:2005 要求进行危害分析(7.4),危害分析可以任何书面形式体现。而 GB/T 27341—2009 将危害分析的形式固定为"危害分析工作单"(7.3.4)。

(3)ISO22000:2005 更加详细地规定了不合格品的管理(7.10),这是借鉴了 ISO9001 的经验。而 GB/T 27341—2009 只对已不在企业控制范围内的产品出现不合格需进行召回(6.7.2)进行了规定。

(4)ISO22000:2005 提出了改进(8.5)的要求,这也是参考了 ISO9001 的结构和要求。

(5)食品安全管理体系除 ISO22000:2005 有一个总的通行的要求外,还有针对不同食品企业的 24 个专项要求。例如,CNCA/CTS 0006—2008《食品安全管理体系 谷物磨制品生产企业要求》、CNCA/CTS 0009—2008《食品安全管理体系 制糖企业要求》、CNCA/CTS

0013—2008《食品安全管理体系 烘培食品生产企业要求》。而 GB/T 27341—2009 没有专项。

应说明的是,食品安全管理体系的建立、文件编写、内部审核和管理评审的程序和方法与 GB/T 27341—2009 基本一致,只需在上述不同点予以注意。

8.2.2 实训任务

实训组织:上网查找有关 ISO22000:2005"7.10.4 撤回"和"7.9 可追溯性系统"的案例各 3 个,并进行分析,判断是否符合标准及其原因。

实训成果:案例综述类文章,总字数 500 字以上。

实训评价:5 分制。要求内容全面,条理清楚。

任务 3　危害分析与关键控制点体系、食品安全管理体系认证实施规则比较

8.3.1　基础知识

本部分内容对 CNCA-N-008:2011《危害分析与关键控制点(HACCP)体系认证实施规则》(简称 HACCP 认证规则)和 CNCA-N-007:2010《食品安全管理体系认证实施规则》(简称 FSMS 认证规则,国家认监委 2010 年第 5 号公告)进行对比,从中学习食品安全管理体系认证公司的选择、认证实施过程和认证后管理的相关知识。

食品安全管理体系认证规则和危害分析与关键控制点体系认证规则不仅在结构上完全相同,在内容上也具有很强的一致性。其实,危害分析与关键控制点体系认证规则是在食品安全管理体系认证规则的基础上编写的,而且大部分认证机构是在首先获得食品安全管理体系认证资格的基础上申请危害分析与关键控制点体系认证资格。为避免内容拖沓,本部分只对食品安全管理体系认证规则和危害分析与关键控制点体系认证规则的不同部分进行解释。

1. 人员要求

食品安全管理体系对认证人员有了更详细的要求。通常,获得食品安全管理体系注册审核员资格需要经过培训→中国认证认可协会(CCAA)全国统考→积攒审核经历→在中国认证认可协会注册。而要获得危害分析与关键控制点体系注册审核员资格,其申请人应首先取得食品安全管理体系审核员注册资格(含实习)。也就是说,只有先获得食品安全管理体系注册审核员资格的人,才可能成为危害分析与关键控制点体系注册审核员。如只在企业内部实施内审,其审核员应经过培训,需要获得内审员证书方可从事相关工作。

2. 专项技术要求

食品安全管理体系在审核依据,也就是审核遵循的标准部分,"认证机构实施食品安全管理体系认证时,在以上基本认证依据要求的基础上,还应将本规则规定的专项技术规范作为认证依据同时使用(见附件)"(详见条款 4.2)。也就是说,在审核食品安全管理体系时不仅需要遵守 ISO22000:2005 标准的要求,对于不同的食品企业还应遵循专项技术规范。而危害分析与关键控制点体系审核只遵循 GB/T 27341—2009 这个通用的标准。

3. 申请评审

食品安全管理体系要求"认证机构应根据认证依据、程序等要求,在 15 个工作日对申请人提交的申请文件和资料进行评审并保存评审记录"(详见条款 5.2.2)。而危害分析与关键控制点体系要求"认证机构应在申请人提交材料齐全后 10 个工作日内对其提交的申请文件和资料进行评审并保存评审记录"(详见条款 5.2.2)。食品安全管理体系要求 15 天,危害分析与关键控制点体系要求 10 天。

4. 认证机构应根据审核需要组成审核组

食品安全管理体系要求"审核组成员身体健康,并有健康证明","审核组如果需要技术专家提供支持,技术专家应具有大学本科以上的学历,身体健康具有健康证明"(详见条款 5.3.1)。食品安全管理体系要求审核员和技术专家有健康证,而危害分析与关键控制点体系规定得很笼统。

5. 阶段审核重点

食品安全管理体系认证规则没有对第一阶段和第二阶段审核的审核重点进行阐述,而危害分析与关键控制点体系则有具体的说明,详见条款 5.4.1.1 和 5.4.1.2。

6. 是否通知现场审核

食品安全管理体系认证规则没有"不通知现场审核"的要求,而危害分析与关键控制点体系有要求,详见条款 5.6.2.2。

7. 信息报告

食品安全管理体系要求审核计划的通报:当受审核方为出口企业时,向受审核方所在地的直属出入境检验检疫局通报;非出口企业向省级质量技术监督局通报。而危害分析与关键控制点体系要求无论任何类型的企业,向中国食品农产品认证信息系统通报。同时,食品安全管理体系要求每年 11 月底向认监委通报工作报告,而危害分析与关键控制点体系要求每年 2 月底报告。

8.3.2 实训任务

实训组织:将食品安全管理体系认证规则和危害分析与关键控制点体系认证规则的区别以表格的形式进行总结,文字要求简练,每个空格字数少于 30 字。

食品安全管理体系认证规则和危害分析与关键控制点体系认证规则对比模板

食品安全管理体系认证规则		危害分析与关键控制点体系认证规则	
	条款		条款

实训成果:食品安全管理体系认证规则和危害分析与关键控制点体系认证规则对比表格。

实训评价:5 分制。要求内容全面,条理清楚。

思考题

ISO22000 食品安全管理体系的作用有哪些?

拓展学习网站

1. 食品伙伴网(http://www.foodmate.net)
2. 中国质量认证中心(http://www.cqc.com.cn)
3. ISO22000:2005 食品安全管理体系(http://www.iso.org/standard/35466.html)

项目 9 绿色食品与有机食品的认证

项目概述

本项目分别从申报绿色食品、有机食品的条件、申报程序、所需上报材料及各种申报表的填写方法等方面详细介绍一个企业如何进行绿色食品及有机食品的认证。

任务 1 绿色食品认证

9.1.1 基础知识

1. 绿色食品认证的机构和性质

1) 绿色食品认证机构

绿色食品认证是农产品质量认证体系的组成部分。中国绿色食品发展中心成立于1992年,是负责绿色食品标志许可、有机农产品认证、农产品地理标志登记保护、协调指导地方无公害农产品认证工作的"三品一标"专门机构,同时负责农产品品质规格、营养功能评价鉴定,协调指导名优农产品品牌培育、认定和推广等工作。中心为隶属于农业农村部的正局级事业单位,与农业农村部绿色食品管理办公室合署办公,内设办公室、财务处、体系标准处、审核评价处、标识管理处、地理标志处、品牌发展处、基地建设处、国际合作与信息处、中绿华夏有机食品认证中心等10个处室和部门。中国绿色食品协会秘书处和优质农产品开发服务协会秘书处挂靠中心,中心内部还设有北京中绿田源农业发展有限公司和后勤服务中心。

中国绿色食品发展中心职能包括:①参与绿色食品、有机农产品和地理标志农产品发展有关规章制度、规划计划、政策措施的拟订及实施;②负责绿色食品、有机农产品认证及农产品地理标志登记审查;③承担绿色食品、有机农产品标志授权管理和产品质量跟踪检查的具体工作;④参与拟订绿色食品、有机农产品和地理标志农产品相关质量标准、技术规范并组织实施,开展有关国际交流与技术合作;⑤开展绿色食品、有机农产品和地理标志农产品生产基地创建、技术推广和宣传培训,承办农产品品牌培育、市场发展和展示推介等工作;⑥组织实施绿色食品、有机农产品和农产品地理标志登记相关检验检测工作;⑦组织开展农产品品质规格、营养功能评价鉴定等相关工作;⑧协调指导名优农产品品牌培育、认定和推广等工作;⑨协调指导地方无公害农产品认证相关工作;⑩承担农业农村部交办的其他工作。

2) 中国绿色食品认证的性质

中国绿色食品实行统一、规范的标志管理,即通过对合乎特定标准的产品发放特定的标志,用以证明产品的特定身份以及与一般同类产品的区别。从形式上看,绿色食品标志管理是一种质量认证行为,但绿色食品标志是在国家工商行政管理局(现为国家市场监督管理局)注册的一个商标,受《中华人民共和国商标法》严格保护,在具体运作上完全按商标性质处理。因此,绿色食品标志管理实现了质量认证和商标管理的结合。

2. 绿色食品相关知识介绍

1) 绿色食品概念

绿色食品是指遵循可持续发展原则，按照特定生产方式生产，经专门机构认定，许可使用绿色食品标志商标的无污染的安全、优质、营养类食品。简单说就是无污染的安全、优质、营养类食品。

2) 绿色食品特征

绿色食品与普通食品相比有三个显著特征。

(1) 强调产品出自最佳生态环境。绿色食品生产从原料产地的生态环境入手，通过对原料产地及其周围的生态环境因子严格监测，判定其是否具备生产绿色食品的基础条件。

(2) 对产品实行全程质量控制。绿色食品生产实施"从土地到餐桌"全程质量控制。通过产前环节的环境监测和原料检测，产中环节具体生产、加工操作规程的落实，以及产后环节产品质量、卫生指标、包装、保鲜、运输、储藏、销售控制，确保绿色食品的整体产品质量，并提高整个生产过程的标准化水平和技术含量，而不是简单地对最终产品的有害成分和卫生指标进行测定。

(3) 对产品依法实行标志管理。绿色食品标志是一个质量证明商标，属知识产权范畴，受《中华人民共和国

Ⅱ. Introduction of Knowledge on Green Food

A. The Concept of Green Food

Green food is a safe, high-quality and nutritious food that is produced in accordance with the principle of sustainable development, produced in accordance with specific production methods and approved by specialized agencies. To put it simply, it is a safe, high-quality and nutritious food without pollution.

B. The Characteristics of Green Food

Green food has three distinct characteristics compared with common food.

1. Emphasize that the product comes from the best ecological environment. The production of green food starts with the ecological environment of a raw material producing area and determines whether it has the basic condition of producing green food by strictly monitoring the ecological environment factors of the raw material producing area and its surrounding.

2. Carry out the whole process quality control to the product. Green food production implement the whole process of quality control "from the land to the table". Through the environmental monitoring and raw material detection of the pre-production links, the implementation of the specific production and processing operation rules in the mid-production links, and the control of the product quality, hygiene indicators, packaging, preservation, transportation, storage and sales control in the post-production links, and to ensure the overall product quality of green food and improve standardization and technical content throughout the production process, rather than simply measuring harmful components and hygiene indicators of the final product.

3. Carry out mark management of products according to law. The green food logo is a quality certification trademark, which belongs to the

商标法》保护。绿色食品标志管理的手段包括技术手段和法律手段。技术手段是指按照绿色食品标准体系对绿色食品产地环境、生产过程及产品质量进行认证,只有符合绿色食品标准的企业和产品才能使用绿色食品标志商标。法律手段是指对使用绿色食品标志的企业和产品实行商标管理。绿色食品标志商标已由中国绿色食品发展中心在国家工商行政管理局注册,专用权受《中华人民共和国商标法》保护。

category of intellectual property rights and is protected by the *Trademark Law of the People's Republic of China*. The means of green food label management include technical means and legal means. The technical means is to certify the environment, production process and product quality of green food production area according to the green food standard system. Only enterprises and products that meet the green food standard can use green food trademark. Legal means refers to the use of green food logo enterprises and products to implement trademark management. The trademark of green food has been registered with the State Administration for Industry and Commerce by the China Green Food Development Center and the exclusive right is protected by the *Trademark Law of the People's Republic of China*.

3) 绿色食品分级

从 1996 年起,绿色食品采用分级制度,分为 A 级绿色食品和 AA 级绿色食品两类。

(1) A 级绿色食品是指生产产地的环境符合 NY/T 391—2000 的要求,生产过程中严格按照绿色食品生产资料使用准则和生产操作规程要求,限量使用限定的化学合成生产资料,产品质量符合绿色食品产品标准,经专门机构认定,许可使用 A 级绿色食品标志的产品。

(2) AA 级绿色食品是指生产产地的环境符合 NY/T 391—2000 的要求,生产过程中不使用化学合成的肥料、农药、兽药、饲料添加剂、食品添加剂和其他有害于环境和身体健康的物质,按有机生产方式生产,产品质量符合绿色食品产品标准,经专门机构认

C. The Grading of Green Food

Since 1996, green food has been classified into Class A green food and Class AA green food.

1. Class A green food means that the environment of the producing area conforms to the requirements of NY/T 391—2000. In the process of production, the limited chemical synthetic means of production is used in strict accordance with the standards for the use of the means of production of green food and the requirements of the operating rules of production. The product quality conforms to the green food product standard, by the specialized agency confirmation, and the products licensed to use Class A green food logo.

2. Class AA green food refers to the environment of the producing area that conforms to the requirements of NY/T 391—2000 and does not use chemical synthetic fertilizers, pesticides, veterinary drugs, feed additives, food additives and other substances harmful to the environment and health during the production process.

定,许可使用 AA 级绿色食品标志的产品。

A 级绿色食品与 AA 级绿色食品的出发点不同,发展 A 级绿色食品的目的是为广大人民群众提供安全、优质食品,满足国内消费的需要;发展 AA 级绿色食品是为了与国际有机食品接轨,促进我国农产品的出口创汇。

A 级绿色食品与 AA 级绿色食品的主要区别在于生产技术标准的不同,AA 级绿色食品要求完全按有机农业生产方式生产;A 级绿色食品要求基本按有机农业生产方式,但可适当保留常规生产方式。

AA 级和 A 级绿色食品的区别如表 9-1 所示。

According to the organic production mode, the product quality conforms to the green food product standard, through the specialized agency confirmation, and the products licensed to use Class AA green food logo.

The starting point of Class A green food is different from Class AA green food. The purpose of developing Class A green food is to provide safe and high quality food for the broad masses of people and to meet the needs of domestic consumption. The development of Class AA green food is in line with the international organic food, and promote the export of Chinese agricultural products to earn foreign exchange.

The main difference between Class A green food and Class AA green food is the difference of production technology standard. Class AA green food should be produced according to organic agricultural production mode and Class A green food should be based on organic agricultural production mode, but the conventional production mode could be kept properly.

The differences between Class AA and Class A green food are shown in Table 9-1.

表 9-1 绿色食品分级标准的区别

评价体系	AA 级绿色食品	A 级绿色食品
环境评价	采用单项指数法,各项数据均不得超过有关标准	采用综合指数法,各项环境监测的综合污染指数不得超过 1
生产过程	生产过程中禁止使用任何化学合成肥料、化学农药及化学合成食品添加剂	生产过程中允许限量、限时间、限定方法使用限定品种的化学合成物质
产品	各种化学合成农药及合成食品添加剂均不得检出	允许限量使用的化学合成物质的残留量低于国家标准或达到发达国家普通食品标准,其他禁止使用的化学物质残留不得检出
包装标识与标志编号	标志和标准字体为绿色,底色为白色,防伪标签的底色为蓝色,标志编号以 AA 结尾	标志和标准字体为白色,底色为绿色,防伪标签底色为绿色,标志编号以 A 结尾

4) 绿色食品标志

D. Logo of Green Food

绿色食品标志是指中文"绿色食品"、英文"Green Food"、绿色食品标志图形及这三者相互组合等四种形式。图 9-1 为绿色食品标志与文字组合商标。

The green food sign is made up of four forms, such as Chinese Green Food, English Green Food, green food logo, and these three combinations. For example, Figure 9-1 is a green food logo and word combination trademark.

图 9-1 绿色食品标志、文字组合商标

绿色食品标志图形由三部分组成,即上方的太阳、下方的叶片和中心的蓓蕾,象征自然生态;颜色为绿色,象征着生命、农业、环保;图形为正圆形,意为保护。整个图形描绘了一幅阳光照耀下的和谐生机,告诉人们绿色食品正是出自纯洁、良好生态环境中的安全无污染食品,能给人们带来蓬勃的生命力。绿色食品的标志还提醒人们要保护环境,通过改善人与环境的关系,创造自然界新的和谐。

The green food logo is composed of three parts, namely, the sun above, the leaf below and the bud in the center, which symbolize natural ecology; the color is green, symbolizing life, agriculture and environmental protection; the figure is positive circle, meaning protection. The whole picture depicts the harmonious vitality of the sun and tells people that green food is a safe and pollution-free food in a pure and good ecological environment, which can bring people vigorous vitality. Green food logo also remind people to protect the environment by improving the relationship between people and the environment to create a new harmony in nature.

A 级绿色食品标志和字体为白色,底色为绿色(图 9-2);AA 级绿色食品标志和字体为绿色,底色为白色(图 9-3)。

Class A green food logo and font is white, the background color is green, as shown in Figure 9-2; Class AA green food logo and font is green, the background color is white, as shown in Figure 9-3.

图 9-2 A 级绿色食品标志　　图 9-3 AA 级绿色食品标志

绿色食品标志的使用是产品通过专门机构认证,许可企业依法使用。凡从事食品生产、加工的企业,需要在某项产品上使用"绿色食品"标志的,必须依照《绿色食品标志管理办法》的有关规定提出申请,经审查符合标准的,授予"绿色食品标志使用证书",准其使用,企业方可在指定的产品上使用"绿色食品"标志。

The use of the green food logo is that the product is certified by a specialized agency and licensed to be used by the enterprise in accordance with the law. Where an enterprise engaged in food production or processing needs to use the "Green Food" logo on a certain product, it must submit an application in accordance with the relevant provisions of the *Measures for the Administration of Green Food Logo*. After examination, if the application meets the standards, the "Green Food Logo Use Certificate" shall be granted, and the enterprise may only use the logo of "Green Food" on the designated product.

3. 绿色食品标准体系

我国绿色食品标准是应用科学技术原理,结合绿色食品生产实践,借鉴国内外相关标准所制定的,是在绿色食品生产中必须遵守,在绿色食品质量认证时必须依据的技术性文件。

绿色食品标准体系以全程质量控制为核心,对绿色食品产前、产中和产后全过程质量控制技术和指标作了全面的规定,构成了一个科学、完整的标准体系(图9-4),包括绿色食品产地环境质量标准、生产技术标准、产品标准,以及包装、贮藏、运输标准四部分。现行有效标准共126项,其中通用准则类标准16项,产品标准110项。绿色食品的标准为原农业部发布的推荐性行业标准,但是对于绿色食品生产企业来说,为强制性执行标准。

Ⅲ. Green Food Standards System

The green food standard of our country is a technical document that must be observed in the production of green food and must be based on when attestation of the quality of green food by applying the principle of science and technology, combining the practice of green food production, drawing lessons from the relevant standards at home and abroad.

The green food standard system takes the whole process quality control as the core, and makes the overall stipulation to the whole process quality control technology and the index before, during and after the green food production. It constitutes a scientific and complete standard system (Figure 9 - 4), including four parts of environmental quality standards of green food production area, production technology standards, product standards and packaging, storage and transportation standards. There are 126 effective standards, of which 16 are generic standards and 110 are product standards. The green food standard is the recommended industry standard issued by the Ministry of Agriculture, but for the green food production enterprises, it is mandatory to enforce the standard.

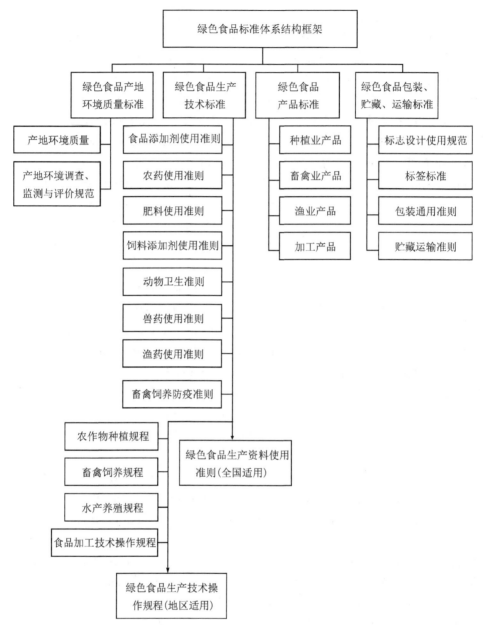

图9-4 绿色食品标准体系结构框架

1) 绿色食品产地环境质量标准

产地环境是绿色食品生产的基本条件,产地环境质量状况直接影响绿色食品质量,是绿色食品可持续发展的先决条件。绿色食品产地环境质量是指绿色食品植物生长地和动物养殖地的生态环境、空气质量、水环境和土

A. Environmental Quality Standard of Green Food Producing Area

The environment of producing area is the basic condition of green food production. The environmental quality of producing area directly affects the quality of green food and is the precondition of sustainable development of green food. The environmental quality of green food producing area refers to the ecological

壤环境质量。《绿色食品 产地环境质量》(NY/T 391—2013)是在遵循自然规律和生态学原理,强调农业经济系统和自然生态系统的有机循环的基本原则指导下,充分依据国内外各类环境标准,结合绿色食品生产实际情况,辅以大量科学实验验证,确定不同产地环境的监测项目及限量值。该标准规定了绿色食品产地的生态环境、空气质量、水质和土壤质量要求以及各项指标的检测方法。

为了规范绿色食品产品产地环境质量调查、监测、评价的原则、内容和方法,科学、正确地评价绿色食品产地环境质量,为绿色食品认证提供科学依据,原农业部发布了标准《绿色食品产地环境调查、监测与评价规范》(NY/T 1054—2013),规定了绿色食品产地环境调查、监测和评价的具体要求。

2)绿色食品生产技术标准

绿色食品生产过程的控制是绿色食品质量控制的关键环节。绿色食品生产技术标准是绿色食品标准体系的核心,它包括绿色食品生产资料使用准则和绿色食品生产技术操作规程两部分。

environment, air quality, water environment and soil environmental quality of green food plant growing area and animal breeding area. *Environmental Quality of Green Food Producing Areas* (NY/T 391—2013) is guided by the basic principles of natural laws and ecological principles, emphasizing the organic circulation of agricultural economic systems and natural ecosystems, and fully based on all kinds of environmental standards at home and abroad. Combined with the actual situation of green food production, with a large number of scientific experiments to determine the monitoring of the environment of different production areas and limited value. This standard stipulates the ecological environment, air quality, water quality and soil quality requirements of green food producing areas, as well as the testing methods of each index.

In order to standardize the principles, contents and methods of environmental quality investigation, monitoring and evaluation of green food production area, the environmental quality of green food producing area is evaluated scientifically and correctly, which provides scientific basis for green food certification. The Ministry of Agriculture has issued the standard *Norms for Environmental Investigation, Monitoring and Evaluation of Green Food Producing Areas* (NY/T 1054—2013), which specifies the specific requirements for environmental investigation, monitoring and evaluation of green food producing areas.

B. Technical Standard for Green Food Production

The control of the green food production process is the key link of green food quality control. The green food production technical standard is the core of the green food standard system. It includes two parts: the green food production material usage standard and the green

绿色食品生产资料使用准则是对生产绿色食品过程中物质投入的一个原则性规定,由原农业部发布,在全国范围内适用。它包括生产绿色食品的肥料、农药、兽药、渔药、食品添加剂和饲料添加剂的使用准则,以及动物卫生准则和畜禽饲养防疫准则等。绿色食品生产技术操作规程是以绿色食品生产资料使用准则为依据,按不同农业区域的生产特性、作物种类、畜禽种类分别制定的,用于指导绿色食品生产活动,规范绿色食品生产技术的技术规定,只在地区范围内适用,具体包括农作物种植、畜禽饲养、水产养殖和食品加工技术操作规程。

3)绿色食品产品标准

绿色食品产品标准是绿色食品标准体系的重要组成部分,是衡量绿色食品最终产品质量的指标尺度。它反映了绿色食品生产、管理及质量控制水平,突出了绿色食品产品无污染、安全、优质、营养的特征。绿色食品产品标准跟普通食品的标准一样,规定了产品的质量要求、各项指标的检验方法、检验规则以及标签、包装、贮藏和运输要求,其重点是产品品质要求。产品品质包括外观品质、营养品质和卫生品质,具体体现在对原料、感官、理化、卫生和微生物学五个方面的要求。绿色食品的卫生品质要求高于普通食品的国家现行标准,主要表现在对农药残留和重金属的检测项目种类

food production technical operation rule.

The guidelines for the use of green food production data are a principle of material input in the process of producing green food, which is issued by the Ministry of Agriculture and is applicable throughout the country. It includes guidelines for the use of fertilizers, pesticides, veterinary drugs, fish medicines, food additives and feed additives for the production of green food, as well as guidelines for animal health and animal breeding and prevention. The operation rules for green food production are technical provisions for guiding green food production activities and standardizing green food production technologies, and which are based on the guidelines for the use of green food production materials, according to the production characteristics of different agricultural regions, crop types, livestock and poultry species are formulated respectively. It is only applicable in the regional scope, including crop cultivation, livestock and poultry breeding, aquaculture and food processing technical operating rules.

C. Green Food Product Standards

The green food product standard is an important part of the green food standard system, and it is an indicator to measure the quality of the final product of green food. It reflects the level of green food production, management and quality control, and highlights the features of pollution-free, safe, high-quality and nutritious green food products. The standards for green food products are the same as those for ordinary food products, which stipulate the quality requirements of products, the inspection methods and rules for various indicators, as well as the requirements for labels, packaging, storage and transportation, with the focus on product quality requirements. Product quality includes appearance quality, nutrition quality and hygiene quality, which are

多、指标严。而且绿色食品的主要原料必须是来自绿色食品产地的,是按绿色食品生产技术操作规程生产出来的产品。为了适应绿色食品迅速发展的形势,中国绿色食品发展中心在全国范围内组织有关技术力量,有计划、有步骤地制(修)定了一批绿色食品产品标准,目前已经颁布实施的绿色食品产品标准达110项。

embodied in five aspects: raw material, sensory, physicochemical, hygiene and microbiology. The hygienic quality requirement of green food is higher than the current national standard of common food, which is mainly reflected in the variety and strict index of pesticide residue and heavy metal detection items. And, the main raw material of green food must come from green food producing area, it is the product that produces according to green food production technology operating regulation. In order to adapt to the rapid development of green food, the China green food development center has organized relevant technical forces nationwide to formulate (revise) a batch of green food product standards in a planned and step-by-step manner. So far, 110 green food product standards have been promulgated and implemented.

4) 绿色食品包装、贮藏、运输标准

《绿色食品 包装通用准则》(NY/T 658—2002)规定了绿色食品的包装必须遵循的原则,以及绿色食品包装的要求、包装材料的选择、包装尺寸、包装检验、抽样、标志和标签、贮存与运输等内容。

绿色食品标签除应符合GB 7718《预包装食品标签通则》的规定外,其外包装上应印有绿色食品标志,绿色食品标志的设计和标识方法应符合《中国绿色食品商标标志设计使用规范手册》的规定;若是特殊营养食品,还应符合GB/T 13432《特殊营养食品标签》的规定。《中国绿色食品商标标志设计使用规范手册》是以绿色食品标志图形为核心,对绿标、"绿色食品"四个字及英译名及其相互的组合在产品、广告等媒介上的设计、使用进行规范的指导性工具书,绿色食品企业应严格按照该手册的要求,在其获证产

D. Standard for Packing, Storage and Transportation of Green Food

The *General Guidelines for Green Food Packaging* (NY/T 658—2002) stipulate the principles to be followed in the packaging of green foods, the requirements for packaging of green foods, the selection of packaging materials, the size of packaging, packaging inspection, sampling, marking and labeling, storage and transportation.

Except for the compliance with GB 7718 *General Standard for the Labeling of Prepackaged Foods*, the green food label shall be printed on the outer packaging of the green food label. The design and marking methods of green food marks shall conform to the provisions of the *Manual of Specification for the Design and Use of Trademarks of Green Food in China*; if special nutritious food, it should also comply with the requirements of GB/T 13432 *Special Nutritional Food Label*. The *Manual of Specification for the Design and Use of Trademarks of Green Food in China* is a guiding reference book for the design and use of green food logo, which is based on

品包装上设计使用绿色食品标志。

《绿色食品 贮藏运输准则》(NY/T 1056—2006)对绿色食品贮藏、运输的条件、方法、时间作出规定,以保证绿色食品在储运过程中不遭受污染、不改变品质,并有利于环保、节能。

5)绿色食品其他相关标准

绿色食品其他相关标准包括"绿色食品生产资料"认定标准、"绿色食品生产基地"认定标准等,这些标准都是促进绿色食品质量控制管理的辅助标准。

green food logo graphics, and the translation of green food into English and its combination on products, advertisements and other media. Green food enterprises should strictly comply with the requirements of the manual, then design and use green food labels on the packaging of their certified products.

The Standards for the Storage and Transportation of Green Food (NY/T 1056—2006) stipulates the conditions, methods and time of green food storage and transportation, so as to ensure that green food is not polluted, does not change the quality, and is conducive to environmental protection and energy saving.

E. Other Related Standards for Green Food

Other related standards of green food include the identification standard of "green food production", the identification standard of "green food production base", these standards are auxiliary standards to promote the quality control and management of green food.

4. 申请绿色食品认证必须具备的条件

1)申请人应当具备的资质条件

(1)能够独立承担民事责任,如企业法人、农民专业合作社、个人独资企业、合伙企业、家庭农场等,国有农场、国有林场和兵团团场等生产单位。

(2)具有稳定的生产基地。

(3)具有绿色食品生产的环境条件和生产技术。

(4)具有完善的质量管理体系,并至少稳定运行一年。

(5)具有与生产规模相适应的生产技术人员和质量控制人员。

(6)申请前三年内无质量安全事故和不良诚信记录。

(7)与绿色食品工作机构或检测机构不存在利益关系。

2)申请使用绿色食品标志的产品应当具备的条件

申请使用绿色食品标志的产品应当符合《中华人民共和国食品安全法》和《中华人民共和国农产品质量安全法》等法律法规规定,在原国家工商行政管理总局商标局核定的绿色食品标志商标涵盖商品范围内,并具备下列条件:

(1)产品或产品原料产地环境符合绿色食品产地环境质量标准;

(2)农药、肥料、饲料、兽药等投入品使用符合绿色食品投入品使用准则;

(3)产品质量符合绿色食品产品质量标准;

(4)包装储运符合绿色食品包装储运标准。

5. 绿色食品标志许可审查程序

绿色食品标志许可审查程序如图 9-5 所示,可分为认证申请、受理及文审、现场检查、产地环境及产品检测和评价、审核和评审、颁证等六个程序。

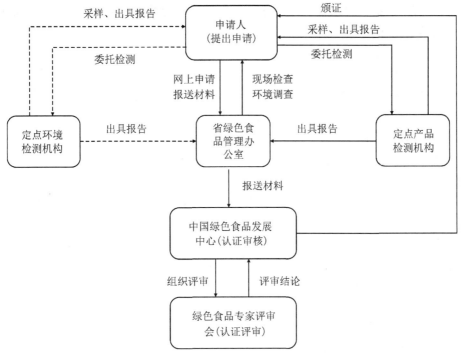

图 9-5 绿色食品认证程序

1) 认证申请

申请人至少在产品收获、屠宰或捕捞前三个月,向所在省级绿色食品工作机构提出申请,完成网上在线申报并提交下列文件:

（1）"绿色食品标志使用申请书"及"调查表";

（2）资质证明材料,如营业执照、全国工业产品生产许可证、动物防疫条件合格证、商标注册证等证明文件复印件;

（3）质量控制规范;

（4）生产技术规程;

（5）基地图、加工厂平面图、基地清单、农户清单等;

（6）合同、协议、购销发票,生产、加工记录;

（7）含有绿色食品标志的包装标签或设计样张（非预包装食品不必提供）;

（8）应提交的其他材料。

2) 受理及文审

省级工作机构应当自收到上述规定的申请材料之日起十个工作日内完成材料审查。符合要求的,予以受理,向申请人发出"绿色食品申请受理通知书",执行现场检查程序;不符合要求的,不予受理,书面通知申请人本生产周期不再受理其申请,并告知理由。

3）现场检查

（1）省级工作机构应当根据申请产品类别，组织至少两名具有相应资质的检查员组成检查组，在材料审查合格后四十五个工作日内组织完成现场检查（受作物生长期影响可适当延后）。

现场检查前，应提前告知申请人并向其发出"绿色食品现场检查通知书"，明确现场检查计划。现场检查工作应在产品及产品原料生产期内实施。

（2）现场检查要求。

①申请人应当根据现场检查计划作好安排。检查期间，要求主要负责人、绿色食品生产负责人、内检员或生产管理人员、技术人员等在岗，开放场所设施设备，备好文件记录等资料。

②检查员在检查过程中应当收集好相关信息，作好文字、影像、图片等信息记录。

（3）现场检查程序。

①召开首次会议。由检查组组长主持会议，明确检查目的、内容和要求，申请人、主要负责人、绿色食品生产负责人、技术人员和内检员等参加。

②实地检查。检查组应当对申请产品的生产环境、生产过程、包装储运、环境保护等环节逐一进行严格检查。

③查阅文件、记录。核实申请人全程质量控制能力及有效性，如质量控制规范、生产技术规程、合同、协议、基地图、加工厂平面图、基地清单、记录等。

④随机访问。在查阅资料及实地检查过程中随机访问生产人员、技术人员及管理人员，收集第一手资料。

⑤召开总结会。检查组与申请人沟通现场检查情况，并交换现场检查意见。

（4）现场检查完成后，检查组应当在十个工作日内向省级工作机构提交"绿色食品现场检查报告"。省级工作机构依据"绿色食品现场检查报告"向申请人发出"绿色食品现场检查意见通知书"。现场检查合格的，执行产地环境、产品检测和评价程序；不合格的，通知申请人本生产周期不再受理其申请，告知理由并退回申请。

4）产地环境、产品检测和评价

（1）申请人按照"绿色食品现场检查意见通知书"的要求委托检测机构对产地环境、产品进行检测和评价。

（2）检测机构接受申请人委托后，应当分别依据《绿色食品 产地环境调查、监测与评价规范》（NY/T 1054）和《绿色食品 产品抽样准则》（NY/T 896）及时安排现场抽样，并自环境抽样之日起三十个工作日内、产品抽样之日起二十个工作日内完成检测工作，出具"环境质量监测报告"和"产品检验报告"，提交省级工作机构和申请人。

（3）申请人如能提供近一年内绿色食品检测机构或国家级、部级检测机构出具的"环境质量监测报告"，且符合绿色食品产地环境检测项目和质量要求的，可免做环境检测。

经检查组调查确认产地环境质量符合《绿色食品 产地环境质量》（NY/T 391）和《绿色食品 产地环境调查、监测与评价规范》（NY/T 1054）中免测条件的，省级工作机构可作出免做环境检测的决定。

5）审核和评审

（1）省级工作机构应当自收到"绿色食品现场检查报告""环境质量监测报告"和"产品检验报告"之日起二十个工作日内完成初审。初审合格的，将相关材料报送中心，同时完成网上报送；不合格的，通知申请人本生产周期不再受理其申请，并告知理由。

(2)中心应当自收到省级工作机构报送的完备申请材料之日起三十个工作日内完成书面审查,提出审查意见,并通过省级工作机构向申请人发出"绿色食品审查意见通知书"。

①需要补充材料的,申请人应在"绿色食品审查意见通知书"规定的时限内补充相关材料,逾期视为自动放弃申请。

②需要现场核查的,由中心委派检查组再次进行检查核实。

③审查合格的,中心在二十个工作日内组织召开绿色食品专家评审会,并形成专家评审意见。

(3)中心根据专家评审意见,在五个工作日内作出是否颁证的决定,并通过省级工作机构通知申请人。同意颁证的,进入绿色食品标志使用证书(以下简称证书)颁发程序;不同意颁证的,告知理由。

6)颁证

认证评审合格后,由省绿色食品管理办公室负责组织企业签订"绿色食品标志商标使用许可合同",中心统一向省绿色食品管理办公室寄发"绿色食品标志使用证书",经其转发企业。

6. 绿色食品认证申请材料的编制

1)绿色食品认证申请材料编制的总体要求

(1)要求申请人用钢笔、签字笔正楷如实填写"绿色食品标志使用申请书",或用 A4 纸打印,字迹整洁、术语规范、印章(签名)端正清晰。

(2)所有表格栏目不得空缺,如不涉及本项目,应在表格栏目内注明"无";如表格栏目不够,可附页,但附页必须加盖公章。

(3)申请认证材料应装订成册,编制页码,并附目录。所有材料一式三份。

2)申请材料的封面和目录

(1)材料封面如图 9-6 所示。

图 9-6 绿色食品申请材料封面

(2)材料目录。中国绿色食品发展中心将绿色食品分为种植产品、畜禽产品、加工产品、水产品、食用菌、蜂产品六大类,分别发布了申请材料清单。企业应根据清单要求准备相应的材料,目录内容按清单要求的顺序进行编排。

以加工产品为例,申请材料清单如下。

①"绿色食品标志使用申请书"和"加工产品调查表"。

②营业执照复印件。

③商标注册证复印件(有必要的应提供续展证明、商标转让证明、商标使用许可证明等)。

④食品生产许可证、食盐定点生产许可证、定点屠宰许可证、饲料生产许可证等其他国家强制要求办理的资质证书复印件(适用时)。

⑤工厂所在地行政区域图(市、县或乡的行政图,标明加工厂位置)。

⑥加工厂区平面布局图(包括厂区各建筑物、设备和周围土地利用情况)。

⑦加工厂所使用的证明文件(如为委托加工,提供委托加工合同书、委托加工厂的营业执照、食品生产许可证)。

⑧质量管理手册:

A. 绿色食品生产、加工、经营者的简介;

B. 绿色食品生产、加工、经营者的管理方针和目标;

C. 管理组织机构图及其相关岗位的责任和权限;

D. 可追溯体系;

E. 内部检查体系;

F. 文件和记录管理体系。

⑨生产加工管理规程,需申请人盖章:

A. 加工规程,技术参数;

B. 产品的包装材料、方法和贮藏、运输环节规程;

C. 污水、废弃物的处理规程;

D. 防止绿色食品与非绿色食品交叉污染的规程(存在平行生产的企业须提交);

E. 运输工具、机械设备及仓储设施的维护、清洁规程;

F. 加工厂卫生管理与有害生物控制规程;

G. 生产批次号的管理规程。

⑩配料固定来源和购销证明:

A. 对于购买绿色食品原料标准化生产基地原料的,申请人需提供基地证书复印件、购销合同和发票复印件;

B. 对于购买绿色食品产品或其副产品的,申请人需提供有效期内的证书复印件、购销合同和发票复印件;

C. 对于购买未获得绿色食品认证、原料含量在2%~10%的原料(食盐大于等于5%)的,申请人需提供购销合同和发票复印件,绿色食品检测机构出具的符合绿色食品标准的检测报告;

D. 对于购买未获得绿色食品认证、原料含量小于2%的原料(食盐小于5%)的,申请人需提供固定来源的证明文件。

⑪生产加工记录(能反映产品生产过程和投入品使用情况)。

⑫预包装食品标签设计样张(非预包装食品不必提供)。

⑬加工水监测报告。

⑭产品检验报告。

3)"绿色食品标志使用申请书"的填写

"绿色食品标志使用申请书"包括封面、填写说明、保证声明,以及"表一 申请人基本情况""表二 申请产品情况""表三 原料供应情况""表四 申请产品统计表"七个部分。详见绿色食品标志使用申请书填写要求及范本。

绿色食品标志使用申请书填写要求及范本
(申请书封面)

绿色食品标志使用申请书
初次申请□　续展申请□
(根据实际申请的需要在□打"√")

申请人(盖章)　(公司名称应与营业执照、公章一致)

申请日期　××××年　××月　××日

中国绿色食品发展中心

填写说明

一、本申请书一式三份,中国绿色食品发展中心、省级工作机构和申请人各一份。

二、本申请书无签名、盖章无效。

三、申请书的内容可打印或用蓝、黑钢笔或签字笔填写,语言规范准确、印章(签名)端正清晰。

四、申请书可从http://www.moa.gov.cn/sydw/lssp/下载,用A4纸打印。

五、本申请书由中国绿色食品发展中心负责解释。

保证声明

我单位已仔细阅读《绿色食品标志管理办法》有关内容,充分了解绿色食品相关标准和技术规范等有关规定,自愿向中国绿色食品发展中心申请使用绿色食品标志。现郑重声明如下。

1. 保证"绿色食品标志使用申请书"中填写的内容和提供的有关材料全部真实、准确,如有虚假成分,我单位愿承担法律责任。

2. 保证申请前三年内无质量安全事故和不良诚信记录。

3. 保证严格按《绿色食品标志管理办法》、绿色食品相关标准和技术规范等有关规定组织生产、加工和销售。

4. 保证开放所有生产环节,接受中国绿色食品发展中心组织实施的现场检查和年度检查。

5. 凡因产品质量问题给绿色食品事业造成的不良影响,愿接受中国绿色食品发展中心所作的决定,并承担经济和法律责任。

法定代表人(签字):(应与营业执照上一致)　申请人(盖章):(与营业执照、公章一致)

　　　　　　　　　　　　　　　　　　　　　＿＿＿＿＿ ××××年 ×× 月 ×× 日

表一　申请人基本情况

申请人(中文)	××××××××公司			
申请人(英文)	(有就填写,没有填"无")			
联系地址	××省××市××县××街道××号 (确保能接收到绿色食品商标使用许可合同)		邮　编	
网址				
营业执照注册号	(按营业执照证书上的编号填写)		首次获证时间	
企业法定代表人	(与营业执照一致)	座机	手机	
联系人	(负责办理认证申请的人员)	座机	手机	
传真		E-mail		
龙头企业	国家级□　　省(市)级□　　地市级□　　其他□ (根据企业实际情况在□打"√")			
年生产总值 /万元		年利润 /万元		
申请人简介	(主要介绍与生产加工有关的事项,如企业规模、生产能力、主要产品、申报产品项目的发展情况等,以上表格中已有的内容不要重复介绍) 例如,以下为绿色食品苹果生产企业的简介: 　　本企业于2005年开始建设,至2010年底共计开发土地面积10128亩,有效苹果种植面积7000亩;分布在三个片区(命名为1—3号基地)。栽植苹果树48.2万株;栽植防风林4.57万株;共修建灌溉主渠、支渠和毛渠33公里;共修建果园主道路和环园路65公里;兴建了5000立方米的水库三座,配套节水灌溉系统全覆盖;建设办公、生活设施共计3200平方米,各种基础设施建设完善。			

内检员(签字):

注:(1)内检员适用于已有中心注册内检员的申请人。
(2)首次获证时间仅适用于续展申请。

表二　申请产品情况

产品名称	商标	产量/吨	是否有包装	包装规格	备注
商品名（即产品包装上的名称，不能填写系列产品名称或集合名词）	（与商标注册证明上一致）	（按实际填写）	有/无	（按实际填写）	

注：(1)续展产品名称、商标变化等情况需在备注栏说明。
(2)若此表不够，可附页。

表三　原料供应情况

原料来源	原料供应情况		
	生产商	产品名称	使用量/吨
绿色食品	（填写原料本身为绿色食品产品的供应商名称，与购销合同上一致）	（与购销合同上一致）	（与购销合同上一致）
	基地名称	使用面积/万亩	使用量/吨
全国绿色食品标准化原料生产基地	（填写主要原料的全国绿色食品原料生产基地名称或自建基地名称）	（填写种养殖面积，依据"关于种养殖面积的有关说明"填写）	（按实际填写）

关于种养殖面积(单位：万亩)的有关说明

1. 初级产品

(1)种植业产品：直接填报种植面积(食用菌不需填报)。

(2)畜禽产品：牛、羊肉产品既要填报放牧草场面积，又要填报主要饲料原料(如玉米、小麦、大豆等)的种植面积；猪肉、禽肉与禽蛋类产品只填报饲料主要原料种植面积。

(3)水产品(包括淡水、海水产品)：填报水面养殖面积。

2. 加工产品

主要原料是绿色食品产品的，不需要填报种养殖面积；主要原料来自全国绿色食品标准化原料生产基地或申报单位自建基地的，需要填报种养殖面积。

(1)需要填报主要原料(或饲料)种养殖面积的加工产品。

①农林类加工产品：小麦粉、大米、大米加工品、玉米加工品、大豆加工品、食用植物油、机制糖、杂粮加工品、冷冻保鲜蔬菜、蔬菜加工品、果类加工品、山野菜加工品、其他农林加工

产品。

②畜禽类加工产品:蛋制品、液体乳、乳制品、蜂产品。

③水产类加工产品:淡水加工品、海水加工品。

④饮料类产品:果蔬汁及其饮料、固体饮料(果汁粉、咖啡粉)、其他饮料(含乳饮料及植物蛋白饮料、茶饮料及其他软饮料)、精制茶、其他茶(如代用茶)、白酒、啤酒、葡萄酒、其他酒类(黄酒、果酒、米酒等)。

⑤其他加工产品:方便主食品(米制品、面制品、非油炸方便面、方便粥)、糕点(焙烤食品、膨化食品、其他糕点)、果脯蜜饯、淀粉、调味品(味精、酱油、食醋、料酒、复合调味料、酱腌菜、辛香料、调味酱)、食盐(海盐、湖盐)。

(2)不需要填报主要原料(或饲料)种养殖面积的加工产品。

①农林类加工产品:食用菌加工品。

②畜禽类加工产品:肉食加工品(包括生制品、熟制品、畜禽副产品加工品、肉禽类罐头、其他肉食加工品)。

③饮料类产品:瓶(罐)装饮用水、碳酸饮料、固体饮料(乳精、其他固体饮料)、冰冻饮品、其他酒类(露酒)。

④其他加工产品:方便主食品(包括速冻食品、其他方便主食品)、糖果(包括糖果、巧克力、果冻等)、食盐(包括井矿盐、其他盐)、调味品(包括水产调味品、其他调味品、发酵制品)、食品添加剂。

表四　申请产品统计表

产品名称	年产值/万元	年销售额/万元	年出口量/吨	年出口额/万美元	绿色食品包装印刷数量
商品名(与表二中的产品名称一致)	产品年产值＝申报产量×当年产品平均出厂价格	(申报产品上年度国内销售额)	(申报产品上年度的出口量)	(申报产品上年度的出口额)	(根据产品的申报产量确定)

注:表三、表四可根据需要增加行数。

4)调查表的填写

根据申请产品的类别,调查表分为"种植产品调查表""畜禽产品调查表""加工产品调查表""水产品调查表""食用菌调查表""蜂产品调查表"六种,企业根据实际情况选择填写。

以"加工产品调查表"为例,除封面和填表说明以外,调查表内容包括以下九个方面:

(1)加工产品基本情况;

(2)加工厂环境基本情况;

(3)加工产品配料情况;

(4)加工产品配料统计表;

(5)产品加工情况;

(6)包装、储藏、运输;

(7)平行加工;

(8)设备清洗、维护及有害生物防治;

(9)污水、废弃物处理情况及环境保护措施。

5)生产技术规程(种植规程、养殖规程、加工规程)的编制

(1)种植规程的编写。农作物种植规程要根据企业实际情况制定,一般包括适用范围、环境要求、种植技术、田间管理、施肥、病虫害防治、收获及包装运输等部分。有以下几点要求:

①应根据申报产品或产品原料的特点,因地制宜地编制具备科学性、可操作性的种植规程;

②规程的编制应体现绿色食品生产的特点,病虫草害的防治应以生物、物理、机械防治为主,施肥应以有机肥为基础,以维持或增加土壤肥力为核心;

③规程编制的内容应包括立地条件(环境质量、肥力水平等)、品种与茬口、育苗与移栽、种植密度、田间管理(包括肥、水等)、病虫草鼠害的防治、收获等;

④对病虫草鼠害的防治,应根据近三年的植保概况制定较全面的防治措施,表三的内容在规程中应全部体现;

⑤农药的使用应注明农药名称、剂型规格、使用目的、使用方法、全年使用次数、安全间隔期等内容;

⑥正式打印文本,并加盖种植单位或技术推广单位公章。

(2)养殖规程的编写。养殖规程的编写要求如下:

①规程的制定要因地制宜,具有科学性和可操作性;

②规程的制定应体现绿色食品生产特点,以预防为主,优先建立严格的生物安全体系,改善饲养环境,加强饲养管理,增强动物自身的抗病力;

③内容应详细,包括养殖场所卫生环境条件、环境消毒、饲料、饲料添加剂、饲料加工、防疫、体内外寄生虫及疾病防治、屠宰、检疫、仓储、运输、包装及生产管理等环节;

④饲料及饲料添加剂使用应根据动物各生长阶段营养需要合理调配,药剂的使用应注明品种、剂型、使用方法、使用剂量及停药期;

⑤需正式打印件,并加盖公章。

(3)加工规程的编写。绿色食品加工操作规程应根据产品加工实际情况编写,包括以下内容:

① 生产工艺流程；

② 对原、辅料的要求，包括原、辅料来源，原、辅料进厂验收（感官指标、理化指标），进厂后的储存、预处理等；

③ 生产工艺，应根据生产工艺流程将加工的每一环节用简要的文字表述出来，其中有关温度、浓度、杀菌的方法、添加剂的使用等应详细说明；

④ 主要设备及清洗方法；

⑤ 成品检验制度；

⑥ 储藏（储藏的方法、地点等）。

6) 公司对"基地＋农户"的质量控制体系的编制

(1) 基地使用协议：公司应与各基地签订合同（协议），合同（协议）有效期应为三年；合同（协议）条款中应明确双方职责，明确要求严格按绿色食品生产操作规程及标准进行生产，并明确监管措施；合同（协议）中应标明基地（农户）名称、作物（动物）品种、种植面积（养殖规模）、预计收购数量及质量要求等。

(2) 基地图：在当地行政区划图基础上绘制，应清楚标明各基地方位及周边主要标志物方位。

(3) 基地清单：列出各基地名称、地址、负责人、电话、作物（或动物）品种、种植面积（养殖规模）、预计产量；基地要求具体到最小单元村（场）。

(4) 基地农户清单：公司应建立详细农户清单，包括所在基地名称、农户姓名、作物（动物）品种、种植面积（养殖规模）、预计产量；对于基地农户数超过1000户的申请企业，可以只提供一个基地的农户清单样本，但企业必须以文字形式声明已建立了农户清单。

(5) 基地管理制度：公司应建立一套详细的管理制度，确保基地（农户）严格按绿色食品要求进行生产；基地的生产管理可由公司委托当地（乡、镇）农技推广部门负责，并签订委托管理协议。

7) 产品执行标准

产品执行标准按绿色食品标准、国家标准、行业标准、企业标准的顺序采标，即国家已经发布了产品绿色食品标准的，必须执行该标准；国家没有发布绿色食品标准的，可以执行现行国家标准、行业标准、企业标准。

8) 企业资质材料（复印件）

(1) 企业的营业执照：经年检过的营业执照副本，企业名称应与申请人对应，经营范围应涵盖所申报产品。

(2) 生产许可证：产品实行生产许可证的企业提供，证书在有效期内。

(3) 卫生许可证：需要卫生许可证的行业提供，证书在有效期内。

(4) 职工健康证：有效期内5~8人的样本。

(5) 商标注册证或受理证明：商标注册人与绿色食品申报主体一致，如有主体变更、许可使用、续展等情况，需有文件说明。

(6) 其他需要提供的资质材料按要求提供。

9) 企业质量管理手册

如果企业通过国际标准化组织认证，提供质量管理手册复印件；如没有，质量管理手册应包含质量方针、质量目标、基地管理、组织机构图、质量工作职责等内容。

公司应建立一套科学合理的组织机构,明确组织、管理绿色食品生产的机构、职责及主要负责人;所设机构应全面,包括基地管理、技术指导、生产资料供应、监督、收购、加工、仓储、运输、销售等各个环节的部门。

公司应建立一套详细的培训制度,加强对干部、主要技术人员、基地农户有关绿色食品的知识培训。

要求公司对基地和农户进行统一管理(即统一供应品种、统一供应生产资料、统一技术规程、统一指导、统一监督管理、统一收购、统一加工、统一销售),各管理措施要求详细,符合实际情况,并具可操作性;如公司委托第三方技术服务部门进行管理,需签订有效期为三年的委托管理合同,受托方按上述要求制定具体的管理制度。

10)要求提供的其他材料

如通过体系认证的,附证书复印件;购买原料证明(合同、购货发票、原料绿色食品证书等);包装标签或设计样张;当地政府环保规定或规划;企业和产品获奖证明;检查员现场检查、培训及其他有关企业产品的照片;有关肥料、农药的标签、说明书;绿色食品知识培训登记表等。

7. 绿色食品标志的使用管理

未经中国绿色食品发展中心许可,任何单位和个人不得使用绿色食品标志。已获准使用绿色食品标志的企业(简称标志使用人)必须严格依照《商标法》《绿色食品标志管理办法》和"绿色食品标志商标使用许可合同"的要求正确使用绿色食品商标标志,并接受绿色食品监管部门的监督检查,保证使用标志产品的质量,维护"绿色食品"标志的商标信誉。

1)绿色食品标志使用证书

绿色食品标志使用申请人在签订"绿色食品标志商标使用许可合同"并缴纳相关费用后,即可获得由中国绿色食品发展中心颁发的"绿色食品标志使用证书",该证书上标明获证产品名称、产品编号、企业信息码、生产商名称、批准产量、许可期限、颁证机构等内容。

产品编号样式如下:

LB	—	××	—	××	××	××	××××	A(AA)
绿标		产品类别		认证年份	认证月份	省份(国别)	产品序号	产品级别

编号形式规定:产品类别代码为两位数,产品分类为5大类57小类,并按小类编号;认证时间包括年份和月份;省份按行政区划的序号编码,国外产品从第51号开始,中国不编国别代码。

企业信息码的编码形式为GF××××××××××××。GF是绿色食品英文"Green Food"首字母的缩写组合,后面为12位阿拉伯数字,其中一到六位为地区代码(按行政区划编制到县级),七到八位为企业获证年份,九到十二位为当年获证企业序号。样式如下:

GF	××××××	××	××××
绿色食品 英文缩写	地区 代码	获证 年份	企业 序号

绿色食品标志使用证书有效期三年。证书有效期满,需要继续使用绿色食品标志的,标志使用人应当在有效期满三个月前向省级工作机构提出续展申请,同时完成网上在线申报。标志使用人逾期未提出续展申请,或者续展未通过的,不得继续使用绿色食品标志。

2)绿色食品标志设计使用规范

(1)获证企业应按照中国绿色食品发展中心所发布的《绿色食品标志设计使用规范手册》的要求在产品包装和产品宣传材料上印制绿色食品标志。绿色食品标志商标(图形商标、文字商标组合图形)和企业信息码同时使用,如图9-7所示。

(a)　　　　　　　　　　　　　　　　　　　　(b)

图9-7　绿色食品标志使用示例

(2)绿色食品企业应在其认证产品上使用绿色食品标志防伪标签。绿色食品标志防伪标签由中国绿色食品发展中心统一委托定点专业生产单位印刷。企业不得自行生产或从其他渠道获取防伪标签,也不可直接向中心委托的防伪标签生产企业定货。绿色食品企业应根据认证产品的核准产量订制和使用绿色食品标志防伪标签,不得超出认证核准范围订制和使用。每种产品只能使用对应的防伪标签(印有该产品的编号)。防伪标签应贴于食品标签或包装正面显著位置,不能掩盖原有绿标、企业信息码等绿色食品标志整体形象。防伪标签粘贴位置应固定,不能随意变化。

3)标志使用人的权限和要求

(1)标志使用人使用标志的权限：

①在获证产品及其包装、标签、说明书上使用绿色食品标志；

②在获证产品的广告宣传、展览展销等市场营销活动中使用绿色食品标志；

③在农产品生产基地建设、农业标准化生产、产业化经营、农产品市场营销等方面优先享受相关扶持政策；

④禁止将绿色食品标志用于非许可产品及其经营性活动；

⑤禁止转让绿色食品标志使用证书。

(2)对标志使用人的要求：

①严格执行绿色食品标准,保持绿色食品产地环境和产品质量稳定可靠；

②健全和实施产品质量控制体系,对其生产的绿色食品质量和信誉负责；

③遵守标志使用合同及相关规定,规范使用绿色食品标志；

④积极配合县级以上人民政府农业行政主管部门的监督检查及其所属绿色食品工作机构的跟踪检查；

⑤在证书有效期内,标志使用人的单位名称、产品名称、产品商标等发生变化的,应当经省级工作机构审核后向中国绿色食品发展中心申请办理变更手续；

⑥产地环境、生产技术等条件发生变化,导致产品不再符合绿色食品标准要求的,标志使用人应当立即停止标志使用,并通过省级工作机构向中国绿色食品发展中心报告。

4)绿色食品及其标志使用的监督检查

为了加强对绿色食品企业产品质量和绿色食品标志使用的监督管理,《绿色食品标志管理办法》规定：由县级以上农业行政主管部门依法对辖区内绿色食品产地环境、产品质量、包装标识、标志使用等情况进行监督检查；中国绿色食品发展中心和省级工作机构组织对绿色食品及标志使用情况进行跟踪检查;省级工作机构应当组织对辖区内标志使用人使用绿色食品标志的情况实施年度检查;对检查不合格或违反规定的标志使用人,应取消其标志使用权。

绿色食品监管工作形式主要有年度检查、产品抽检和市场监察三种,中国绿色食品发展中心分别制定了相关工作制度予以规范,即《绿色食品年度检查工作规范》《绿色食品产品质量年度抽检管理办法》和《绿色食品标志市场监察实施办法》。

5)绿色食品标志使用权的取消

标志使用人有下列情形之一的,由中国绿色食品发展中心取消其标志使用权,收回标志使用证书,并予公告:

(1)生产环境不符合绿色食品环境质量标准的;
(2)产品质量不符合绿色食品产品质量标准的;
(3)年度检查不合格的;
(4)未遵守标志使用合同约定的;
(5)违反规定使用标志和证书的;
(6)以欺骗、贿赂等不正当手段取得标志使用权的。

标志使用人依照规定被取消标志使用权的,三年内中国绿色食品发展中心不再受理其申请;情节严重的,永久不再受理其申请。

9.1.2 实训任务

填写果汁饮料生产企业猕猴桃汁产品"绿色食品标志使用申请书"和"调查表"。

实训组织:对学生进行分组,每个组参照"基础知识"中的内容并利用网络资源,填写一份果汁饮料生产企业猕猴桃汁产品"绿色食品标志使用申请书"和"调查表"。

实训成果:果汁饮料生产企业猕猴桃汁产品"绿色食品标志使用申请书"和"调查表"。

实训评价:由获证企业绿色食品内检员或主讲教师进行评价。

绿色食品认证申请技能考核评分表

项目	考核内容	参考标准	分值	得分
申请书的填写	申请书文本格式和完整性	申请书文本格式正确;申请书内容完整	2	
	封面的填写	填写完整、规范、正确;申请单位全称与公章全称一致	2	
	保证声明的填写	填写完整、规范、正确;有公章和法人签字	2	
	表一 申请人基本情况	填写完整、规范、正确,无漏项;申请人简介语言简练、条理清晰、抓住要点	10	
	表二 申请产品情况	填写完整、规范、正确,无漏项	4	
	表三 原料供应情况	填写完整、规范、正确,无漏项	10	
	表四 申请产品统计表	填写完整、规范、正确,填写数据合理	3	
	整洁性	无随意涂改现象	2	

续表

项目	考核内容	参考标准	分值	得分
调查表的填写	调查表文本格式和完整性	调查表文本格式选择正确;内容完整	2	
	调查表封面的填写	填写完整、规范、正确;申请单位全称与公章全称一致	2	
	(1)加工产品基本情况	填写完整、规范、正确,无漏项	3	
	(2)加工厂环境基本情况	填写完整、规范、正确;内容符合绿色食品生产环境要求	5	
	(3)加工产品配料情况	填写完整、规范、正确;内容符合绿色食品生产对于原料、添加剂、加工助剂的使用要求;数据合理	10	
	(4)加工产品配料统计表	填写完整、规范、正确;数据合理,与材料中的相关内容一致	4	
	(5)产品加工情况	产品加工工艺流程图绘制正确,工艺条件描述清楚;其他内容填写完整、规范、正确	8	
	(6)包装、储藏、运输	填写内容完整、规范、正确;符合绿色食品的包装储运标准要求	8	
	(7)平行加工	填写内容完整、规范、正确;符合绿色食品生产加工的相关要求	8	
	(8)设备清洗、维护及有害生物防治	填写完整、规范、详尽;内容符合绿色食品加工、存储、运输相关要求	8	
	(9)污水、废弃物处理情况及环境保护措施	填写完整、规范、详尽;内容符合环境保护相关要求;有填表人、内检员签字	5	
	整洁性	无随意涂改现象	2	
合 计			100	
总体评价				

思考题

1. 什么是绿色食品,绿色食品如何分级?
2. 绿色食品标准体系包括哪些内容?
3. 简述绿色食品认证程序。

4. 以绿色食品乳制品为例,申请绿色食品认证时需要提交哪些材料?
5. 申请绿色食品认证的产品和企业需要满足哪些基本条件?

任务 2 有机食品认证

9.2.1 基础知识

1. 有机食品的概念与优点

1)有机食品的概念

有机食品是一种国际通称,是从英文 Organic Food 直译过来的。这里所说的"有机"不是化学上的概念,而是指采取一种有机的耕作和加工方式。有机食品是指按照这种方式生产和加工的,产品符合国际或国家有机食品要求和标准,并通过国家认证机构认证的一切农副产品及其加工品,包括粮食、蔬菜、水果、奶制品、禽畜产品、蜂蜜、水产品、调料等。

除有机食品外,国际上还把经认证后的一些派生的产品如有机化妆品、纺织品、林产品,或由有机产品生产而提供的生产资料,包括生物农药、有机肥料等,统称为有机产品。

我国对有机产品的定义通常为来自于有机农业生产体系,按照 GB/T 19630 相关生产要求和标准生产、加工,并通过独立的有机产品认证机构认证的供人类消费、动物食用的产品。有机产品中绝大部分为有机食品类。

2)有机食品的优点

(1)用自然、生态平衡的方法从事

I. The Concept and Advantages of Organic Food

A. The Concept of Organic Food

Organic food is an international term that is literally translated from the English Organic Food. The term "organic" is not a chemical concept, but an organic form of farming and processing. Organic food means production and processing in this manner; products that meet international or national requirements and standards for organic food; and all agricultural by-products and processed products certified by national certification bodies, including food, vegetables, fruits, dairy products, livestock products, honey, aquatic products, spices and so on.

In addition to organic foods, the international production of some derived products, such as organic cosmetics, textiles, forest products or organic products, including biological pesticides and organic fertilizers, which are referred to as organic products after certification.

The definition of organic products in China refers to the production and processing of organic products from the organic agricultural production system in accordance with the relevant production requirements and standards of GB/T 19630, as well as the products that are produced, processed and certified by independent organic product certification bodies for human consumption and animal consumption. Most of the organic products are organic foods.

B. Advantages of Organic Food

1. Agricultural production and management

农业生产和管理,保护环境,满足人类需求,实现可持续发展。

(2)顺应国际市场潮流,扩大有机农业生产及有机食品出口,提高产品市场竞争力。

(3)满足国内对"绿色""环保"的消费需求。

(4)保护生产者,特别是通过有机食品的增值来提高生产者的收益,同时有机认证是消费者可以信赖的重要证明。

2. 有机食品的特征

有机食品生产于生态良好的有机农业生产体系,在生产和加工过程中不使用化学农药、化肥、化学防腐剂等化学合成物质,并杜绝基因工程生物及其产物。

(1)原料必须来自已经建立或正在建立的有机农业生产体系(又称有机农业生产基础),或采用有机方式采集的野生天然产品。

(2)产品在整个生产过程中必须严格遵循有机食品的生产、加工、包装、贮藏、运输等要求。

(3)生产者在有机食品的生产和流通过程中,有完善的跟踪检查体系和完整的生产、销售的档案记录。

(4)必须通过独立的有机产品认证机构的认证检查。

3. 有机食品标准体系

有机食品标准发展至今,已在世界范围内初步形成了不同层次的标准

by natural and ecological balance, so as to protect the environment, meet human needs and achieve sustainable development.

2. Comply with the international market trend, expand organic agricultural production and organic food exports, improve the market competitiveness of products.

3. Meet the domestic consumption demand of "green" and "environmental".

4. Protect the producers, especially through the value-added of organic food to improve the income of producers, and organic certification is an important proof that consumers can be trusted.

II. The Characteristics of Organic Food

Organic food is produced in the organic agricultural production system with good ecology. Chemical synthetic substances such as chemical pesticides, chemical fertilizers, chemical preservatives and genetic engineering organisms and their products are not used in the production and processing procedures.

1. Raw materials must come from organic agricultural production systems that have been established or are being established (also known as organic agricultural production bases), or wild natural products collected by organic means.

2. In the whole production process, the products must strictly follow the production, processing, packaging, storage, transportation and other requirements of organic food.

3. Producers in the production and circulation of organic food, there is a sound tracking and inspection system and complete production, sales records.

4. Must pass the certification inspection of independent organic products certification body.

III. Standard System for Organic Food

Since the development of organic food standards, different levels of standard system have been formed in

体系,主要表现在国际水平、地区水平、国家水平和认证机构水平等四个方面。

1)国际水平

从国际水平上看,有机食品标准有国际有机农业运动联盟(IFOAM)的基本标准。该标准和准则作为国际标准已在国际标准化组织注册,是地区标准、国家标准和认证机构自身标准的基础,是有机食品标准的标准。

国际食品法典委员会(CAC)的有机农业和有机农产品标准是由联合国粮农组织(FAO)与世界卫生组织(WHO)制定的,是《食品法典》的一部分,属于建议性标准。《食品法典》作为联合国协调各个成员国食品卫生和质量标准的跨国性标准,一旦成为强制性标准,就可以作为世界贸易组织仲裁国际食品生产和贸易纠纷的依据。

2)地区水平

欧盟(EU)标准属于地区水平标准。1991年欧盟有关有机农业的规则被发表于欧盟的官方刊物。1999年12月,欧盟委员会决定通过有机产品的标志,这个标志可以由 EU 2092/91 规则下的生产者使用。欧盟关于有机生产的 EU 2092/91 规则中有很多对消费者和生产者的保护。欧盟的 EU 2092/91 是 1991 年 6 月制定的,对有机农业和有机农产品的生产、加工、贸易、检查、认证以及物品使用等全过程进行了具体规定,共分 16 个条款和 6 份附件。1991 年制定的时候,标准只包括植物生产的内容,1998 年完成了动物标准的制定,2000 年 8 月 24 日正

the world, mainly in four aspects: international level, regional level, national level and certification institutional level.

A. International Level

From the international level, the organic food standards have the "basic standards" of the International Federation of Organic Agricultural Movements (IFOAM). As an international standard, the basic standards and guidelines of IFOAM have been registered in ISO, which is the basis of regional standards, national standards and certification bodies' own standards, and also the standard of organic food standards.

The organic agriculture and organic agricultural products standards of CAC are formulated by FAO and WHO, which is a part of *Codex Alimentarius* and belong to recommended standards. *Codex Alimentarius*, as a transnational standard for coordinating food hygiene and quality standards in various member countries of the CAC, can be used as the basis for WTO to arbitrate the international food production and trade disputes once it becomes a mandatory standard.

B. Regional Level

The EU standard belongs to the regional level standard. EU rules on organic agriculture were published in the EU's official publication in 1991. In December 1999, the European Union Commission decided to adopt the logo for organic products, which could be used by producers under the EU 2092/91 rule. EU's EU 2092/91 rules on organic production have a lot of protection for consumers and producers, it was formulated in June 1991, it specifies the whole process of production, processing, trade, inspection, certification and use of organic agriculture and organic agricultural products, which is divided into 16 articles and 6 annexes. When it was established in 1991, the standard included only plant

式生效。

欧盟标准适用于其15个成员国的所有有机食品的生产、加工、贸易（包括进口和出口）。也就是说，所有进口到欧盟的有机食品的生产过程应该符合欧盟的有机农业标准。因此，欧盟标准制定完成后，对世界其他国家的有机食品的生产、管理，特别是出口产生了很大影响。

3）国家水平

在国家水平上，除了15个欧盟成员国外，日本、阿根廷、巴西、澳大利亚、美国、智利、匈牙利、以色列、瑞士等国家都有自己的标准。不同国家的有机食品标准的发展历程各异，但共同的特点是发展历史短，主要集中在近10年左右。现举几例说明如下。

（1）美国：1990年通过联邦法《有机食品生产法案》；1992年成立国家有机食品标准委员会（NOSB）；1994年该委员会提交有机食品标准建议稿；1997年美国农业部（USDA）制定有机食品规章提案；1998年美国农业部着手修改有机食品规章；1999年有机贸易协会（OTA）发布美国民间有机食品标准；2000年3月美国农业部第二次提交有机食品规章提案；2001年夏天公布并开始执行美国的有机食品标准。

（2）日本。早在1935年，日本宗教和哲学领袖冈田茂吉就倡导"建立一个不依赖人造化学品和保护稀有资源的农业生态系统"。进入20世纪60、70年代，日本民间一些人士纷纷

production. The animal standard was completed in 1998 and came into effect on August 24, 2000.

EU standards apply to the production, processing, trade, import and export of all organic foods in its 15 member countries. In other words, all organic foods imported into the EU should be produced in accordance with EU organic agricultural standards. As a result, the completion of EU standards has had a significant impact on the production, management and export of organic food in other countries of the world.

C. National Level

At the national level, Japan, Argentina, Brazil, Australia, the United States, Chile, Hungary, Israel and Switzerland all have their own standards, in addition to the 15 EU member states. The development process of organic food standards varies from country to country, but the common characteristic is that the development history is short, mainly concentrated in the past 10 years or so. Here are a few examples.

1. The United States: passed the federal law of *Organic Food Production Act* in 1990; National Organic Food Standards Board (NOSB) established in 1992; NOSB submitted proposals for organic food standards in 1994; United States Department of Agriculture (USDA) developed the organic food regulations proposal in 1997; in 1998, the USDA began to revise organic food regulations; in 1999, the Organic Trade Association (OTA) issued the United States folk organic food standards; in March 2000, USDA submitted its second proposal for organic food regulations; in the summer of 2001, it published and began to implement United States organic food standards.

2. Japan. As early as 1935, Japanese religious and philosophical leader Maoki Okada advocated the "establishment of an agro ecosystem without relying on artificial chemicals and protecting rare resources", and entered the 1960s and 1970s, some

探索保护环境的农业生产体系,并相继产生了一批有机农业的民间交流和促进组织,如自然农法国际基金会、日本有机农业研究会、日本有机农业协会等。1992年日本农林水产省制定了《有机农产品、蔬菜、水果生产准则》和《有机农产品生产管理要点》,并于1992年将以有机农业为主的农业生产方式列入保护环境型农业政策。2001年正式出台的日本有机农业标准(JAS)标志着在日本有机农业生产的规范化管理已完全纳入政府行为。

(3)中国。按照国际有机食品标准和管理要求,1995年原国家环保局制定了《有机食品标志管理章程》和《有机食品生产和加工技术规范》,初步形成了较为健全的有机食品生产标准和认证管理体系。《有机产品认证标准》是目前生态环境部有机食品发展中心进行有机食品认证的基本依据。

4)认证机构水平

从认证机构水平上看,基本上每一个认证机构都建立了自己的认证标准。这里需要说明的是,一个国家可以有1个认证机构,也可以有多个认

folk people in Japan have explored the agricultural production system to protect the environment, and have produced a number of folk exchanges and promotion organizations for organic agriculture one after another. Such as the Natural Agriculture Law International Foundation, Japan Organic Agriculture Research Society, Japan Organic Agriculture Association, and so on. In 1992, the Ministry of Agriculture, Forestry and Fisheries of Japan formulated *Guidelines for the Production of Organic Agricultural Products*, *Vegetables and Fruits* and *Key Points in the Production and Management of Organic Agricultural Products*. In 1992, the agricultural production mode, which is based on organic agriculture, was included in the environmental protection agricultural policy. The Japanese Agriculture Standard (JAS), which was formally issued in 2001, indicates that the standardized management of organic agricultural production in Japan has been fully incorporated into the government behavior.

3. China. In accordance with international organic food standards and management requirements, in 1995, the former State Environmental Protection Bureau formulated the *Regulations for the Management of Organic Food Marks* and the *Specification for the Production and Processing of Organic Food*, which relatively sound organic food production standards and certification management system was initially formed. *Organic Product Certification Standard* is the basic basis for organic food certification by the Organic Food Development Center of the Ministry of Ecology and Environment at present.

D. Institutional Level

From the level of certification bodies, almost every certification body has established its own certification standards. It needs to be explained here that a country can have 1 certification body

证机构(比如美国境内就有40多个认证机构)。这些认证机构多数是民间的(如德国的Naturland、英国的Soil Associatin和美国国际有机作物改良协会等),也可以是官方的。不同认证机构执行的标准都是在国际有机农业运动联盟基本标准的基础上发展起来的,但侧重点有所不同。比如欧洲一些认证机构的有机食品标准,其主要内容多是围绕畜禽饲养,包括了牲畜、家禽饲养、牧草、饲料生产,肉、奶制品加工等。而中国以及其他一些亚洲国家的认证机构,其标准则多集中在大田作物(蔬菜水果)生产、野生产品开发、茶叶以及水产等方面。这也从一个侧面反映了不同国家或地区不同的资源特色。此外,根据不同地区的特征和需要,不同认证机构对标准的发展也有所不同。这其中多数认证机构仍以国际有机农业运动联盟基本标准的内容为主,标准比较原则化;也有一部分认证机构已根据本地区或本国实际,进一步发展了国际有机农业运动联盟标准,使之更具体化,便于操作,比如德国的Bioland已经建立了针对不同产品的标准系列。

4. 中国有机食品国家标准

2004年,国家标准化管理委员会和国家认证监督管理委员会组织有机农业研究机构、认证机构、认可机构等单位组成标准起草工作组,并完成了《有机产品》国家标准的制定。国家质量监督检验检疫总局、国家标准化管

and also have multiple certification bodies (for example, there are more than 40 certification bodies in the United States), most of these institutions are folk (such as Naturland in Germany, Soil Associatin in the UK, and Organic Crop Improvement Association in the United States), and they are also official. The standards implemented by different certification bodies are developed on the basis of the basic standards of IFOAM, but the emphasis is different. For example, the organic food standards of some certification bodies in Europe mainly focus on raising livestock and poultry, including livestock, poultry raising, forage, feed production, meat, dairy processing, etc., and certification bodies in China and some other Asian countries, the standards mainly focus on the production of field crops (vegetables and fruits), wild product development, tea and aquatic products. This also reflects the different resource characteristics of different countries or regions from one aspect. In addition, according to the characteristics and needs of different regions, the development of standards is also different among different certification bodies. Most of them still focus on the content of the basic standards of IFOAM, and the standards are more principled. Some certification bodies have further developed IFOAM standards to make them more specific and easy to operate based on local or national realities. For example, Bioland in Germany has established a series of standards for different products.

IV. National Standards for Organic Food in China

In 2004, the National Standardization Management Committee and the National Certification Supervision and Administration Committee organized a standard drafting working group composed of organic agricultural research institutions, certification institutions, and accredited institutions and completed the formulation of

理委员会和国家认证认可监督管理委员会分别发布了 GB/T 19630—2011《有机产品》和 CNCA-N-009:2011《有机产品认证实施规则》,并于 2012 年 3 月 1 日代替 GB/T 19630—2005《有机产品》和 CNCA-OG-001:2005《有机产品认证实施规则》。

1)制定的原则

(1)以国际有机农业标准(国际有机农业运动联盟和国际食品法典委员会)为基础。国际有机农业运动联盟是有机食品生产的民间机构,其基本标准的目的在于促进全球有机农业的发展,从技术和可操作性角度提出了有机食品生产的最低标准,作为各国制定有机农业标准的基础。国际食品法典委员会发布的《有机食品生产、加工、标识和销售指南》(GL 32—1999)是规范全球有机农业生产、加工、标志和销售的综合标准。为促进中国有机农业的发展与国际接轨,促进有机产品和体系的国际认证,《有机产品》国家标准采用国际和国外先进标准为基础。

(2)参考其他国家的标准。为了促进和加快我国有机农业标准与其他国家的互认,在具体指标和物质投入表中,参考了欧盟委员会有机食品

the national standards for *Organic Products*. The State General Administration of Quality Supervision, Inspection and Quarantine, the State Standardization Administration and the National Certification and Accreditation Regulatory Commission have recently issued GB/T 19630—2011 *Organic Products* and CNCA-N-009:2011 *Organic Products Certification Implementation Rules* respectively, and replace the GB/T 19630—2005 *Organic Products* and CNCA-OG-001:2005 *Organic Product Certification Implementation Rules* on March 1, 2012.

A. Principles Established

1. Based on International Organic Agriculture Standards (IFOAM and CAC). IFOAM is a non-governmental organization of organic food production, whose purpose is to promote the development of global organic agriculture. From the point of view of technology and maneuverability, IFOAM puts forward the minimum standard of organic food production, which is the basis of formulating organic agriculture standards in various countries. The GL 32—1999 *Guidelines for the Production, Processing, Labelling and Marketing of Organically Produced Foods* issued by CAC is a comprehensive standard for regulating the production, processing, marking and marketing of organic agriculture in the world. In order to promote the development of Chinese organic agriculture in line with the international standards and promote the international certification of organic products and systems, the national standards of *Organic Products* are based on international and foreign advanced standards.

2. Reference to standards in other countries. In order to promote and accelerate the mutual recognition between China's organic agriculture standards and other countries, in the specific

(EEC) No. 2092/2091法规、欧盟委员会有机食品（EC) No. 1788/2001法规、美国有机食品生产法规（NOP）、德国Naturland有机水产养殖标准等，保证了标准的衔接性和先进性。

（3）符合我国有机农业生产的实际。为结合我国有机农业生产的实际情况，《有机产品》国家标准在国家环保局发布的《有机（天然）食品标志管理章程》、《有机（天然）食品生产和加工技术规范》、国家环保总局有机食品发展中心发布的《OFDC有机认证标准》以及中国认证机构国家认可委员会发布的《有机产品生产和加工认证规范》等基础上，进行完善和修改。

2) 标准的内容

中国《有机产品》标准分为四部分。

（1）第一部分是生产，主要包括作物种植、食用菌栽培、野生植物采集、畜禽养殖、水产养殖、蜜蜂养殖及其产品的运输、贮藏和包装，是农作物、食用菌、野生植物、畜禽、水产、蜜蜂及其未加工产品的有机生产通用规范和要求。

（2）第二部分是加工，主要包括有机产品加工的通则，根据第一部分生

indicators and material input tables, reference is made to the European Commission's organic food (EEC) No. 2092/2091 regulation, the European Commission organic food (EC) No. 1788/2001 regulation, the American organic food production regulation (NOP), the German Naturland organic aquaculture standard, etc., which ensures the convergence and advanced nature of the standards.

3. Conform to the practice of organic agricultural production in China. In order to combine the actual situation of organic agriculture production in China, on the basis of the *Organic (Natural) Food Label Management Regulations* and the *Organic (Natural) Food Production and Processing Technical Specifications* issued by the State Environmental Protection Administration, and the *OFDC Organic Certification Standard* issued by the Organic Food Development Center of the State Environmental Protection Administration, as well as the *Certification Specification for the Production and Processing of Organic Products* issued by the National Accreditation Committee of China's certification body, the national standards for *Organic Products* have been improved and revised.

B. Content of the Standard

The standard of *Organic Products* in China is divided into four parts.

1. Part one Production: mainly including crop cultivation, edible fungus cultivation, wild plant collection, livestock and poultry breeding, aquaculture, bee breeding and transportation, storage and packaging of their products. It is a general standard and requirement for the organic production of crops, edible fungi, wild plants, livestock and poultry, aquatic products, bees and their unprocessed products.

2. Part two Processing: mainly including general rules for the processing of organic products, unprocessed

产标准生产的未加工产品为原料进行加工及包装、贮藏和运输，是有机产品加工通用规范和要求。

(3) 第三部分是标志与销售，是按第一部分和第二部分生产或加工并获得认证的产品的标志和销售，是有机产品标志和销售的通用规范及要求。

(4) 第四部分是管理体系，主要包括有机产品生产、加工、经营过程中必须建立和维护的管理体系，是有机产品的生产者、加工者、经营者及相关的供应环节质量管理的通用规范和要求。

3) 主要特点

(1) 涵盖一个大农业体系。《有机产品》国家标准不仅涵盖了种植业方面的要求，还包括了畜牧和家禽饲养业、水产品、蜜蜂等方面的要求，不仅包括生产过程的要求，还延伸到收获、加工和包装、标签等方面的要求。所以，《有机产品》国家标准是控制从农业（包括粮食、饲料和纤维）、畜牧和家禽、水产的田间生产到加工成最终消费产品的一个完整的、基础性的指导法规，也可以说是有机农业的根本法则。

(2) 将技术和管理融为一体。《有机产品》国家标准的框架分为有机生产过程控制和管理体系控制两大部分，明确了管理体系对有机产品生产的统领作用，从而丰富了世界有机农

products produced according to the first part of the production standard are processed and packaged, stored and transported as raw materials, which is a general specification and requirement for the processing of organic products.

3. Part three Logo and sales: marks and sales of products manufactured or processed and certified as part Ⅰ and part Ⅱ. It is a general specification and requirement for the marking and sale of organic products.

4. Part four Management system: it includes the management system which must be established and maintained in the process of production, processing and management of organic products. It is the general standard and requirement of the quality management of organic products' producers, processors, operators and related supply links.

C. Main Features

1. Covering a large agricultural system. The national standards for *Organic Products* cover not only the requirements of the farming industry, but also the requirements of livestock and poultry, aquatic products, bees, etc., and not only the requirements of the production process, it also extends to the requirements of harvesting, processing and packaging, labeling, etc. Therefore, the national standard for *Organic Products* is a complete and fundamental guideline for controlling the production of agricultural (including food, feed and fiber), livestock and poultry, and aquatic products, to processing them into final consumer products. It can also be said to be the fundamental principle of organic agriculture.

2. Integration of technology and management. The frame of the national standard of *Organic Products* is divided into two parts: organic production process control and management system control. It clarifies the leading role of management

业的内容。

（3）将健康、环保和安全有机结合。《有机产品》国家标准强调以健康的生态系统为基础，以良好的操作为规范；提倡在保证有机生产基地基本的本底（土壤、水等）环境条件下，重视生产过程中对环境的保护；并设定了评估由于不可避免的因素（土壤本身和外来物质的漂移）对最终产品安全指标造成风险的最低限值。

（4）符合国情，国际接轨。《有机产品》国家标准充分考虑到中国几千年农业发展的传统经验和实践，遵循国家相关法律法规，保证标准的实用性；在国际标准有争议和敏感的具体环节（如人粪尿、集约化生产的畜禽粪便、烟草、食品添加剂等）保持与国际标准的一致性。

system on organic production and enriches the content of organic agriculture in the world.

3. An organic combination of health, environmental protection and safety. The national standard of *Organic Products* emphasizes on the basis of healthy ecosystem and on the basis of good operation, advocates to attach importance to the protection of the environment in the production process under the basic background (soil, water, etc.) in the organic production base, and sets a minimum value to evaluate the risk caused by the inevitable factors (drift of the soil itself and foreign matter) to the final product safety indicators.

4. In line with national conditions and in line with international standards. The national standard of *Organic Products* fully takes into account the traditional experience and practice of thousands of years of agricultural development in China, follows the relevant laws and regulations of the country, and ensures the practicability of the standard. The controversial and sensitive aspects of international standards (such as human excrement, intensive production of animal manure, tobacco, food additives, etc.) are consistent with international standards.

5. 有机食品的法律法规

1）认证机构的授权和认可

（1）认证机构的授权。

①授权机构。有机食品认证机构的授权目前主要有两种形式：政府授权和非政府组织授权。目前欧盟、美国以及日本等国家和地区既可通过政府组织（如农业部等）进行授权，又可通过私人组织进行授权，如私人机构按照ISO65的规定进行第三方授权。国际有机农业运动联盟认证授权属于非政府组织，其活动主要通过总部设在美国的国际有机认可公司（IOAS）完成。国际有机认可公司称，其活动是严格按照国际标准ISO65的原理进行操作的。由于国际有机农业运动联盟在世界上影响非常大，尽管它是非政府组织，但很多国家和地区都承认国际有机农业运动联盟授权的机构。我国有机食品认证机构的授权机构为国家认证认可监督管理委员会（CNCA）。

②授权依据。我国认证机构的授权依据为《中华人民共和国认证认可条例》。该条例对国内认证机构资格的规定包括：有固定的场所和必要的设施；有符合认证认可要求的管理制度；注册资本不得少于人民币300万元；有10名以上相应领域的专职认证人员。从事产品认证活动的认证机构，还应当具备与从事相关产品认证活动相适应的检测、检查等技术

能力。

《中华人民共和国认证认可条例》对国外认证机构资格的规定包括：设立外商投资的认证机构除应当符合上述条件外，还应当符合：外方投资者取得其所在国家或者地区认可机构的认可；外方投资者具有3年以上从事认证活动的业务经历；设立外商投资认证机构的申请、批准和登记，按照有关外商投资法律、行政法规和国家有关规定办理。

(2) 认证机构的认可。

①认可机构。认可是国家依法设立的权威机构对认证机构实施认证的能力进行评定和承认。我国有机产品认证机构的认可机构为中国认证机构国家认可委员会(CNAB)。

②认可依据。CNAB-AC 23:2003《认证机构实施有机产品生产和加工认证的认可基本要求》为我国认证的认可依据。该要求是依据 CNAB-AC 21(ISO/IEC 导则 65)、国际认可论坛(IAF)发布的相应指南文件和《IFOAM 有机生产和加工认证标准》，并结合中国的具体情况而制定的。

2) 有机产品认证管理办法

根据《中华人民共和国认证认可条例》，国家质量监督检验检疫总局于2013年11月颁布了新修订的《有机产品认证管理办法》(总局令第155号)，并于2014年4月1日起施行。新《有机产品认证管理办法》共分七章六十三条。

第一章为总则(共六条)，主要规定了立法目的和立法依据，明确了有机产品的定义、有机产品标准和合格评定程序的制定规范，并对有机产品的监督管理体制、本办法的适用范围等进行了具体规范。

第二章为认证实施(共十条)，规定了从事有机产品认证活动的认证机构及其人员的具体要求；受理有机产品认证的条件；对有机产品产地(基地)环境检测、产品样品检测活动的机构的资质要求作出了规定；对有机加工品的有机配料含量作出了规定；规定了有机产品认证的具体过程，并对有机产品认证机构的跟踪检查等义务性要求作出了规定。

第三章为认证产品进口(共八条)，规定了向中国出口有机产品的国家或者地区的有机产品主管机构应采取的措施和要求；规定了进口有机产品认证委托人应完成的工作事项；对进口有机产品入境检验检疫涉及材料等作出了规定。

第四章为认证证书和认证标志(共十二条)，规定了有机产品认证证书的基本格式、内容以及标志的基本式样，并明确了有机产品认证证书和标志在使用中的具体要求。

第五章为监督管理(共十条)，规定了国家认证认可监督管理委员会、地方质监部门和出入境检验检疫机构对有机产品监督检查工作中的具体监管方式，以及对获得有机产品认证的生产、加工、经营单位、个人和进口有机产品进行了具体规范，并对有机产品认证认可活动中的申诉、投诉制度作出了层级规定。

第六章为罚则(共十一条)，规定了对有机产品认证认可活动中的各类违法行为的处罚。

第七章为附则(共五条)，规定了对有机产品认证认可活动的收费要求和加工配料含义，并明确了本办法的具体运行时间和解释权。

新《有机产品认证管理办法》在旧版的基础上有了更新和改进，对于促进有机产品产业发展、保护消费者健康和生态环境有着重要意义。

3) 有机产品认证实施规则

为规范有机产品认证活动，根据《中华人民共和国认证认可条例》《有机产品认证管理办法》等有关规定,新修订了《有机产品认证实施规则》(CNCA-N-009:2011)。该规则于2011年12月2日由中国国家认证认可监督管理委员会以2011年第34号公告公布,自2012年3月1日起实施。

新《有机产品认证实施规则》分为目的和范围,认证机构要求,认证人员要求,认证依据,认证程序,认证后管理,再认证,认证证书、认证标志的管理,信息报告,认证收费等十部分。新《有机产品认证实施规则》在旧版本的基础上新增规定每件达到新国标的有机产品加贴17位的唯一编码,"一品一码"、不可二次包装,产品质量可全程追溯等内容;还明确了新的《有机产品认证目录》。总之,《有机产品认证实施规则》是对认证机构开展和实施有机产品认证程序的统一要求。

4)有机产品认证目录

根据《有机产品认证管理办法》《有机产品认证实施规则》规定,国家认证认可监督管理委员会在各认证机构已认证产品的基础上,按照风险评估的原则,组织相关专家制定了《有机产品认证目录》(2018)。目前国家认监委共六次发布有机产品认证增补目录,表9-2是截止到2018年6月19日更新后的有机产品认证目录。

表9-2 有机产品认证目录(2018)

序号	产品名称	产品范围
生产		
植物类(含野生植物采集)		
谷物		
1	小麦	小麦
2	玉米	玉米、鲜食玉米、糯玉米
3	水稻	稻谷
4	谷子	谷子
5	高粱	高粱
6	大麦	大麦、酿酒大麦、饲料大麦、青稞
7	燕麦	莜麦、燕麦
8	杂粮	黍、粟、苡仁、荞麦、花豆、泥豆、鹰嘴豆、饭豆、小扁豆、羽扁豆、瓜尔豆、利马豆、木豆、红豆、绿豆、青豆、黑豆、褐红豆、油莎豆、芸豆、糜子、苦荞麦、藜麦、穄子、红稗
蔬菜		
9	薯芋类	马铃薯、木薯、甘薯、山药、葛类、芋、魔芋、菊芋、蕉芋(旱藕)
10	豆类蔬菜	蚕豆、菜用大豆、豌豆、菜豆、刀豆、扁豆、长豇豆、黎豆、四棱豆
11	瓜类蔬菜	黄瓜、冬瓜、丝瓜、西葫芦、节瓜、菜瓜、笋瓜、越瓜、瓠瓜、苦瓜、中国南瓜、佛手瓜、蛇瓜
12	白菜类蔬菜	白菜、菜薹

续表 9-2

序号	产品名称	产品范围
13	绿叶蔬菜	散叶莴苣、莴笋、苋菜、茼蒿、菠菜、芹菜、苦菜、菊苣、苦苣、芦蒿、蕹菜、苜蓿、紫背天葵、罗勒、荆芥、乌塌菜、荠菜、茴香、芸薹、叶甜菜、猪毛菜、寒菜、番杏、灰灰菜、榆钱菠菜、木耳菜、落葵、紫苏、莳萝、芫荽、水晶菜、菊花脑、珍珠菜、养心菜、帝王菜、芦荟、海篷子、碱蓬、冰菜、人参菜、马兰头
14	新鲜根菜类蔬菜	芜菁、萝卜、牛蒡、芦笋、甜菜、胡萝卜、鱼腥草
15	新鲜甘蓝类蔬菜	芥蓝、甘蓝、花菜
16	新鲜芥菜类蔬菜	芥菜
17	新鲜茄果类蔬菜	辣椒、西红柿、茄子、人参果、秋葵
18	新鲜葱蒜类蔬菜	葱、韭菜、蒜、姜、圆葱、岩葱
19	新鲜多年生蔬菜	笋、鲜百合、金针菜、黄花菜、朝鲜蓟、香椿、辣木、沙葱、荨麻、椒蒿
20	新鲜水生类蔬菜	莲藕、茭白、荸荠、菱角、水芹、慈菇、豆瓣菜、莼菜、芡实、蒲菜、水芋、水雍菜、莲子
21	新鲜芽苗类蔬菜	苗菜、芽菜
22	食用菌类	菇类、木耳、银耳、块菌类、北虫草
水果与坚果		
23	柑桔类	桔、橘、柑类
24	甜橙类	橙
25	柚类	柚
26	柠檬类	柠檬
27	葡萄类	鲜食葡萄、酿酒葡萄
28	瓜类	西瓜、甜瓜、厚皮甜瓜、木瓜
29	苹果	苹果、沙果、海棠果
30	梨	梨
31	桃	桃
32	枣	枣
33	杏	杏
34	其他水果	梅、杨梅、草莓、黑豆果、橄榄、樱桃、李子、猕猴桃、香蕉、椰子、菠萝、芒果、番石榴、荔枝、龙眼、杨桃、菠萝蜜、火龙果、红毛丹、西番莲、莲雾、面包果、榴莲、山竹、海枣、柿、枇杷、石榴、桑椹、酸浆、沙棘、山楂、无花果、蓝莓、黑莓、树莓、高钙果、越橘、黑加仑、雪莲果、诺尼果、黑果腺肋花楸、黑老虎（布福娜）、蓝靛果、神秘果、番荔枝
35	核桃	核桃
36	板栗	板栗

续表 9-2

序号	产品名称	产品范围
37	其他坚果	榛子、瓜籽、杏仁、咖啡、椰子、银杏果、芡实(米)、腰果、槟榔、开心果、巴旦木果、香榧、苦槠果、栝蒌、澳洲坚果、角豆、可可
豆类与其他油料作物		
38	大豆	大豆
39	其他油料作物	油菜籽、芝麻、花生、茶籽、葵花籽、红花籽、油棕果、亚麻籽、南瓜籽、月见草籽、大麻籽、玫瑰果、琉璃苣籽、苜蓿籽、紫苏籽、翅果油树、青刺果、线麻、南美油藤、元宝枫
花卉		
40	花卉	菊花、木槿花、芙蓉花、海棠花、百合花、茶花、茉莉花、玉兰花、白兰花、栀子花、桂花、丁香花、玫瑰花、月季花、桃花、米兰花、珠兰花、芦荟、牡丹、芍药、牵牛、麦冬、鸡冠花、凤仙花、百合、贝母、金银花、荷花、藿香蓟、水仙花、腊梅、霸王花、紫藤花、金花葵
香辛料作物产品		
41	香辛料作物产品	花椒、青花椒、胡椒、月桂、肉桂、丁香、众香子、香荚兰豆、肉豆蔻、陈皮、百里香、迷迭香、八角茴香、球茎茴香、孜然、小茴香、甘草、薄荷、姜黄、红椒、藏红花、芝麻菜、山葵、辣根、草果、甘菊、神香草、猫薄荷、啤酒花、山苍子
制糖植物		
42	制糖植物	甘蔗、甜菜、甜叶菊
其他类植物		
43	青饲料植物	苜蓿、黑麦草、芜菁、青贮玉米、绿萍、红萍、羊草、皇竹草、甜象草、老芒麦、构树
44	纺织用的植物原料	棉、麻、桑、竹、木棉
45	调香的植物	香水莲、薰衣草、迷迭香、柠檬香茅、柠檬马鞭草、藿香、鼠尾草、小地榆、天竺葵、紫丁香、艾草、佛手柑

续表 9-2

序号	产品名称	产品范围
46	野生采集的植物	蕨菜、刺嫩芽、山芹、山核桃、松子、沙棘、蓝莓、羊肚菌、松茸、牛肝菌、鸡油菌、板蓝根、月见草、蒲公英、红花、贝母、灰树花、当归、葛根、石耳、榛蘑、草蘑、松蘑、栗蘑、红蘑、小麦草、塔花、水飞蓟、益母草、茯苓、高良姜、接骨木、蒺藜、天门冬、积雪草、蔓荆子、独活、葫芦巴、苦橙、缬草、车前草、远志、山葡萄、红树莓、雪菊、罗布麻、橡籽、刺五加、华西银腊梅、笋、刺梨、沙葱、荨麻、椒蒿、鹅绒委陵菜、山苦茶（鸥鸪茶）、青钱柳、毛建草（岩青兰）、地耳、鹿角菜、霞草（麻杂菜）、猕猴桃、黑果枸杞、毛豹皮樟（老鹰茶）、鸡血藤、龙胆草、夏枯草、香樟、滇重楼、白及、山刺玫（刺玫）、杜鹃、蹄盖蕨菜（猴腿菜）、荚果蕨（黄瓜香）、黄芩、金莲花、柳蒿、香青兰（山薄荷）、山菠菜、小根蒜、鸭舌草（鸭嘴菜）、马齿苋、蘋（四叶菜）、花脸香菇（花脸蘑）、滑子菇（珍珠菇）、双孢菇（内蒙白蘑）、亚侧耳（元蘑）、黄花菜、木耳、荠菜、苋菜、榛子、白柳、决明子、芦苇、胖大海、砂仁、凉粉草（仙草）、栀子、刺槐（洋槐）、紫花碎米荠、元宝枫、油茶籽、火棘、野草莓、山苍子、乌头（川乌、附子）、皂荚（皂角）、银柴胡、虎杖、天南星（天南星、异叶天南星、东北天南星）、藁本（藁本、辽藁本）、木贼、半边莲、百部（直立百部、蔓生百部、对叶百部）、葶苈子（播娘蒿、独行菜）、祁州漏芦（漏芦）、白前（柳叶白前、芫花叶白前）、大血藤、款冬、泽泻、光叶菝葜（土茯苓）、白花前胡（前胡）、射干、半夏、北乌头（草乌）、桑寄生、蝙蝠葛（北豆根）、独角莲（白附子）、升麻（大三叶升麻、兴安升麻、升麻）、苍术（茅苍术、北苍术）、茵陈（滨蒿、茵陈蒿）、木通（木通、三叶木通、白木通）、地肤、半枝莲、苍耳、小通草（喜马山旌节花、中国旌节花、青荚叶）、谷精草、白术、木香、玄参、莎草（香附）、益智、乌药、川芎、五倍子（盐肤木、青麸杨、红麸杨叶上的虫瘿）、郁李（欧李、郁李、长柄扁桃）、高良姜、吴茱萸（吴茱萸、石虎、疏毛吴茱萸）、莪术（蓬莪术、广西莪术、温郁金）、毛叶地瓜儿苗（泽兰）、延胡索（元胡）、麦蓝菜（王不留行）、槲蕨（骨碎补）、石菖蒲、阿尔泰银莲花（九节菖蒲）、蛇床、槐、密蒙花、茜草、粗茎鳞毛蕨（绵马贯众）、马勃（脱皮马勃、大马勃、紫色马勃）、山银花（灰毡毛忍冬、红腺忍冬、华南忍冬、黄褐毛忍冬）
47	茶	茶
种子与繁殖材料		
48	种子与繁殖材料	种子、繁殖材料（仅限本目录列出的植物类种子及繁殖材料）

续表 9-2

序号	产品名称	产品范围
	植物类中药	
49	植物类中药	三七、大黄、婆罗门参、人参、西洋参、土贝母、黄连、黄芩、菟丝子、牛蒡根、地黄、桔梗、槲寄生、钩藤、通草、土荆皮、白鲜皮、肉桂、杜仲、牡丹皮、五加皮、银杏叶、石韦、石南叶、枇杷叶、苦丁茶、柿子叶、罗布麻、枸骨叶、合欢花、红花、辛夷、鸡冠花、洋金花、藏红花、金银花、大草蔻、山楂、女贞子、山茱萸、五味子、巴豆、牛蒡子、红豆蔻、川楝子、沙棘、大蓟、广藿香、小蓟、马鞭草、龙葵、长春花、仙鹤草、白英、补骨脂、羊栖菜、海蒿子、冬虫夏草、茯苓、灵芝、石斛、除虫菊、甘草、罗汉果、巴戟天、黄荆、何首乌、川芎、天麻、厚朴、柴胡、莞香、苁蓉、锁阳、蝉花、玛咖、玉竹、连翘、金线莲、绞股蓝、当归、丹参、党参、黄芪、扯根菜、黄精、巴拉圭冬青、苦参、萝芙木、牛大力、黑果枸杞、枸杞*(试点)、猫尾草(石参)、平卧菊三七、牛皮消、红豆杉、大白茅(白茅根)、白芷、破布叶(布渣叶)、穿心莲、菘蓝(大青叶、板蓝根)、淡竹叶、秤星树(岗梅根)、鸡蛋花、橘红、决明子、莲(莲子心)、芦苇、胖大海、忍冬(忍冬藤)、砂仁、夏枯草、凉粉草(仙草)、栀子、鸡血藤、辽细辛、滇重楼、白及、淫羊藿(淫羊藿、巫山淫羊藿)、三叶崖爬藤(三叶青)、构树、山桃、山苍子、大花红景天(红景天)、乌头(川乌、附子)、皂荚(皂角)、银柴胡、虎杖、天南星(天南星、异叶天南星、东北天南星)、藁本(藁本、辽藁本)、木贼、半边莲、百部(直立百部、蔓生百部、对叶百部)、葶苈子(播娘蒿、独行菜)、祁州漏芦(漏芦)、麻黄(草麻黄、中麻黄、木贼麻黄)、白前(柳叶白前、芫花叶白前)、大血藤、款冬、泽泻、光叶菝葜(土茯苓)、白花前胡(前胡)、射干、半夏、北乌头(草乌)、防风、桑寄生、蝙蝠葛(北豆根)、独角莲(白附子)、升麻(大三叶升麻、兴安升麻、升麻)、秦艽(秦艽、麻花秦艽、粗茎秦艽、小秦艽)、苍术(茅苍术、北苍术)、羌活(羌活、宽叶羌活)、茵陈(滨蒿、茵陈蒿)、黄皮树(黄柏)、木通(木通、三叶木通、白木通)、地肤、半枝莲、苍耳、小通草(喜马山旌节花、中国旌节花、青荚叶)、明党参、谷精草、知母、白术、木香、珊瑚菜(北沙参)、太子参、藤茶(显齿蛇葡萄)、赤小豆(赤小豆、赤豆)、玄参、莎草(香附)、益智、乌药、川芎、五倍子(盐肤木、青麸杨、红麸杨叶上的虫瘿)、郁李(欧李、郁李、长柄扁桃)、高良姜、吴茱萸(吴茱萸、石虎、疏毛吴茱萸)、莪术(蓬莪术、广西莪术、温郁金)、毛叶地瓜儿苗(泽兰)、延胡索(元胡)、麦蓝菜(王不留行)、槲蕨(骨碎补)、石菖蒲、阿尔泰银莲花(九节菖蒲)、蛇床、槐、密蒙花、茜草、粗茎鳞毛蕨(绵马贯众)、马勃(脱皮马勃、大马勃、紫色马勃)、猪苓、山银花(灰毡毛忍冬、红腺忍冬、华南忍冬、黄褐毛忍冬)
	畜禽类	
	活体动物	
50	肉牛(头)	肉牛
51	奶牛(头)	奶牛
52	乳肉兼用牛(头)	乳肉兼用牛
53	绵羊(头)	绵羊

续表 9-2

序号	产品名称	产品范围
54	山羊(头)	山羊
55	马(头)	马
56	驴(头)	驴
57	猪(头)	猪
58	鸡(只)	鸡
59	鸭(只)	鸭
60	鹅(只)	鹅
61	其他动物(头/只)	兔、羊驼、鹌鹑、火鸡、鹿、蚕、鹧鸪、骆驼、鸵鸟、黄粉虫
动物产品或副产品		
62	牛乳	牛乳
63	羊乳	羊乳
64	马乳	马乳
65	其他动物产品	驴奶、骆驼奶
66	鸡蛋(枚)	鸡蛋
67	鸭蛋(枚)	鸭蛋
68	其他禽蛋(枚)	鹌鹑蛋、鸵鸟蛋、鹅蛋
69	动物副产品	毛、绒、蚕蛹、蚕茧
水产类		
鲜活鱼		
70	海水鱼(尾)	文昌鱼、鳗、鲱鱼、鲇鱼、鲑、鳕鱼、鲉、鲈、黄鱼、鲷、鳗鲡、鲷、鲀、鲈鱼、鲆、鲽鱼、鳟、军曹鱼
71	淡水鱼(尾)	青鱼、草鱼、鲢鱼、鳙鱼、鲤鱼、鳜鱼、鲟鱼、鲫鱼、鲶鱼、鲌鱼、黄鳝、鳊鱼、罗非鱼、鲂鱼、鲴鱼、乌鳢、鲳鱼、鳗鲡、鳟鱼、鲮、鲴鱼、鮠、鲇、梭鱼、餐条鱼、狗鱼、雅罗鱼、池沼公鱼、武昌鱼、黄颡鱼、泥鳅、亚东鱼(鲑)、银鱼、丁鲹、梭鲈、河鲈、江鳕、东方欧鳊、银鲫、欧鲇、鳡浪白鱼
甲壳与无脊椎动物		
72	虾类(吨)	虾
73	蟹类(只)	绒螯蟹、三疣梭子蟹、红螯相手蟹、锯缘青蟹
74	无脊椎动物	牡蛎、鲍、螺、蛤类、蚶、河蚬、蛏、西施舌、蛤蜊、河蚌、海蜇、海参、卤虫、环刺螠、海胆、扇贝
其他水生脊椎动物		
75	两栖和爬行类动物	鳖、中华草龟、大鲵

续表 9-2

序号	产品名称	产品范围
水生植物		
76	藻类	海带、紫菜、裙带菜、麒麟菜、江蓠、羊栖菜、螺旋藻、蛋白核小球藻
加工		
肉制品及副产品加工		
77	冷鲜肉和冷冻肉	猪、牛、羊、鸭、鸡、鹅、鹿、驴、兔、鸵鸟、骆驼、羊驼、马、鹌鹑、鹧鸪、火鸡
78	加工肉制品和可食用屠宰副产品	肉制品(以第50~61项中的动物为原料加工的制品)、可食用屠宰副产品(第50~61项中的动物内脏、骨骼、血、皮、油脂及其制品)
水产品加工		
79	冷鲜鱼和冷冻鱼	海水鱼(文昌鱼、鳗、鲱鱼、鲇鱼、鲑、鳕鱼、鲉、鲈、黄鱼、鲷、鳗鲡、鲷、鲀、鲈鱼、鲆、鲽鱼、鳟)、淡水鱼(青鱼、草鱼、鲢鱼、鳙鱼、鲤鱼、鳜鱼、鲟鱼、鲫鱼、鲶鱼、鲌鱼、黄鳝、鳊鱼、罗非鱼、鲂鱼、鲷鱼、乌鳢、鲳鱼、鳗鲡、鳡鱼、鲮、鲴鱼、鲍、鲇、梭鱼、餐条鱼、狗鱼、雅罗鱼、池沼公鱼、武昌鱼、黄颡鱼、丁鲅、梭鲈、河鲈、江鳕、东方欧鳊、银鲫、欧鲇、鳡浪白鱼)
80	加工鱼制品	加工鱼制品
81	其他水产加工制品	海参、海胆、扇贝、小龙虾、海带、紫菜、裙带菜、麒麟菜、江蓠、羊栖菜、海苔、螺旋藻(粉、片)、鲍鱼、虾
加工或保藏的蔬菜		
82	冷冻蔬菜	速冻蔬菜
83	保藏蔬菜	保藏蔬菜
84	腌渍蔬菜	盐渍菜、糖渍菜、醋渍菜、酱渍菜
85	脱水蔬菜	蔬菜干制品
86	蔬菜罐头	蔬菜罐头
饮料		
87	果蔬汁及其饮料	果蔬汁及其饮料
88	其他植物饮料	杏仁露、菊粉、核桃露(乳)、豆奶
加工和保藏的水果和坚果		
89	保藏的水果和坚果	保藏的水果和坚果(限于以本目录生产/植物类中第23~37项为原料的)
90	冷冻水果	冷冻水果
91	冷冻坚果	冷冻板栗
92	果酱	果酱
93	烘焙或炒的坚果	松籽、核桃(仁)、杏(仁)、葵花籽(仁)、五香瓜子、榛子(仁)、花生、澳洲坚果(仁)

续表 9-2

序号	产品名称	产品范围
94	其他方法加工及保藏的水果和坚果	坚果粉(粒、片)、水果干制品(限于以本目录生产/植物类中第23~34项为原料加工的)
植物油加工		
95	食用植物油	食用植物油(限于以第1~43、45~49项中的产品或其植株的其他部分作为原料加工的)
植物油加工副产品		
96	植物油加工副产品	植物油加工副产品
经处理的液体乳或奶油		
97	经处理的液体乳和奶油	黄油、巴氏杀菌乳(含调制乳)、灭菌乳(含调制乳)、乳脂(奶油)、乳清液、含乳饮料
其他乳制品		
98	乳粉及其制品	乳粉、乳清粉、乳糖、乳清蛋白粉、含乳固态成型制品
99	发酵乳	发酵乳、干酪、再制干酪
谷物磨制		
100	小麦(粉)	小麦、小麦粉、麦麸
101	玉米(粉)	玉米、玉米粉
102	大米(粉)	大米、米粉、米糠
103	小米(粉)	小米、小米粉
104	其他谷物碾磨加工品和副产品	其他谷物去壳产品及副产品、其他谷物磨制粉、其他植物磨制粉、碾压的片、藜麦
淀粉与淀粉制品		
105	淀粉	淀粉
106	淀粉制品	粉丝、其他淀粉制品
107	豆制品	豆制品
加工饲料		
108	加工的植物性饲料	植物性饲料
109	加工的动物性饲料	动物性饲料
烘焙食品		
110	饼干、面包及其他烘焙产品	饼干、面包、月饼
面条等谷物粉制品		
111	米面制品	面制品、米制品

续表 9-2

序号	产品名称	产品范围
112	方便食品	粮食制品
不另分类的食品		
113	茶	红茶、黑茶、绿茶、花茶、乌龙茶、白茶、黄茶、速溶茶、茶粉
114	代用茶	苦丁茶、杜仲茶、柿叶茶、桑叶茶、银杏叶茶、野菊花茶、野藤茶、菊花茶、薄荷、大麦茶、其他代用茶(仅限以本目录生产/植物类中第1~49项为原料加工的)
115	其他食品	咖啡、巧克力、1,3-二油酸-2-棕榈酸甘油三酯(仅作为营养强化剂)、溶豆
116	保藏的禽蛋及其制品	禽蛋及其制品
117	调味品	糖、酱油、食醋、芝麻盐、酱、香辛料、低聚半乳糖、低聚果糖
118	植物类中草药加工制品(仅限于经切碎、烘干等物理工艺加工的产品)	三七、大黄、人参、西洋参、菟丝子、牛蒡根、地黄、桔梗、槲寄生、肉桂、杜仲、牡丹皮、五加皮、银杏叶、苦丁茶、罗布麻、红花、藏红花、金银花、山楂、女贞子、山茱萸、五味子、牛蒡子、沙棘、大蓟、广藿香、小蓟、补骨脂、冬虫夏草、茯苓、灵芝、松花粉、大花红景天(红景天)、泽泻、光叶菝葜(土茯苓)、升麻(大三叶升麻、兴安升麻、升麻)、苍术(茅苍术、北苍术)、知母、白术、木香、珊瑚菜(北沙参)、太子参、玄参、莎草(香附)、益智、川芎、郁李(欧李、郁李、长柄扁桃)、高良姜、吴茱萸(吴茱萸、石虎、疏毛吴茱萸)、毛叶地瓜儿苗(泽兰)、槲蕨(骨碎补)、茜草
白酒		
119	白酒和配制酒	白酒、食用酒精、配制酒(限于以白酒为配基,以第1~49项中的植物为原料生产的)
葡萄酒和果酒等发酵酒		
120	葡萄酒	红葡萄酒、白葡萄酒、桃红葡萄酒
121	果酒	果酒、水果红酒/冰酒/干酒
122	黄酒	黄酒
123	米酒	米酒
124	其他发酵酒	红曲酒
啤酒		
125	啤酒	啤酒
纺纱用其他天然纤维		
126	纺纱用其他天然纤维	竹纤维、蚕丝、皮棉、麻、木棉
纺织品		
127	纺织制成品	纱及其制成品、线及其制成品、丝及其制成品

* 限于"宁夏枸杞"种。

注:(1)获得有机产品认证的植物类产品可包括该产品的整个植株或者植株的某一部分。例如葡萄获

得有机产品认证,其葡萄籽和葡萄叶无需另外申请认证。

(2)认证委托人在生产基地外对新鲜蔬菜、水果、杂粮产品进行包装的,认证机构可根据《有机产品 第1部分:生产》标准的5.10条款对包装场所进行检查,符合要求后颁发"生产"范围认证证书。

(3)第92项"果酱"中含"果泥"。

(4)第112项"方便食品"的"粮食制品"中含"糊类食品"。

6. 有机食品的认证程序

有机产品认证,是指经过授权的认证机构按照有机产品国家标准和相关规定对有机产品生产和加工过程进行评价的活动,以规范化的检查为基础,包括实地检查、质量保证体系的审核和最终产品的检测,并以有机产品认证证书的文件形式予以确认。

1)有机产品认证主管单位

不同于无公害农产品以及绿色食品的认证,有机产品认证属于独立第三方认证。我国的有机产品认证开始于20世纪90年代初,最初由国家环境保护(总)局"国家有机食品认证认可委员会"负责有机产品认证机构的管理与认可。以2002年11月1日《中华人民共和国认证认可条例》的正式颁布实施为起点,有机产品认证工作由国家认证认可监督管理委员会统一管理,进入规范化阶段。

到目前为止,经国家认监委认可的专职或兼职有机产品认证机构总共有31家(每年数据会有变化)。如中国质量认证中心、中绿华夏有机食品认证中心、南京国环有机产品认证中心等。

2)有机产品认证流程

有机产品认证属于产品认证的范畴,虽然各认证机构的认证程序有一定差异,但根据《中华人民共和国认证认可条例》、国家质量监督检验检疫总局《有机产品认证管理办法》、国家认证认可监督管理委员会《有机产品认证实施规则》和中国认证机构国家认可委员会《产品认证机构通用要求:有机产品认证的应用指南》的要求以及国际通行做法,有机产品认证的模式通常为"过程检查+必要的产品和产地环境检测+证后监督"。认证程序一般包括认证申请与受理、现场检查准备与实施、认证决定、认证后管理这些主要流程。

(1)认证申请与受理。

①申请。对于申请有机产品认证的单位或者个人,根据有机产品生产或者加工活动的需要,可以向有机产品认证机构申请有机产品生产认证或者有机产品加工认证。根据《有机产品认证管理办法》和《有机产品认证实施规则》等的规定,认证委托人应当向有机产品认证机构提出书面申请,并提交下列材料。

A. 认证委托人的合法经营资质文件复印件,如营业执照副本、组织机构代码证、土地使用权证明及合同等。

B. 认证委托人及其有机生产、加工、经营的基本情况,包括认证委托人名称、地址、联系方式;当认证委托人不是产品的直接生产、加工者时,生产、加工者的名称、地址、联系方式;生产单元或加工场所概况;申请认证产品名称、品种及其生产规模,包括面积、产量、数量、加工量等;同一生产单元内非申请认证产品和非有机方式生产的产品的基本信息;过去3年间的生产历史,如植物生产的病虫草害防治、投入物使用及收获等农事活动描述;野生植物采集情况的描述;动物、水产养殖的饲养方法,疾病防治,投入物使用,动物运输和屠宰等情况的描述;申请和获得其他认证的情况。

C.产地(基地)区域范围描述,包括地理位置、地块分布、缓冲带及产地周围临近地块的使用情况等;加工场所周边环境描述、厂区平面图、工艺流程图等。

D.申请认证的有机产品生产、加工规划,包括对生产、加工环境适宜性的评价,对生产方式、加工工艺和流程的说明及证明材料,农药、肥料、食品添加剂等投入物质的管理制度,以及质量保证、标识与追溯体系建立、有机生产加工风险控制措施等。

E.本年度有机产品生产、加工计划,上一年度销售量、销售额和主要销售市场等。

F.承诺守法诚信,接受行政监管部门及认证机构的监督和检查,保证提供材料真实,执行有机产品标准、技术规范的声明。

G.有机生产、加工的管理体系文件。

H.有机转换计划(适用时)。

I.当认证委托人不是有机产品的直接生产、加工者时,认证委托人与有机产品生产、加工者签订的书面合同复印件。

J.其他相关材料。

在此期间,认证机构应当对申请者提出的认证申请进行评审,重点关注申请是否符合有机产品认证基本要求以及相关文件和资料是否齐全,明确该申请是否符合申请条件。另一方面,明确该申请是否处在本认证机构的认可范围、能力范围或资源范围之内,以及完成该项认证所需的资源和时间等,并在规定的时间内作出是否受理的决定。在此基础上,认证机构和申请者之间应当签订正式的书面认证协议,明确认证依据、认证范围、认证费用、现场检查日期、双方责任、证书使用规定、违约责任等事项。

②受理。认证机构应当自收到申请人书面申请之日起10个工作日内,完成对申请材料的评审,并作出是否受理的决定。

申请材料齐全、符合要求的,予以受理认证申请,认证机构与申请人签订认证合同;不予受理的,应当书面通知认证委托人,并说明理由。认证机构的评审过程应确保:认证要求规定明确,形成文件并得到理解;和认证委托人之间在理解上的差异得到解决;对于申请的认证范围、认证委托人的工作场所和任何特殊要求有能力开展认证服务,认证机构应保存评审过程的记录。

(2)现场检查准备与实施。认证协议签订后,认证机构即安排相关人员对该项认证进行策划,根据申请者的专业特点和性质确定认证依据,选择并委派进行现场检查的检查员组成检查组,每个检查组应至少有一名相应认证范围注册资质的专业检查员。对同一认证委托人的同一生产单元不能连续3年以上(含3年)委派同一检查员实施检查。

认证机构应向检查员提供充分的信息,以便检查员为实施检查作适当准备。针对申请者递交的有机产品认证所需要的文件资料,认证机构或检查组一般要对其符合性、完整性和充分性进行审核和基本判定。文件审核时应重点关注有机生产技术规程、有机加工操作规程、与保持有机完整性有关的基本情况及其控制程序、产品检测报告,以及法律法规的基本要求等。文件审核结束后,应将审核意见编制成文件审核报告,并提交给申请者。若申请者提交的文件不能完全符合要求时,一般要求申请者在双方确定的现场检查日期前将文件审核报告中提出的不符合内容全部纠正完毕,也可安排检查员在现场检查中进行验证。

现场检查分为例行检查和非例行检查。例行检查包括首次认证检查和例行换证检查,也称监督检查,例行检查每年至少一次。非例行检查是在获证者中按一定比例随机抽取检查对象或对被举报对象进行的不通知检查,也称飞行检查。对于产地(基地)的首次检查,检查范围应不少于2/3的生产活动范围。对于多农户参加的有机生产,访问的农户数不少于农户总数的平方根。

检查组根据文件审核评审的结果和相关信息,对现场检查进行策划,并与受检查方保持密切的沟通,确定检查的范围、场所、日期及检查组的分工等。现场检查计划一般以书面形式通知受检查方并获得确认。

对受检查方的有机生产或加工场所进行现场检查是有机产品认证的核心环节。检查时间应当安排在申请认证产品的生产、加工的高风险阶段,通常在认证产品收获前或加工期间进行。特别是对农产品的检查,应在作物和畜禽的收获或屠宰以前进行。

现场检查的主要工作内容是对受检查方的有机生产和加工、包装、仓储、运输、销售等过程及其场所进行检查和核实,评价这些过程是否符合认证依据的要求,技术措施和管理体系能否保证有机产品的质量,评估是否存在破坏有机完整性的风险,审核记录保持系统是否具有可追溯性,收集与支持认证决定有关的证据和材料等。

现场检查的另一项重要工作是对受检查方的有机生产或加工的能力和规模进行核实,核算认证年度中有机作物、畜禽等生产或加工产品的种类及其数量,以便在有机产品证书上予以明确界定。

现场检查包括对转换期的追溯核查、平行生产、转基因产品的核查,也包括对特殊情况和范围的检查,如小农户的检查、投入物的核查等,以确认生产、加工过程与认证依据标准的符合性。检查过程至少应包括:

①对生产、加工过程和场所的检查,如生产单元存在非有机生产或加工时,也应对其非有机部分进行检查;

②对生产或加工管理人员、内部检查员、操作者的访谈;

③对 GB/T 19630.4 所规定的管理体系文件与记录进行审核;

④对认证产品的产量与销售量的汇总核算;

⑤对产品和认证标志追溯体系、包装标识情况的评价和验证;

⑥对内部检查和持续改进的评估;

⑦对产地和生产加工环境质量状况的确认,并评估对有机生产、加工的潜在污染风险;

⑧样品采集;

⑨适用时,对上一年度认证机构提出的不符合项采取的纠正和/或纠正措施进行验证。

在结束检查前,对检查情况进行总结,向受检查方及认证委托人明确并确认存在的不符合项,对存在的问题进行说明。在完成现场检查后,根据现场检查发现,编制并向认证机构递交公正、客观和全面的关于认证要求符合性的检查报告。

(3)合格评定与认证决定。认证机构应根据评价过程中收集的信息、检查报告和其他有关信息,评价所采用的标准等认证依据及法律法规的适用性和符合性,现场检查的合理性和充分性,检查报告及证据和材料的客观性、真实性和完整性等,重点判定有机生产和加工过程的符合性、产品安全质量符合性及产品质量是否符合执行标准的要求,并最终作出能否发放证书的决定。

申请人的生产活动及管理体系符合认证标准的要求,认证机构予以批准认证。生产活动、管理体系及其他相关方面不完全符合认证标准的要求,认证机构提出整改要求。申请人已经在规定的期限内完成整改或已经提交整改措施并有能力在规定的期限内完成整改以满足认证要求的,认证机构经过验证后可以批准认证。

申请人的生产活动存在以下情况之一,认证机构不予批准认证:

①提供虚假信息,不诚信的;
②未建立管理体系,或建立的管理体系未能有效实施;
③生产加工过程使用了禁用物质或者受到禁用物质污染的;
④产品检测发现存在禁用物质的;
⑤申请认证的产品质量不符合国家相关法规和/或标准强制要求的;
⑥存在认证现场检查场所外进行再次加工、分装、分割情况的;
⑦一年内出现重大产品质量安全问题或因产品质量安全问题被撤销有机产品认证证书的;
⑧未在规定的期限完成不符合项纠正和/或纠正措施,或者提交的纠正和/或纠正措施未满足认证要求的;
⑨经监(检)测产地环境受到污染的;
⑩其他不符合本规则和/或有机标准要求,且无法纠正的。

认证机构应当按照认证依据的要求及时作出认证结论,并保证认证结论的客观、真实。对不符合认证要求的,应当书面通知申请人,并说明理由。根据相关认可准则的规定,认证决定可以由认证机构委托的一组人(一般称作颁证委员会、技术委员会)或某个人作出。认证机构应当对其作出的认证结论负责。

对符合有机产品认证要求的,认证机构应当向申请人出具有机产品认证证书,并准许其使用有机产品认证标志。从2014年4月1日起正式实施的新版《有机产品认证管理办法》明确规定,即日起取消"有机转换"认证标志,今后市场上只会出现有机产品。

证书的主要内容应当包括以下几个方面:获证单位名称、地址;获证产品的数量、产地面积和产品种类;有机产品认证的类别;依据的标准或者技术规范;有机产品认证标志的使用范围、数量、使用形式或方式;颁证机构、颁证日期、有效期和负责人签字;属于有机产品转换期间的,注明"转换"字样和转换期限。

(4)监督和管理。有机产品认证证书有效期为一年。获证者应当在有效期期满前向认证机构申请年度换证,认证机构将由此启动监督换证检查程序。认证机构应当按照规定对获证单位和个人、获证产品、生产及变更情况等进行有效跟踪检查,即年度换证例行检查。例行检查至少一年一次。

申请人应及时就产品更改、生产过程更改或区域扩大、管理权或所有权等更改通知认证机构。

监督检查还包括非例行检查,非例行检查不应事先通知。非例行检查的对象和频次等可基于有关认可规则和认证机构对风险的判断及来源于社会、政府、消费者对获证产品的信息反馈。

根据需要定期或不定期进行产地(基地)环境检测和产品样品检测,保证认证、检查结论能够持续符合认证要求。

根据有关规定,认证机构在发放证书时应当告知获证者有关证书变更、重新申请、撤销、注销、暂停等的管理规定或事项。

①证书变更。获证者在有机产品认证证书有效期内,发生下列情形之一的,应当向认证

机构办理变更手续：

　　A.获证单位或者个人、有机产品生产、加工单位或者个人发生变更的；

　　B.产品种类变更的；

　　C.有机产品转换期满，需要变更的。

　②重新申请证书。获证者在有机产品认证证书有效期内，发生下列情形之一的，应当向有机产品认证机构重新申请认证：

　　A.产地（基地）、加工场所或者经营活动场所发生变更的；

　　B.其他不能持续符合有机产品标准、相关技术规范要求的。

　③证书撤销、注销、暂停的规定。获证者发生下列情形之一的，认证机构应当及时作出暂停、注销、撤销认证证书的决定：

　　A.获证产品不能持续符合有机产品标准、相关技术规范要求的；

　　B.获证单位或者个人、有机产品生产、加工单位发生变更的；

　　C.产品种类与证书不相符的；

　　D.证书超过有效期的；

　　E.未按规定加施或者使用有机产品标志的。

　对于撤销和注销的证书，有机产品认证机构应当予以收回。

3）有机产品认证方法

有机产品认证机构在接到申请者的书面申请后，应当对申请材料进行评审，如申请材料齐全符合受理条件的，认证机构可直接受理，否则可不受理或要求申请者补充、修改，直至符合要求为止。在正式受理认证申请之后，认证机构就会安排相关人员对该项认证进行策划，并及时安排现场检查。

企业和个人要想成功申报有机食品认证，主要应注意三件事：①有机食品认证材料要符合要求；②基地和加工厂要能接受检查官的现场检查；③要有有机生产管理资质的相关人员。

（1）熟悉标准。了解标准中允许使用的物质和禁止使用的物质以及技术方面的要求。我国已于2012年3月1日起实施有机产品新标准（GB/T 19630—2011）。管理者、基地负责人和内部检查员应参加有机培训，以便系统了解有机生产的技术要求和相关知识。有机生产者尤其是有机生产管理者和内部检查员必须充分理解和透彻掌握有机标准条款和具体要求，并随时接受检查组的询问。

（2）文件材料的准备。申请者在接受认证机构的检查之前，要积极配合认证机构提供有关的文件材料。根据质量管理的理念，制定并实施有机质量管理体系文件。文件包括四个方面，第一是有机转换证明材料，说明你已经进行过相应的有机生产。例如，一年生作物在播种前至少需要两年的转换期，而多年生作物（牧草除外）在第一次收获有机产品之前至少需要三年的转换期。也就是说，如果希望产品获得有机认证，必须确保在转换期内的生产要求符合有机标准并有足够的证据来证明。第二是质量手册，说明企业内部整体的管理方式，如管理方针、组织机构、内部审核等。第三是程序文件或作业指导书，比如"××作物的有机生产规程"，是指导企业进行有机生产的内部标准。第四是记录，即所有生产、加工、经营活动的记录，比如有机肥的购买单据等。文件没有统一规定的格式，企业可以根据自己的实际情况灵活编制实施。检查员将在现场核实文件内容是否与实际生产活动一致。

申请者最好在接受检查组(员)检查前进行一次全面的自检。

(3)设施、人员等方面的准备:有机生产地点;有机作物、动物的生长状况;有机和常规的隔离措施;有机生产对周围环境的影响,以及环境对于有机生产的影响等;管理人员、内部检查员和生产加工人员对有机农业和标准的理解,以及根据标准活动的实施情况。

(4)检查现场的准备。对受检查的现场和设施进行检查,确保一切处于受控状态。应在作物收获前接受检查组的现场检查;加工厂的检查应安排在产品加工期间进行。

①农场:地块、作物、种子/种苗、培育种苗的设施、温棚、灌溉水、灌溉设施、土壤培肥物质(叶面肥料、外购商品有机肥、自制堆肥、生物活性制剂等)、病虫草害控制物质(植物源农药、微生物农药、矿物性农药、植物生长调节剂、杀虫剂等)、畜禽、缓冲带及缓冲带内种植的作物、农场生产设备(收割机、喷雾器等)、运输工具、农场内简易加工设备、包装材料、仓库和用于仓库有害生物控制的物质等。

②加工厂:原(辅)料(有机配料、非有机配料、添加剂和加工助剂)、加工设备、水源、水质报告、锅炉房、锅炉水处理剂、锅炉除垢剂、有害生物防治设施(诱捕器、杀虫灯)和控制物质、卫生清洁消毒物质、原料收货区、产品搬运区、生产区、包装区、装货区、包装材料仓库、原料仓库、半成品仓库、成品仓库、厂外仓库、运输工具、废料(废渣、废水和废气)处理场所、分析化验室等。

(5)落实受访谈人员。要求在检查组(员)进行现场检查时相关人员在场,以便检查组能够与其进行充分交流,能够获得客观、真实的信息、事实及证据。

(6)落实后勤安排。申请者应指定陪同人员,并对办公、交通和就餐作出相应安排。

9.2.2 实训任务

编制有机食品标准认证内审资料。

实训组织:根据实际情况,将全班学生分为若干小组,每个小组参照"基础知识"中的内容并利用网络资源,结合任务内容具体分工,共同完成实训任务。实训任务含三项内容:有机食品认证申请书的编制、有机食品认证内部检查报告的编制、有机食品认证调查表的填写。

实训成果:以每一小组为单位,每一项任务出一份材料。实训课时结束时,上交材料。

实训评价:由有机食品生产企业负责人或主讲教师进行评价。

有机食品标准认证内审资料评价表

姓名		班级		任务	编制有机食品标准认证内审资料
评价项目	评价内容	评价方法	分值	自评	小组评价	教师评价
学习态度	自学充分,不迟到、早退,上课认真		10			
学习方法	设计的方案可行,操作安排统筹设计合理,材料准备充分有序		20			

续表

评价项目	评价内容	评价方法	分值	自评	小组评价	教师评价
教学参与	能积极参与各项教学活动,与老师、同学密切配合,步调一致		30			
目标1	能正确、规范地填写有机食品认证申请书		10			
目标2	能完整、规范地编制有机食品认证内部检查报告		10			
目标3	能正确、规范地填写有机食品认证调查表		10			
团队协作	善于沟通,积极与他人合作完成任务		10			
合计			100			
权重				20%	30%	50%
得分						

思考题

1. 有机食品认证的流程是怎样的?
2. 有机食品认证需要提交哪些材料?
3. 有机产品认证现场检查的主要环节有哪些?需要注意些什么?

拓展学习网站

1. 中国绿色食品发展中心(http://www.greenfood.org.cn)
2. 中国绿色食品网(http://www.zhongguolsspw.roboo.com)
3. 各省绿色食品网
4. 中国有机食品网(http://www.ujget.com)